Lecture Notes in Physics

Volume 1007

The series Lecture Notes in Physics (LNP), founded in 1969, reports new developments in physics research and teaching - quickly and informally, but with a high quality and the explicit aim to summarize and communicate current knowledge in an accessible way. Books published in this series are conceived as bridging material between advanced graduate textbooks and the forefront of research and to serve three purposes:

- to be a compact and modern up-to-date source of reference on a well-defined topic;
- to serve as an accessible introduction to the field to postgraduate students and non-specialist researchers from related areas;
- to be a source of advanced teaching material for specialized seminars, courses and schools.

Both monographs and multi-author volumes will be considered for publication. Edited volumes should however consist of a very limited number of contributions only. Proceedings will not be considered for LNP.

Volumes published in LNP are disseminated both in print and in electronic formats, the electronic archive being available at springerlink.com. The series content is indexed, abstracted and referenced by many abstracting and information services, bibliographic networks, subscription agencies, library networks, and consortia.

Proposals should be sent to a member of the Editorial Board, or directly to the responsible editor at Springer:

Dr Lisa Scalone
Springer Nature
Physics
Tiergartenstrasse 17
69121 Heidelberg, Germany
lisa.scalone@springernature.com

Andrea Di Vita

Non-equilibrium Thermodynamics

 Springer

Andrea Di Vita
Dipartimento di Ingegneria Civile
Chimica e Ambientale
University of Genoa
Genoa, Italy

ISSN 0075-8450 ISSN 1616-6361 (electronic)
Lecture Notes in Physics
ISBN 978-3-031-12220-0 ISBN 978-3-031-12221-7 (eBook)
https://doi.org/10.1007/978-3-031-12221-7

*Et ignem regunt numeri (Numbers rule
fire too)*

*Jean Baptiste Joseph Fourier (quoting Plato)
Théorie analytique de la chaleur (1822)*

Preface

This book offers a panoramic view of non-equilibrium thermodynamics. Since the golden age of steam engines, thermodynamics has been the undisputed queen of applied research, including not only engineering science but also—indeed, from the very beginning—the sciences of life. The importance of thermodynamics to all branches of science in which systems with very large numbers of particles are involved cannot be overstated.

The Second Principle of thermodynamics—which prescribes maximization of entropy—describes the stability of isolated systems at thermodynamic equilibrium without the need for any detailed knowledge of the microscopic processes underlying the dynamics of the system. Thus, applications of thermodynamics still spread across all fields of science, from astrophysics to physiology to Earth sciences.

This success is even more amazing, as most steady, stable macroscopic systems in our everyday life are far from being isolated from their environment—think, for example, of a current-carrying street lamp, or of a mighty, calm, flowing river—so that their obvious stability does not correspond to a maximum of their entropy. Investigation of stability of these systems relies on the solutions of the relevant equations of dynamics on a case-by-case basis, and lacks therefore the generality of thermodynamics.

For a long time, many researchers have tried to describe such stability in the same way that the Second Principle describes the stability of thermodynamic equilibrium—and failed. Most of them invoked either entropy, or its production rate, or some modified version of it. In their hapless efforts, however, those researchers have found a lot of useful stability criteria for far-from-equilibrium states.

These criteria usually take the form of variational principles, in terms of the minimization or maximization of some quantity. Most of them are convincing rules-of-thumb in their proponents' research field; their effectiveness usually relies on considerable empirical evidence rather than on rigorous justification based on first principles. When more configurations are accessible, these stability criteria act as selection rules. Quite surprisingly, criteria independently formulated in different research fields may resemble each other.

In this book, we discuss these variational principles by both highlighting the role of macroscopic quantities, outlining their limits of validity and investigating their connection with other principles. Its audience includes STEM undergraduates as well as the professionals involved in biological and medical physics, physics, chemistry and engineering sciences.

The book contains many results that are not commonly covered in textbooks written for these communities and in large parts are generally not even known in these communities. Its aim is to help the reader to take full advantage of the variational approaches to non-equilibrium thermodynamics with the help of skilled use of those thermodynamic variables which are relevant to the particular problems of interest.

This book could not exist without the continuous encouragement and the staunch support of Prof. A. Bottaro, DICCA, University of Genova.

The unending patience of my wife Ania—who silently endured my grunt— and the warm enthusiasm of my daughter Susanna—who also drew all figures very patiently—deserve gratitude beyond my words. My dedication of this book to them is definitely too small a prize.

Genoa, Italy   Andrea Di Vita

Contents

List of Main Variables and Acronyms

Latin[1]

a	Generic quantity (2.2)
a_0	Value of a in the environment (2.2), unperturbed quantity (4.3)
a_1	Perturbation of a (5.8)
a_{behind}	Value of a behind a discontinuity surface (5.6)
a_{eq}	Value of a at thermodynamic equilibrium (5.2)
a_{front}	Value of a in front of a discontinuity surface (5.6)
a_{max}	Maximum amount of a (5.7)
a_{min}	Minimum amount of a (2.2)
a_{thr}	Threshold value of a (5.3)
a_{TOT}	Total value of a (3.1)
$\bar{a}$	Time-average of a (3.4)
$[a]$	$a_{behind} - a_{front}$ (5.6)
$\langle a \rangle$	Statistical average of a (4.1)
$\{a\}$	Spatial average of a over the horizontal plane (5.6)
A_r^0	Affinity of the r-th reaction (4.2)
A_s	Shock wavefront area (5.6)
A_{wire}	Cross section of a wire (4.1)
$\mathbf{a}$	Generic vector quantity (4.2)
$\mathbf{A}$	Vector potential (6.1)
b_i	i-th component of a generic array (4.3)
b_n	Component of $\mathbf{b} \perp$ wavefront (5.6)

---

[1] Some symbols have different meanings in different Chapters, due to the different formalism of different bibliographical references. In order to help further bibliographical enquiries, we aim at reproducing the original symbols we found in the literature as far as possible, even at the cost of some redundancy. In case, all meanings are listed here, to remove all ambiguities. However, it turns out that the meaning of each variable is always obvious from the context. For simplicity, no dummy index is included. Each row displays the number of the Section where the item appears in the text for the first time with the highlighted meaning.

$\mathbf{b}$	Generic vector quantity (4.2)
$\mathbf{B}$	Magnetic induction (2.4)
c	Speed of light in vacuum $= 2.99792458 \cdot 10^8 m \cdot s^{-1}$ (4.2), heat capacity (5.6)
c_k	Mass fraction of the k-th species (3.3)
c_p	Specific heat at constant pressure per unit mass (5.8)
c_s	Speed of sound (5.2)
c_p	Specific heat at constant volume per unit mass (5.8)
c_{ij}	Effort required by a travel between the i-th and the j-th city (5.5)
$\mathbf{c}$	Generic vector quantity (4.2)
C	Total effort for all travels (5.5)
C_p	Specific heat at constant pressure (3.2)
C_v	Specific heat at constant volume (3.2)
d	Distance between parallel plates (5.6)
d_i	i-th distance (5.5)
d_{ij}	Distance between the i-th and the j-th city (5.5)
d_{AB}	Distance between points A and B (5.5)
D_j	Number of arrivals to the j-th city (5.5)
D_D	Dufour coefficient (4.3)
D_T	Thermodiffusion coefficient (4.3)
D_{ij}	$(ij)-$ the component of the matrix of particle diffusion coefficients (4.3)
D_{ikpq}	Coefficient in the relationship $J_{ip} = D_{ikpq}X_{kq}$ (4.3)
$\frac{d_X P}{dt}$	$\int d\mathbf{x} Y_m \frac{dX_m}{dt}$ (4.3)
$\frac{d_Y P}{dt}$	$\int d\mathbf{x} X_m \frac{dY_m}{dt}$ (4.2)
$\frac{d_i S}{dt}$	Amount of entropy produced per unit time in the bulk (4.2)
$\frac{d_e S}{dt}$	Amount of entropy produced per unit time in the bulk (4.2)
E	Energy (2.2)
e_k	Electric charge of one particle of the k-th species (4.2)
$\mathbf{E}$	Electric field (4.2)
f	Rayleigh's dissipation function (4.1)
F	Helmholtz's free energy (2.3)
g	Gibbs' free energy per unit mass (3.3), modulus of the acceleration of gravity (5.3)
G	Gibbs' free energy (2.3), universal gravitational constant (3.4)
h	Enthalpy per unit mass (3.3)
h_k	Enthalpy per unit mass of the k-th species (3.3)
H	Enthalpy (3.3), elevation above the base level (6.2)
H_w	Hydraulic head (5.3)
H_{mag}	$\frac{1}{2} \int |\mathbf{H}|^2 d\mathbf{x}$ (2.4)
$\mathbf{H}$	Magnetic intensity (2.4)
i	Rank (5.5)

i_{arc}	Arc current (5.3)
I_e	Electric current (4.1)
I_S	Entropy current (4.1)
$I_{streaming}$	Streaming current (4.3)
J_k	k-th thermodynamic flow (3.5)
J_{kp}	p-th component of J_k (4.3)
$J_{streaming}$	Thermodynamic flow conjugated to $I_{streaming}$ (4.3)
$\mathbf{j}_{el}$	Electric current density (4.2)
$\mathbf{j}_k$	Mass flow of the k-th species (4.2)
$\mathbf{j}_s$	Entropy flow (4.2)
$\mathbf{j}'_s$	$\mathbf{j}_s + \rho s \mathbf{v}$ (4.2)
k_B	Boltzmann's constant $= 1.380649 \cdot 10^{-23} \, J \cdot K^{-1}$ (4.1)
$k_{-3,-2-,1-,+1,+2,+3}$	Time constants of chemical reactions (4.3)
K	Kinetic energy (3.4), number of thermodynamic quantities (3.5)
K_w	Hydraulic conductivity (5.3)
l_{wire}	Length of a wire (4.1)
L	Lagrangian (4.3), scalar function of ϕ (4.3), length (5.6)
L_{ij}	(i, k)-th component of Onsager's coefficient matrix (4.1)
m	Traveller's mass (6.2)
m_{el}	Mass of charge carriers (5.3)
m_{ion}	Ion mass (6.1)
m_k	Mass of one particle of the k-th species (3.3)
m_m	Molar mass (4.2)
M	Common destination of many travellers (5.4)
Ma	Mach number (5.6)
$\mathbf{M}$	Magnetization (2.4)
n	Number of moles (4.1), density of travellers (6.2)
n_{el}	Density of charge carriers (5.3)
n_k	Particle density of the k-th species (4.2)
$\mathbf{n}$	Unit normal vector (5.6)
N	Number of particles (1)
N_i	Number of trips on the i-th distance (5.5)
N_j	Population of the j-th city (5.5)
N_k	Number of particles of the k-th species (3.3)
N_l	$N_{TOT} \cdot p_l$ (5.5)
$N_{A,B}$	Population of A, B (5.5)
N_{TOT}	Maximum rank (5.5)
$N_>(Y)$	$N_{TOT} \cdot \int_Y^{+\infty} p_l \, dl$ (5.5)
$\dot{N}$	Number of travellers who reach M per unit time (5.4)
O_i	Number of departures from the i-th city (5.5)
p	Pressure (2.2)
p_c	Capillary pressure (5.3)
p_l	Distribution function (5.5)
P	$\frac{d_i S}{dt}$ (4.2)

Pe	Péclet number (5.8)
P_{ext}	Amount of mechanical energy supplied per unit time by the external world (5.3)
P_h	Heating power density (4.2)
P_{st}	Value of P in steady state (4.3)
P_L	Power dissipated in the load (5.7)
P_{TOT}	Total heating (dissipated) power (5.6)
q_{el}	Electric charge of charge carriers (5.3)
q_{ion}	Ion electric charge (6.1)
q_1	P_{h1} (5.8)
$\mathbf{q}$	Heat flux (4.1)
$\mathbf{q}_w$	Volumetric water flux (5.3)
Q	Amount of heat per unit time (4.1)
Q_j	Ohmic power dissipated in the j-th resistor (6.2)
Q_{loss}	Amount of heat lost across the boundaries per unit time (5.6)
Q_C	Amount of heat per unit time to the cold heat exchanger (5.8)
Q_H	Amount of heat per unit time from the hot heat exchanger (4.1)
Q_{Joule}	Ohmic heating power (5.8)
$Q_{I,II,III,l}$	Amount of exchanged resources (5.5)
$Q_{Peltier}$	Peltier heating power (4.1)
$Q_{Thomson}$	Thomson heating power (4.1)
Q_{thr}	Threshold value of Q_H (5.8)
$\dot{Q}$	Amount of heat per unit time (6.2)
r	$c_p \cdot \frac{\gamma-1}{\gamma}$ (5.8), density of power supplied to matter by electromagnetic radiation (6.2)
R	Amount of work made by an external source (2.2), radius (3.4), number of regions (3.5), constant of perfect gases (4.1), number of chemical reactions (4.2), electrical resistance (5.3)
R_k	Overall amount of work spent at the k-th attempt (5.5)
R_L	Load resistance (5.7)
R_S	Inner resistance (5.7)
Ra	Rayleigh number (5.3)
Re	Reynolds' number (5.3)
Re_M	Magnetic Reynolds' number (5.3)
R_Ω	Electric resistance (4.1)
s	Entropy per unit mass (3.3)
s_k	Entropy per unit mass of the k-th species (3.3)
S	Entropy (2.2)
S_g	Entropy generation (6.1)
S_L	Laminar flame speed (5.8)
t	Time (3.5)

T	Absolute temperature (2.2)
$T_{boundary}$	T at the boundary (5.3)
T_c	T at the coolest heat bath (5.8)
T_i	T of the i-th heat bath (6.2)
T_h	T at the hot boundary (5.6)
T_l	T at the cool boundary (5.6)
T_C	T of the cold heat exchanger (5.8)
T_H	T of the hot heat exchanger (5.8)
T_{ij}	Number of travellers between the i-th and the j-th city (5.5)
T_{AB}	Number of travellers between A and B (5.5)
u	Internal energy per unit mass (3.3)
U	Gravitational potential energy (3.4), horizontal velocity (5.3), internal energy (5.6)
U_m	Horizontal velocity averaged along the vertical direction (5.3)
v	Volume per unit mass (3.3)
v_k	Volume per unit mass of the k-th species (3.3)
v_{arc}	Arc voltage drop (5.3)
$\mathbf{v}$	Macroscopic fluid velocity (3.5)
$\mathbf{v}_{el}$	Velocity of charge carriers (5.3)
V	Volume (2.2), voltage (4.1)
V_p	Volume of voids (5.3)
V_S	Volume of shock (5.6)
$\dot{V}$	Volume flow (6.2)
w	Distribution function (4.1)
$\mathbf{V}$	$\mathbf{v} + \frac{q_{ion}\mathbf{A}}{m_{ion}}$ (6.1)
w_i	Amount of work spent on the i-th distance (5.5)
W	Number of possible travellers' combinations (5.5)
W_{diss}	Amount of work dissipated into heat (6.1)
W_{mag}	$\int \frac{|\mathbf{B}|^2}{2\mu_0} d\mathbf{x}$ (2.4)
$W_{I,II,l}$	Amount of supplied resources (5.5)
x	Independent variable (3.1)
x_i	i-th independent variable (2.5), i-th spatial coordinate (4.2)
$\mathbf{x}$	Position vector (2.4)
X	Thermodynamic force (3.1)
X_i	Thermodynamic force (2.5), population of a city with rank i (5.5)
X_{iq}	q-th component of X_i (4.3)
y	Independent variable (3.1)
Y	Thermodynamic flow (3.1)
Y_j	j-th thermodynamic flow (6.1)
Y_{max}	Maximum thermodynamic flow (6.1)
z	Vertical coordinate (5.3)

Greek

α	Point of a wire (4.1), proportionality constant (5.5), thermal diffusivity (5.8)
α_B	Scalar quantity with $\nabla \alpha_B = 0$ (5.3)
β	Point of a wire (4.1), $-\frac{\partial\langle T\rangle}{\partial z}$ (5.6)
β_{ik}	$-\left(\frac{\partial^2 S}{\partial x_i \partial x_k}\right)_{eq}$ (2.5)
γ	$\frac{c_p}{c_v}$ (2.5)
Γ	Ratio of energy flux and particle flux (4.1)
Γ_k	$-\frac{\mu_k^0}{T}$ (4.3), k-th thermodynamic quantity of interest (5.1)
δa	Perturbation of a (4.1)
$\delta^2 a$	Second-order perturbation a (6.1)
Δa	Difference between values of a (2.2), Laplacian of a (5.2)
ΔE_c	Energy expenditure rate per unit volume (5.3)
ΔI	$Q_H - Q_C$ (5.8)
ΔS	Net amount of entropy coming out of the system per unit time (5.8)
ΔU	Increase in internal energy due to $P_{TOT} > 0$ (5.6)
ε	Seebeck coefficient (4.1)
ε_g	Averaged buoyancy power density (5.3)
ε_ν	Averaged viscous power density (5.3)
ε_0	Vacuum electric permittivity $= \mu_0^{-1} c^{-2}$ (4.2)
ζ	Bulk viscosity (5.3)
η	Dynamic viscosity (5.3), efficiency (5.7)
η_l	$\frac{Q_l}{W_l}$ (5.5)
η_{opt}	Optimal efficiency (5.7)
η_C	Carnot efficiency (5.7)
Θ	$\int_a^b \Phi dy$ (5.1)
κ	Scaling factor (5.5)
λ	Lagrange multiplier (4.3)
λ_{ik}	(i, k)-th relaxation coefficient (4.1)
μ_k	Chemical potential of the k-th species (3.3)
μ_k^0	Chemical potential per unit mass of the k-th species (3.3)
μ_0	Vacuum permittivity $= 4\pi \cdot 10^{-7}\ T \cdot A^{-1} \cdot m$ (2.4)
ν	Kinematic viscosity (5.3)
ν_{coll}	Collision frequency (5.3)
ν_{kr}	Stoichiometric coefficient of k-th species in the r-th reaction (4.2)
ξ_r	Reaction degree of the r-th reaction (4.3)
π_{AB}	Peltier coefficient (4.1)
ρ	Mass density (3.3)
ρ_k	Mass density of the k-th species (4.3)
ρ_{el}	Electric charge density (5.3)
σ	Entropy production density (4.2)
σ_A	Thomson coefficient (4.1)
σ_Ω	Electric conductivity (4.1)

σ' Viscous stress tensor (4.2)

σ'_{ik} (i, k)-th component of σ' (4.2)

τ Time shift (4.1)

Υ_r Reaction rate of the r-th reaction (4.2)

ϕ Generic scalar function of spatial coordinates (4.3)

ϕ_g Gravitational potential (3.4)

ϕ_{el} Electrostatic potential (4.2)

ϕ_p Porosity (5.3)

Φ Scalar function of the L_{ij}'S, the X_i'S and the Γ_j'S (5.1)

χ Thermal conductivity (4.1)

χ_{ij} (i, j)-th component of a thermal conductivity tensor (5.2)

Ψ $\frac{d}{dt}\left(\frac{1}{T}\right)\frac{d(\rho u)}{dt} - \rho\frac{d}{dt}\left(\frac{\mu_k^0}{T}\right)\frac{dc_k}{dt} + \frac{\rho}{T}\frac{dp}{dt}\frac{d}{dt}\left(\frac{1}{\rho}\right) - \frac{d}{dt}\left(\frac{1}{T}\right)h\frac{d\rho}{dt}$ (3.6)

ω $2\pi\cdot$ oscillation frequency (5.8)

Ω Geographical region (5.4)

Ω $\nabla \wedge \mathbf{V}$ (6.1)

Acronyms

EIT   Extended irreversible thermodynamics (6.1)

GEC   General evolution criterion (3.6)

LNET   Linear non-equilibrium thermodynamics (4.1)

LTE   Local thermodynamic equilibrium (3.2)

MHD   Magnetohydrodynamics (6.1)

MinEP   Minimum entropy production principle (4.1)

MEPP   Maximum entropy production principle (5.6)

MPP   Maximum power principle (5.7)

PLE   Principle of least effort (5.5)

QTA   Quasi-thermodynamic approach (6.1)

Looking for the Holy Grail?

Abstract

This book is a panoramic view of non-equilibrium thermodynamics. In a macroscopic system at thermodynamic equilibrium the entropy is a maximum, and this maximum property allows us to describe the stability of the system against perturbations regardless of the detailed microscopic dynamics ruling the system. In many problems, however, boundary conditions allow relaxation towards a steady state which is stable but far from thermodynamic equilibrium. It is then only natural to ask if information concerning its stability is available regardless of detailed microscopic dynamics. This is the topic of non-equilibrium thermodynamics. Even if no universal answer exists, answers for selected classes of problems are available in the form of variational principles. The aim of the book is to enable the reader to find a connection between any working hypothesis she/he may want to postulate for her/his own research and some fundamental concepts of non-equilibrium thermodynamics. Hopefully, then, the reader shall be able to formulate a variational principle of her/his own about a particular topic of interest starting from a sound physical basis.

When it comes to describing physical systems near thermodynamic equilibrium, the success of thermodynamics is undisputed. Its Second Principle, which says that the total entropy of an isolated system can never decrease over time, implies that the steady state of thermodynamic equilibrium which isolated system spontaneously evolve ('relaxes') towards is a state with maximum entropy [1,2]. It is difficult to overestimate the impact of the Second Principle on all branches of science, whenever the understanding of systems with a very large number N of particles is at stake.

A feature of thermodynamic equilibrium which makes it so tantalizing is its stability: once an isolated body has achieved thermodynamic equilibrium, should any fluctuation occur it would decrease its entropy, and as a consequence the system would spontaneously relax back to thermodynamic equilibrium because of the Second Principle. Another reason for the success of thermodynamics is that the latter provides us with information which does not rely on the detailed description of the (possibly quite complex) microscopic interactions among the particles the system is made of. For example, thermodynamics tells us that the specific heat at constant volume of a body is always positive but is also always smaller than the specific heat of the same body at constant pressure; and we do not need to solve the equations of motion of the particles the body is made of to get such information. Analogously, we know that the maximum efficiency of a thermal engine is given by Carnot's formula; the validity of this result does not depend on our knowledge of the properties of the metals the piston of our engine is made of.

The work of Gibbs and Boltzmann explained how thermodynamics is rooted in statistical physics. As for steady states in systems where N is not so large, the statistical physics on the nano- and the mesoscale has been enjoying tremendous progress in the last decades—we refer, for example, to Landauer's principle [3], Bochkov-Kuzovlev's theorem [4], Jarzynski's identity [5], Crook's theorem [6] and Evans et al.'s fluctuation theorem [7].[1] Starting from the kinetic theory of gases, moreover, increasingly sophisticated models—including Boltzmann's and Fokker-Planck's kinetic equations—have been describing the unsteady states of systems with $N \gg 1$ and their relaxation to thermodynamic equilibrium.[2]

In many cases, however, suitable boundary conditions keep the steady state a system with $N \gg 1$ relaxes too far from thermodynamic equilibrium. Think, for example, of a copper wire connected to a battery. Definitely, the wire is made of a very large number of particles, i.e. $N \gg 1$. As far as the battery is connected and supplies a constant e.m.f., an electric current flows across the wire, the wire warms up because of Joule heating and radiates heat into the surrounding air. The wire is not isolated, because it is both connected to the battery and in contact with the air. The steady state of the wire is far from thermodynamic equilibrium: the wire is hotter than the surrounding air.

All the same time, this steady state is stable against perturbations. We may, for example, perturb the current flowing across our wire by connecting another wire in parallel to the same battery; alternatively, we may cool the wire locally by putting an ice cube on it, thus reducing its temperature and altering its resistivity. In both cases,

---

[1] Readers familiar with statistical mechanics shall find a comprehensive review in Ref. [8].

[2] Moreover, if N is not so large then the results of thermodynamics which hold for $N \gg 1$ may be suitably generalized with the help, for example, of slightly modified definitions of the thermodynamic quantities. In the words of [9], *macroscopic thermodynamics applies to a large sample of small systems (e.g. a macromolecular solution) [...] we are interested in thermodynamic functions and interrelationships for a 'single' small system, including, in general, variations in the size of the system (e.g. the degree of polymerization or aggregation and the volume) [...] Allowance for these variations in size is, indeed, an important new feature which would not be included in a conventional macroscopic thermodynamic treatment of a large sample of small systems where only the 'number' of small systems (macromolecules, etc.) would be varied [...] Small system effects will be particularly noticeable at phase transitions and in critical regions [...] The special case of surface effects in drops and bubbles has [...] been treated by Gibbs and others.* Moreover, statistical mechanics provides macroscopic thermodynamics with a sound basis even in small systems—see both Ref. [9], notes 26 and 28 of [10] and the note in Sect. 3.3. See also Sect. 5.5.1 for further examples of small systems.

as the perturbation is removed, the wire (and the battery, and the air surrounding it) relaxes back after a while to the initial state. This stability is somehow similar to the stability enjoyed by thermodynamic equilibrium.[3] Only when the battery is switched off, does the electric current go to zero and the wire starts cooling down, until it attains thermodynamic equilibrium with the environment.

Now, it is only natural to ask: in systems far from thermodynamic equilibrium and with $N \to \infty$, is there a way to describe stable relaxed states, which does not rely on the detailed description of the interactions among the particles the system is made of? This is the central question of the so-called 'non-equilibrium thermodynamics'.[4]

The analogy with the thermodynamics of equilibrium provides us with some reasonable suggestion about the answer to this question—if any such answer exists. Firstly, thermodynamic equilibrium corresponds to a maximum of entropy, and entropy is a function of temperature, pressure, etc. We expect that the relaxed state of a system far from equilibrium corresponds to a maximum—or a minimum—of some physical quantity, which depends on the macroscopic quantities which are relevant to the system. Secondly, it is only because of suitable boundary conditions that the system relaxes to a steady state far from thermodynamic equilibrium, rather than to the thermodynamic equilibrium itself. It is reasonable to expect that the impact of these boundary conditions is to put a set of constraints on the maximization (or minimization) which describes the relaxed state; in other words, the latter is likely to satisfy some variational principle subject to some constraint.[5] Finally, just like the Second Principle the desired variational principle should be model-independent, i.e. its validity should not depend on the detailed dynamics of the particular physical system under investigation. This universal variational principle for the stability of systems far from thermodynamic equilibrium, analogous in power and generality to the Second Principle for systems at thermodynamic equilibrium, is the Holy Grail of non-equilibrium thermodynamics.

In spite of the early warnings,[6] the decade-long search for the Holy Grail seemed to be near a glorious, happy end after the astonishing results of Belgian and Dutch researchers: De Donder's chemical affinity, Onsager's symmetry relationships, Prigogine's dissipative structures and the like—see [13] for a historical review. Today, the field is utterly ignored by the vast majority of physicists; most of them whole-

---

[3] A relevant difference is that the wire is no isolated system, then the external world is allowed to interact with it by applying a perturbation. In contrast, an isolated system has no interaction with the external world, and fluctuations are due to purely internal causes.

[4] As for the statistical foundations of non-equilibrium thermodynamics, see Ref. [11].

[5] Admittedly, we are not utilizing the wordings 'variational principle' in a rigorous way (see Sect. 5.1 for a detailed discussion). We should otherwise utilize the word 'extremum property' only. Here and in the following, we neglect these subtleties and keep on speaking of variational principles.

[6] *It is usually found that a [...] description, giving complete information, has an entropy which is constant in time. This is no surprise since the complete specification of each identified particle leaves no room for any disorder [...]. Only an incomplete description will have an entropy which can vary with time (presumably increasing). There is no conflict between these two descriptions; a single physical system can exhibit some features which are apparently reversible and others which require the epithet irreversible* [12].

heartedly agree with Jaynes' sharp judgement [14]: if we have enough information to apply a variational principle far from equilibrium with any confidence, then we have more than enough information to solve the steady-state problem without it. Even if largely overlooked, a nonexistence theorem [15] seems to ring the death knell. Admittedly, the quest for stability remains of utter relevance, because even if we solve a steady-state problem, assessing the stability of the solution against pertur- bations is another matter. However, practically no researcher takes on this problem but with the help of direct (and usually cumbersome) analysis of the equations of motion, on a case-by-case basis. Sadly enough, this is exactly what people hoped to prevent. Nevertheless, people who were not satisfied with this approach have been able to find through trial and error a host of variational principles for relaxed states far from thermodynamic equilibrium in many fields of science, and most of them still need rigorous proof. We guess there is plenty of room for thermodynamics here. By far and large, however, the present mistrust is well justified: attempts to derive the principles under discussion are so far unconvincing since they often require the introduction of additional hypotheses, which by themselves are less evident than the proven statements [16]. The dream came out to be an illusion; but we are rewarded with a host of relevant results. Most of the latter are scarcely known beyond the aca- demic domain of the researchers who have put them forward. Even if these variational principles for relaxed, far-from-equilibrium systems still need a unique, systematic theoretical framework, they have been formulated in order to cope with observations and to describe the outcome of experiments in many fields of science, from acoustics to physiology and astrophysics. References [17, 18] are excellent reviews, which put also in evidence the connection with fundamental advances in statistical physics like the fluctuation theorem.

The aim of this book is rather to discuss these variational principles by highlighting the role of macroscopic quantities. The reason is twofold. Firstly, these principles, even if sometimes stated only vaguely, are of interest per se, and deserve a systematic discussion which highlights their relationships with well-known, generally accepted results of thermodynamics. Secondly, many cans of worms still lurk behind any work which deals with macroscopic systems far from equilibrium, as vague definitions and not-well-established assumptions all too often limit the validity of one's results. In the scientific jargon, few words more than 'entropy' are likely to be invoked in the mounting ocean of pseudoscientific noise which embraces us all. A goal of this work is to help the reader to get rid of as many ill-defined concepts as possible when it comes to far-from-equilibrium problems.[7]

---

[7] Many rigorous discussions in information theory, life sciences and social sciences rely on defini- tions of 'entropy' which formally resemble the familiar entropy of Gibbs and Boltzmann but remain quite separate from it [19]. Unless explicitly stated, in this book we stick to the latter entropy, as *only the Gibbs entropy simultaneously satisfies the Zeroth, First and Second Laws of thermodynamics* [10]. For the First and Second Laws (or 'Principle'), see Sect. 2.1, with the version of Ref. [20] of the Second Law; the Zeroth Law states that *if two systems are both in thermal equilibrium with a third system, then they are in thermal equilibrium with each other* [21]. When we speak of two systems which are in equilibrium with each other, moreover, we recall that *equilibrium is a state-*

The book is organized as follows. In Chap. 2, we start with a review of some well-known features of thermodynamic equilibrium, which turn out to be useful for the following discussion. We introduce in Chap. 3 both Le Châtelier's principle and the all-relevant concept of local thermodynamic equilibrium, together with the related result, the general evolution criterion. We discuss in Chap. 4 the linear non-equilibrium thermodynamics, i.e. the only available theoretical framework where variational principles for far-from-equilibrium relaxed states meet general consensus and find robust physical justification in terms of First Principles even if under quite restrictive assumptions. Up to this point, it is business as usual, and our discussion is basically a collection of results at the textbook level. Now, here comes the fun: many problems of interest lie definitely outside the domain of validity of linear non-equilibrium thermodynamics. Some of them, together with the variational principles proposed in order to cope with them, are listed and discussed in Chap. 5. Admittedly, the list is definitely far from complete. All the same, it gives a hint of the variety of the problems and the fields of research involved; moreover, the listed results suggest a natural generalization of the steady relaxed states to oscillating relaxed states. Till now, Ariadne's thread of our presentation is the wire made of copper. Another familiar (but not trivial) system, a room with a heater and a window, allows us to understand why none of many efforts towards the Holy Grail meets general consensus in Chap. 6. Moreover, it allows us to see the relationships between the variational principles listed in Chap. 5 and the ideas presented in Chap. 3. Conclusions are drawn in Chap. 7 with the help of a third, deceitfully simple system, made of a stick and a river.

This book is based on the lecture notes of a course *Variational approaches to non-equilibrium thermodynamics* held at University of Genova in 2012, and is aimed at undergraduate STEM students. Then, we invoke both statistical physics and kinetic theory as little as possible and discuss the problems from a macroscopic point of view. Moreover, we keep the maths at the simplest level. The relevant relationships of vector analysis invoked in each chapter are systematically listed in the footnotes with no fear of repetition, and the algebraic steps are systematically presented as explicitly as possible. Some fundamental results of variational calculus are briefly explained in Appendices. The reader is supposed to be acquainted with fundamental tools of calculus and fundamental concepts of physics (e.g. energy). SI units are utilized everywhere.

Rather than an encyclopaedia of non-equilibrium thermodynamics, this book is a panoramic view of the topic. Surely, many authors have been omitted or forgotten, and the blame is to be put on the author's ignorance only. All the same, this work is not useless if it enables the reader both to check if linear non-equilibrium thermo-dynamics and its results hold for any physical problem the reader may be concerned

---

ment about the exchange of conserved quantities between systems. To avoid conceptual confusion, one should clearly distinguish between thermal equilibrium (no mean energy transfer), pressure equilibrium (no mean volume transfer), chemical equilibrium (no particle exchange on average), etc. Complete thermodynamic equilibrium corresponds to a state where all the conserved fluxes between two coupled systems vanish [10].

with and, above all, to find a connection between any working hypothesis she/he may want to postulate for her/his own research and some fundamental concepts of thermodynamics. Hopefully, then, the reader shall be able to formulate a variational principle of her/his own about a particular topic of interest starting from a sound physical basis.

References

1. Landau, L.D., Lifshitz, E.: Statistical Physics. Pergamon, Oxford (1960)
2. Callen, H.B.: Thermodynamics and an Introduction to Thermostatistics, 2nd edn. Wiley, New York (1985)
3. Landauer, R.: IBM J. Res. Develop. 5(3), 183–191 (1961)
4. Bochkov, G.N., Kuzovlev, Y.E.: Sov. Phys. JETP 45, 125–130 (1977)
5. Jarzynski, C.: PRL 78, 14 (1997)
6. Crooks, G.E.: J. Stat. Phys. 90, 5/6 (1998)
7. Evans, D.J., Searles, D.J.: Adv. Phys. 51(7), 1529–1585 (2002)
8. Campisi, M., Hänggi, P., Talkner, P.: Rev. Mod. Phys. 83, 771–791 (2011)
9. Hill, T.L.: Thermodynamics of Small Systems. W. A. Benjamin, Inc. Publishing, New York (1964)
10. Hilbert, S., Hänggi, P., Dunkel, J.: Phys. Rev. E 90, 062116 (2014)
11. Kreuzer, H.J.: Nonequilibrium Thermodynamics and its Statistical Foundations. Oxford Science Publications, Clarendon Press, Oxford University Press, Oxford (1983)
12. Grad, H.: Commun. Pure Appl. Math. 14, 234–240 (1961)
13. Kondraputi, D., Prigogine, I.: Modern Thermodynamics. Wiley, New York (1998)
14. Jaynes, E.T.: Ann. Rev. Phys. Chem. 31, 579 (1980)
15. Gage, D.H., Schiffer, M., Kline, S.J., Reynolds, W.C.: The non-existence of a general thermokinetic variational principle. In: Donnelly, R.J. (ed.) Non-Equilibrium Thermodynamics: Variational Techniques and Stability. University of Chicago Press, Chicago (1966)
16. Martyushev, L.M., Seleznev, V.D.: Phys. Rep. 426, 1–45 (2006)
17. Velasco, R.M., Scherer Garcia-Colin, L., Uribe, F.J.: Entropy 13, 82–116 (2011)
18. Bruers, S.: Classification and discussion of macroscopic entropy production principles. arXiv:cond-mat/0604482v2 [cond-mat.stat-mech]
19. Haynes, K.E., Phillips, F.Y., Mohrfeld, J.W.: Socio-Econ. Plan. Sci. 14, 137–145 (1980)
20. Planck, M.: Treatise on Thermodynamics, p. 100, translated by A. Ogg, Longmans Green, London (1903). https://archive.org/details/treatiseonthermo00planrich/page/100/mode/2up?view=theater
21. Guggenheim, E.A.: Thermodynamics. An Advanced Treatment for Chemists and Physicists, 7th edn., p. 8. North-Holland Publishing Company, Amsterdam (1985)

Thermodynamic Equilibrium

2

Abstract

We review some fundamental notions of equilibrium thermodynamics, namely the First and the Second Principle of Thermodynamics, as well as the thermodynamic potentials. As an example, the impact of a magnetic field is discussed. The notion of thermodynamic force is introduced.

2.1 Some Fundamental Concepts

Let us now consider a macroscopic body or system of bodies, and assume that the system is 'isolated'.[1] A part of the system, which is very small compared with the whole system but still macroscopic, may be imagined to be separated from the rest; clearly, when the number of particles in the whole system is sufficiently large, the number in a small part of it may still be very large. Such relatively small but still macroscopic parts will be called 'subsystems'. Generally speaking, the measurement of macroscopic quantities will be affected by fluctuations around some average values; if an isolated macroscopic system is in a state such that in any macroscopic subsystem the macroscopic physical quantities are to a high degree of accuracy equal to their mean values, the system is said to be in a state of 'thermodynamic equilibrium'.[2]

If an isolated macroscopic system is observed for a sufficiently long period of time, it will be in a state of statistical equilibrium for the greater part of this period. If, at any initial instant, this system was not in a state of thermodynamic equilib-

---

[1] Here, we follow the line of reasoning of Sects. 1 and 8 of Ref. [1]. In the following, physical systems are classified according to their interaction with the external world. Usually, the words 'isolated', 'closed' and 'open' refer to systems which exchange no energy and no matter with the external world, which exchange energy but no matter, and which exchange both energy and matter, respectively. A warning: the word 'closed system' in Ref. [1] means *that does not interact with any other bodies*. Here we adopt the usual definitions. Of course, all these labels refer to highly idealized situations. Strictly speaking, indeed, *no physical system is, or ever can be, truly isolated* [2].

[2] And the subsystems are said to be 'at equilibrium with each other'—see note in Sect. 2.1.

© The Author(s), under exclusive license to Springer Nature Switzerland AG 2022   7
A. Di Vita, *Non-equilibrium Thermodynamics*, Lecture Notes in Physics 1007,
https://doi.org/10.1007/978-3-031-12221-7_2

rium (if, for example, it was artificially disturbed from such a state by means of an external interaction and then left to itself, becoming again an isolated system), it will eventually enter a state of thermodynamic equilibrium.

The First Principle and the Second Principle are two fundamental tenets of thermodynamics, involving the energy and the entropy of the system. The First Principle[3] is basically a version of the law of conservation of energy which makes use of macroscopic quantities [4]. The Second Principle of thermodynamics states that if at some instant the entropy of an isolated *system does not have its maximum value, then at subsequent instants the entropy will not decrease; it will increase or at least remain constant* [1].[4] We are mainly going to deal with the consequences of the Second Principle in the following.

2.2   A Minimum Amount of Work

We consider a 'body' in its 'environment'.[5] Body and environment interact with each other. E, V and S are the energy, the volume and the entropy of the body, respectively. V_0 and S_0 are the volume and the entropy of the environment, respectively. Everything is at the same pressure p_0 and at the same temperature T_0—see Fig. 2.1.

Let some external source make a work R upon the body. As a result, the generic physical quantity a undergoes a perturbation Δa. Being the body and the environment coupled with each other, the external source unavoidably perturbs both of them. In particular, the balance of energy reads:

$$\Delta E + \Delta E_0 = R$$

Conservation of total volume reads $V + V_0 = $ const., hence

$$\Delta V + \Delta V_0 = 0$$

Let us apply the First Principle of thermodynamics to the environment[6]:

$$\Delta E_0 + p_0 \Delta V_0 = T_0 \Delta S_0$$

---

[3] *When several systems interact in any way with one another, the whole set of systems being isolated from the rest of the Universe, the sum of the energy of the several systems remains constant* [3].

[4] In other words, *the equilibrium value of any unconstrained internal parameter is such as to maximize the entropy for the given value of the total internal energy* [2]; correspondingly, *every physical or chemical process in nature takes place in such a way as to increase the sum of the entropies of all the bodies taking any part in the process. [...] for reversible processes, the sum of the entropies remains unchanged* [5].

[5] Here, we refer to Sect. 20 of [1].

[6] By writing the First Principle this way, we implicitly assume that only one chemical species is present in the environment. This assumption does not affect the generality of our results below.

Fig. 2.1 A body and its environment

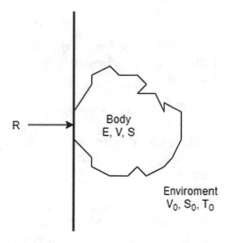

R → Body E, V, S

Enviroment V_0, S_0, T_0

According to the Second Principle of thermodynamics,[7] total entropy cannot decrease:

$$\Delta S + \Delta S_0 \geq 0$$

The relationships above lead to

$$R = \Delta E + T_0 \Delta S_0 - p_0 \Delta V_0 = \Delta E + T_0 \Delta S_0 + p_0 \Delta V \geq R_{min} \equiv \Delta E - T_0 \Delta S + p_0 \Delta V$$

Here, R_{min} is the minimum amount of work required to induce a change of ΔE and ΔS in the energy and the entropy of a body at fixed temperature T_0 and at the same pressure p_0.

We are going to invoke these results again and again in the following.

2.3 Thermodynamic Potentials

We start with the simple case $R = 0$, i.e. no external sources but the environment.[8] Then the inequality $R \geq R_{min} \equiv \Delta E - T_0 \Delta S + p_0 \Delta V$ reduces to

---

[7] Strictly speaking, the Second Principle holds in an isolated system. Admittedly, however, here we write that the total entropy *of the system body+environment* does not decrease. In order to justify this statement in a rigorous way, we may, for example, follow a three-step procedure: (i) suppose that the broader system made of the body, the environment and the external system which acts upon the body with a work R is actually an isolated system, in agreement with the words of Ref. [1], Sect. 20 (*the medium, together with the body in it and the object on which work is done, forms* an isolated system where the words *medium, body* and *object* stand for our *environment, source* and *body*, respectively); (ii) apply the Second Principle to this broader system, and (iii) assume that the contribution of the external system to the entropy balance is negligible. This procedure leaves our results unaffected.

[8] See Sect. 20 of [1].

$$0 \geq \Delta\left(E - T_0 S + p_0 V\right)$$

The symbol $=$ replaces the symbol $\geq$ if and only if $\Delta S + \Delta S_0 = 0$; this occurs at thermodynamic equilibrium between the body and the environment (i.e. $S + S_0 =$ max). Then

$$E - T_0 S + p_0 V = \text{max} \quad ; \quad \text{at thermodynamic equilibrium}$$

We define Gibbs' free energy $G \equiv E - TS + pV$. If the evolution towards thermo dynamic equilibrium occurs at constant T and p then thermodynamic equilibrium satisfies

$$G = \text{min}$$

What if such evolution occurs at constant T and V? If $V = $ const. then $\Delta V = 0$ and the inequality $R \geq R_{\text{min}}$ reduces to $R \geq \Delta E - T_0 \Delta S$. Since $R = 0$, we have $0 \geq \Delta E - T_0 \Delta S$. Since $T = $ const., it is equal to its value at thermodynamic equilib-rium, i.e. $T = T_0$ at all times, and we obtain $0 \geq \Delta(E - TS)$. Again, $=$ replaces $\geq$ at thermodynamic equilibrium only, hence $E - TS = $ min at thermodynamic equi-librium. Introducing the Helmholtz free energy $F \equiv E - TS$, we conclude that if the evolution towards thermodynamic equilibrium occurs at constant T and V then thermodynamic equilibrium satisfies

$$F = \text{min}$$

The quantities F and G are usually referred to as 'thermodynamic potentials'. Sim-ilar arguments allow us to retrieve the Second Principle itself: if the evolution of the system towards thermodynamic equilibrium occurs at constant E and V, then thermodynamic equilibrium satisfies

$$-S = \text{min} \quad i.e. \quad S = \text{max}$$

In all cases, in order to answer the question *What is to be minimized at thermody-namic equilibrium?* we have to know what is kept fixed during the evolution towards thermodynamic equilibrium.

2.4   The Impact of Magnetic Field

What if a magnetic field is present?[9] The relevant equation is

$$\mathbf{B} = \mu_0 \mathbf{H} + \mathbf{M}$$

---

[9] We refer to Sect. 31 of [6].

where the magnetic induction $\mathbf{B}$ is due to all sources, μ_0 is the vacuum permittivity, the magnetic field intensity $\mathbf{H}$ is the contribution of free electric currents and $\mathbf{M}$ is the magnetization.

It is possible to show that if the evolution towards thermodynamic equilibrium occurs at $T = $ const., $p = $ const. and $\mathbf{B}$ const., then thermodynamic equilibrium corresponds to a minimum of the thermodynamic potential $\overline{G}$, where $d\overline{G} = dG + \int_{body} d\mathbf{x}\,(\mathbf{H} \cdot d\mathbf{B})$. Since $T = $ const. and $p = $ const., $G = $ min and

$$0 \leq \int_{body} d\mathbf{x}\,(\mathbf{H} \cdot d\mathbf{B}) = dW_{mag} - \frac{1}{\mu_0} \int_{body} d\mathbf{x}\,(\mathbf{M} \cdot d\mathbf{B}) =$$

$$= dW_{mag} - \frac{1}{\mu_0} \int_{body} d\mathbf{x}d\,(\mathbf{M} \cdot \mathbf{B}) + \frac{1}{\mu_0} \int_{body} d\mathbf{x}\,(\mathbf{B} \cdot d\mathbf{M})$$

where $W_{mag} \equiv \frac{1}{2\mu_0} \int_{body} d\mathbf{x}|\mathbf{B}|^2$ is the magnetic energy. The assumption $\mathbf{B} = $ const. leads to $dW_{mag} = 0$, and we are left with

$$\int_{body} d\mathbf{x}d\,(\mathbf{M} \cdot \mathbf{B}) \leq \int_{body} d\mathbf{x}\,(\mathbf{B} \cdot d\mathbf{M}) \quad ; \quad p \quad , \quad T \quad , \quad \mathbf{B} = \text{const.}$$

Let us apply this inequality to a problem where magnetization $\mathbf{M}$ is constant. Since $\mathbf{B} = $ const., we obtain $\int_{body} d\mathbf{x}d\,(\mathbf{M} \cdot \mathbf{B}) \leq 0$, hence $\int_{body} d\mathbf{x}\,(\mathbf{M} \cdot \mathbf{B}) = $ max at thermodynamic equilibrium, i.e. magnetization aligns itself parallel to the total magnetic field. This is, for example, the case of the compass needle in the Earth's magnetic field.

It is possible to show that if the evolution towards thermodynamic equilibrium occurs at $T = $ const., $p = $ const. and $\mathbf{H}$ const. (i.e. free electric currents are constant), then thermodynamic equilibrium corresponds to a minimum of the thermodynamic potential $\hat{G}$, where $d\hat{G} = dG - \int_{body} d\mathbf{x}\,(\mathbf{B} \cdot d\mathbf{H})$. Since $T = $ const. and $p = $ const. $G = $ min and

$$0 \leq - \int_{body} d\mathbf{x}\,(\mathbf{B} \cdot d\mathbf{H}) = -\mu_0 dH_{mag} - \int_{body} d\mathbf{x}\,(\mathbf{M} \cdot d\mathbf{H}) =$$

$$= -\mu_0 dH_{mag} - \int_{body} d\mathbf{x}d\,(\mathbf{M} \cdot \mathbf{H}) + \int_{body} d\mathbf{x}\,(\mathbf{H} \cdot d\mathbf{M})$$

where $H_{mag} \equiv \frac{1}{2} \int_{body} d\mathbf{x}|\mathbf{H}|^2$. The assumption $\mathbf{H} = $ const. leads to $dH_{mag} = 0$, and we are left with

$$\int_{body} d\mathbf{x}d\,(\mathbf{M} \cdot \mathbf{H}) \leq \int_{body} d\mathbf{x}\,(\mathbf{H} \cdot d\mathbf{M}) \quad ; \quad p \quad , \quad T \quad , \quad \mathbf{H} = \text{const.}$$

Let us apply this inequality to a problem where magnetization $\mathbf{M}$ is constant. Since $\mathbf{H} = $ const., we obtain $\int_{body} d\mathbf{x}d\,(\mathbf{M} \cdot \mathbf{H}) \leq 0$, hence $\int_{body} d\mathbf{x}\,(\mathbf{M} \cdot \mathbf{H}) = $ max at thermodynamic equilibrium, i.e. magnetization aligns itself parallel to the magnetic field created by free currents. This is, for example, the case of a magnet in a solenoid at constant electric current.

2.5   A Symmetry

What if entropy S depends on many variables[10]—say, $x_1, \ldots x_N$ with N integer ≥ 1, so that $S = S(x_1, \ldots x_N)$?

At thermodynamic equilibrium $S = \max \equiv S_{eq}$. With no loss of generality, let us define $x_1, \ldots x_N$ such that $(x_1, \ldots x_N)_{eq} = 0$. Near thermodynamic equilibrium $S \leq S_{eq}$. Since S_{eq} is a maximum, first-order contributions to $S - S_{eq}$ vanish, and we write therefore[11]

$$S - S_{eq} = -\frac{1}{2}\beta_{ik}x_i x_k \leq 0 \quad ; \quad \beta_{ik} = -\left(\frac{\partial^2 S}{\partial x_i \partial x_k}\right)_{eq} \quad ; \quad i, k = 1, \ldots N$$

Its definition implies that

$$\beta_{ik} = \beta_{ki}$$

This symmetry implies $S - S_{eq} \leq 0$. Since S is real, the β_{ik}'s are also real.

Let us define the i-th 'thermodynamic force' X_i as

$$X_i \equiv -\frac{\partial S}{\partial x_i} = \beta_{ik}x_k$$

It follows that $dS = -X_k dx_k = -\beta_{ki}x_i dx_k = -x_i d(\beta_{ki}x_k) = -x_i d(\beta_{ik}x_k) = -x_i dX_i$, i.e.

$$x_i \equiv -\frac{\partial S}{\partial X_i}$$

Then, the permutation $x_i \leftrightarrow X_i$ leaves entropy unaffected. In physicists' jargon, we say that entropy enjoys a *symmetry*. Physically, symmetry means that the selection of a given set of thermodynamic forces is to be given no particular meaning. This selection is rather a matter of mathematical convenience.

References

1. Landau, L.D., Lifshitz, E.: *Statistical Physics* Pergamon, Oxford (1960)
2. Callen, H.B.: Thermodynamics and an Introduction to Thermostatistics, 2nd edn. Wiley, New York (1985)
3. Guggenheim, E.A.: Thermodynamics. An Advanced Treatment for Chemists and Physicists, 7th end., p. 8. North-Holland Publishing Company, Amsterdam (1985)
4. Zemansky, M.W.: Heat and Thermodynamics. McGraw Hill, New York (1968)
5. Planck, M.: Treatise on Thermodynamics, p. 100, translated by A. Ogg, Longmans Green, London (1903). https://archive.org/details/treatiseonthermo00planrich/page/100/mode/2up?view=theater
6. Landau, L.D., Lifshitz, E.: Electrodynamics of Continuous Media. Pergamon, Oxford (1960)

---

[10] We refer to Sects. 111 and 120 of [1].

[11] In order to simplify notation, sum on repeated indices is assumed here and in the following. Here, for example, $\beta_{ik}x_i x_k \equiv \Sigma_{i=1}^{i=N}\Sigma_{k=1}^{k=N}\beta_{ik}x_i x_k$.

Local Thermodynamic Equilibrium

3

Abstract

We discuss the notions of local thermodynamic equilibrium and Le Châtelier's principle, a corollary of the Second Principle of thermodynamics, as well as some of their consequences. The latter include the positiveness of both specific heat at constant volume and of adiabatic compressibility. As a counterexample, we discuss gravitational collapse from the point of view of thermodynamics. If local thermodynamic equilibrium always holds everywhere, then the general evolution criterion follows. The distinction between continuous and discontinuous systems is introduced, as well as the notion of thermodynamic flux.

3.1   Le Châtelier's Principle

What if the body is not at equilibrium with the environment[1] ? In order to obtain an answer, firstly we derive an inequality ('Le Châtelier's principle') from the Second Principle of thermodynamics in a rather abstract way. Then, we introduce the concept of local thermodynamic equilibrium.

We recall that $V_{TOT} = V + V_0 = $ const. Let $E_{TOT} = E + E_0$ and $S_{TOT} = S + S_0$. If the body and the environment are at thermodynamic equilibrium at temperature T_0, then S_{TOT} achieves a maximum[2] value $S_{TOT} = S_{TOT}(E_{TOT})$ and $\frac{1}{T_0} = \left(\frac{\partial S}{\partial E}\right)_{V=V_{TOT}}$ (Fig. 3.1).

---

[1] Here, we refer to Sects. 13, 16, 21, 22 and 103 of [1].

[2] Here, we assume that this is a global maximum, not a local one. Local maxima of S_{TOT} correspond to *states such that the entropy decreases for an infinitesimal deviation from the state and the body then returns to the initial state, whereas for a finite deviation the entropy may be greater than in the original state. After such a finite deviation the body does not return to its original state, but will tend to pass to some other equilibrium state corresponding to a maximum entropy greater than that in the original state. Accordingly, we must distinguish between 'metastable' and 'stable' equilibrium states. A body in a metastable state may not return to it after a sufficient deviation. Although a metastable state is stable within certain limits, the body will always leave it sooner or later for another state which is stable, corresponding to the greatest of the possible maxima of entropy. A*

© The Author(s), under exclusive license to Springer Nature Switzerland AG 2022
A. Di Vita, *Non-equilibrium Thermodynamics*, Lecture Notes in Physics 1007,
https://doi.org/10.1007/978-3-031-12221-7_3

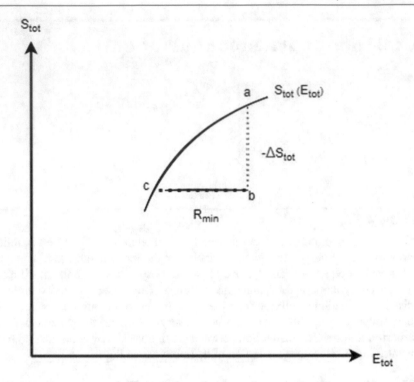

Fig. 3.1 S_{TOT} versus E_{TOT}

At fixed E_{TOT}, let the body be not at equilibrium with the environment. Instead of being located at point a on the $S_{TOT} = S_{TOT}(E_{TOT})$ curve, the point is (say) at point b. How large is the amount of work an external source should do in order to bring the body to the point b? The answer depends on the particular transformation of interest. However, the minimum amount R_{min} of work is obtained at $S_{TOT} =$ const. Accordingly, the minimum amount of work required to bring the body away from thermodynamic equilibrium with the environment up to the point b is cb.

If, furthermore, the body is very small in comparison with the environment then we neglect the impact of the external source on E_{TOT} altogether, i.e. we write $E_{TOT} =$ const. Then, the change ΔS_{TOT} in S_{TOT} which is related to the transition from thermodynamic equilibrium to the point b is equal to $\Delta S_{TOT} = - ab$. The Second Principle of thermodynamics dictates that S_{TOT} decreases as the system goes away from thermodynamic equilibrium. Then, trivially $-\Delta S_{TOT} = \frac{dS_{TOT}}{dE_{TOT}} R_{min} = \frac{R_{min}}{T_0}$, i.e.

$$\Delta S_{TOT} = -\frac{R_{min}}{T_0}$$

---

body which is displaced from this state will always eventually return to it [1]. A problem with local maxima of S_{TOT} is discussed in Sect. 3.4. See also the note in Sect. 3.2.

Indeed, if the body is very small then $E_{TOT} \approx$ const. even if $R \neq 0$; together with $V_{TOT} =$ const., this relationship and the definition $R_{\min} \equiv \Delta E - T_0 \Delta S + p_0 \Delta V$ lead just to $\Delta S_{TOT} = -\frac{R_{\min}}{T_0}$.

Entropy may depend on many variables. For the moment, let us focus on the case where entropy depends just on 2 variables: $S_{TOT} = S_{TOT}(x, y)$. At thermodynamic equilibrium we write $S_{TOT} =$ max, i.e. $-S_{TOT} =$ min. This fact leads to[3]

$$\frac{\partial(-S)}{\partial x} = 0 \quad ; \quad \frac{\partial(-S)}{\partial y} = 0 \quad ; \quad \frac{\partial^2(-S)}{\partial x^2} \geq 0 \quad ; \quad \frac{\partial^2(-S)}{\partial x^2} \geq 0$$

$$\frac{\partial^2(-S)}{\partial x^2} \frac{\partial^2(-S)}{\partial y^2} - \frac{\partial^2(-S)}{\partial x \partial y} \frac{\partial^2(-S)}{\partial y \partial x} \geq 0$$

Thermodynamic forces are $X \equiv -\frac{\partial S}{\partial x}$ and $Y \equiv -\frac{\partial S}{\partial y}$. We rewrite our results as[4]

$$X = 0 \quad ; \quad Y = 0 \quad ; \quad \left(\frac{\partial X}{\partial x}\right)_y \geq 0 \quad ; \quad \left(\frac{\partial Y}{\partial y}\right)_x \geq 0$$

$$\left(\frac{\partial X}{\partial x}\right)_y \left(\frac{\partial Y}{\partial y}\right)_x - \left[\left(\frac{\partial X}{\partial y}\right)_x\right]^2 \geq 0$$

together with the identity $\left(\frac{\partial Y}{\partial x}\right)_y = -\frac{\partial^2 S}{\partial x \partial y} = \left(\frac{\partial X}{\partial y}\right)_x$. Brute force[5] gives therefore

$$\left(\frac{\partial X}{\partial x}\right)_Y = \frac{\partial(X, Y)}{\partial(x, Y)} = \frac{\partial(X, Y)}{\partial(x, y)} \frac{1}{\frac{\partial(x, Y)}{\partial(x, y)}} = \frac{\left(\frac{\partial X}{\partial x}\right)_y \left(\frac{\partial Y}{\partial y}\right)_x - \left(\frac{\partial X}{\partial y}\right)_x \left(\frac{\partial Y}{\partial x}\right)_y}{\left(\frac{\partial Y}{\partial y}\right)_x} =$$

$$= \left(\frac{\partial X}{\partial x}\right)_y - \frac{\left(\frac{\partial X}{\partial y}\right)_x \left(\frac{\partial Y}{\partial x}\right)_y}{\left(\frac{\partial Y}{\partial y}\right)_x} = \left(\frac{\partial X}{\partial x}\right)_y - \frac{\left[\left(\frac{\partial X}{\partial y}\right)_x\right]^2}{\left(\frac{\partial Y}{\partial y}\right)_x}$$

---

[3] The actual definitions of x and y are not relevant here. We drop the subscript $_{TOT}$ in the following for simplicity.

[4] Here, we invoke the identity $\frac{\partial^2}{\partial x \partial y} = \frac{\partial^2}{\partial y \partial x}$

[5] The quantity ('Jacobian') $\frac{\partial(u,v)}{\partial(a,b)} \equiv \frac{\partial u}{\partial a} \frac{\partial v}{\partial b} - \frac{\partial u}{\partial b} \frac{\partial v}{\partial a}$ satisfies the identities $\frac{\partial(u,v)}{\partial(a,v)} = \left(\frac{\partial u}{\partial a}\right)_v$ and $\frac{\partial(u,v)}{\partial(a,b)} = \frac{\partial(u,v)}{\partial(c,d)} \frac{\partial(c,d)}{\partial(a,b)}$ for arbitrary quantities u, v, a, b, c and d [1].

This relationship, together with the inequalities $\left(\frac{\partial X}{\partial x}\right)_y \geq 0$ and $\left(\frac{\partial Y}{\partial y}\right)_x \geq 0$ above, gives

$$\left(\frac{\partial X}{\partial x}\right)_y \geq \left(\frac{\partial X}{\partial x}\right)_Y \geq 0$$

Multiplication of both sides by a perturbation Δx of x leads to 'Le Châtelier's principle':

$$|(\Delta X)_y| \geq |(\Delta X)_Y| \geq 0$$

where $(\Delta X)_y \equiv \Delta x \cdot \left(\frac{\partial X}{\partial x}\right)_y$, $(\Delta X)_Y \equiv \Delta x \cdot \left(\frac{\partial X}{\partial x}\right)_Y$ and the absolute value puts in evidence the fact that Le Châtelier's principle holds regardless of the sign of Δx.

3.2   Local Thermodynamic Equilibrium and Le Châtelier's Principle

Le Châtelier's principle follows from the Second Principle of thermodynamics. Thermodynamic equilibrium corresponds to maximum entropy, i.e. $X = 0$, $Y = 0$. So far, the definitions of x and y (hence of X and Y) are arbitrary. We are therefore free to choose y in such a way that $Y = 0$ describes local thermodynamic equilibrium ('LTE'), i.e. a configuration where the body is in thermodynamic equilibrium with itself but not with the environment. Physically, at LTE all thermodynamic quantities $(E, S, T \ldots)$ are defined within a small mass element just like in thermodynamic equilibrium. Moreover, the relationships among them are the same relationships which would hold, should the whole Universe be at thermodynamic equilibrium with the same values of $E, S, T \ldots$etc. In contrast with the familiar thermodynamic equilibrium, LTE does not require $X = 0$. The word 'local' is justified because we are free to identify with X and Y the thermodynamic force which takes into account the interactions of the different parts of the body with the environment and with each other, respectively; thus, $Y = 0$ is the condition of thermodynamic equilibrium only among the different parts of the body (a 'local' equilibrium, precisely), while $X = Y = 0$ denotes full thermodynamic equilibrium.

LTE is an assumption which is often invoked—either explicitly [2] or implicitly [3]. LTE means that—although the total system is not at equilibrium—the internal energy per unit mass is the same function of the entropy per unit mass, the pressure, the mass density, etc. as in real equilibrium, where all these quantities are defined locally; more generally, the relationships among thermodynamic quantities will be the same as in real equilibrium.[6] In the present discussion, LTE is just a particular state at constant Y, namely $Y = 0$.

---

[6] It is often impossible to write down rigorous criteria a priori for the validity of LTE in a given problem, and the issue is settled by comparison of LTE-based predictions with observations. This is far from surprising, as similar considerations hold for true thermodynamic equilibrium too—see Sect. 1.5 of Ref. [4].

When the body is at LTE, Le Châtelier's principle means just that the absolute value of the perturbation of a thermodynamic force X is minimum. Now, the symmetry $x_i \leftrightarrow X_i$ (Sect. 2.5) allows us to replace the words 'a thermodynamic force' in the last sentence with the words 'a generic quantity'. Minimization of the amplitude of perturbations at LTE implies that a body at LTE counteracts external perturbations: any external operation which pushes the body away from equilibrium triggers backlash, or, in Fermi's words [5]:

If the external conditions of a thermodynamic system are altered, the equilibrium of the system will tend to move in such a direction as to oppose the change in the external conditions.

As a shortcut, with a slight abuse of notation but in agreement with most literature, we are going to refer to this application of Le Châtelier's principle to systems at LTE as 'Le Châtelier's principle' in the following. As an application of Le Châtelier's principle taken from everyday life, imagine a person smoking a cigarette. In a lit cigarette, combustion starts from solid tobacco and produces hot gases, which include the air surrounding the cigarette and the gaseous products of combustion. This production of hot gases tends to raise the pressure, in agreement with the Gay-Lussac law. Admittedly, a lit cigarette is not in an equilibrium state because hot gases continuously escape away from the cigarette, as the smoke shows. However, the process is so slow that we may safely assume that the cigarette is at LTE,[7] even if definitely not at thermodynamic equilibrium with the surrounding environment. By crushing the cigarette into the ashtray, the smoker definitely raises the pressure of the cigarette. The response of the latter is to stop the physical process which raises pressure, i.e. to extinguish combustion: in other words, the smoker puts the cigarette out.

Another example is the synthesis of ammonia, which involves the chemical reaction $N_2 + 3H_2 \leftrightarrow 2NH_3$. Both nitrogen, hydrogen and ammonia are present in gaseous form and can be described as perfect gases. The double arrow means that the reaction acts both ways. LTE is possible when the total number of reactions from left to right (i.e. of synthesis of ammonia) per unit time is equal to the total number of reactions from right to left per unit time. Many different LTE states are allowed for the system made of the molecules of the three gases, as the actual relative abundances of nitrogen, hydrogen and ammonia depend on temperature, pressure, etc. of a particular state. From left to right, the reaction transforms 4 moles of gas (1 mole of nitrogen + 3 of hydrogen) into 2 moles of ammonia, and reduces therefore the total number of moles from 4 to 2. All the rest being unchanged, the equation of state of perfect gases dictates that the lower the total number of moles, the lower the pressure. In other words, the synthesis of ammonia lowers the pressure. Correspondingly, Le Châtelier's principle implies that if the external world raises the pressure then the system reacts by shifting the LTE in the direction of lower pressure, i.e. towards higher abundance of ammonia. In other words, pressurization facilitates the synthesis of ammonia. This conclusion was a breakthrough in the chemical industry and earned Fritz Haber a Nobel Prize in 1918.

---

[7] Or, more precisely, that the evolution of the cigarette is a succession of LTE states.

We obtain many corollaries of Le Châtelier's principle from $\Delta S_{TOT} = -\frac{R_{\min}}{T_0}$, which gives

$$X = -\frac{\partial S}{\partial x} = \frac{1}{T_0}\frac{\partial R_{\min}}{\partial x} \quad ; \quad Y = -\frac{\partial S}{\partial y} = \frac{1}{T_0}\frac{\partial R_{\min}}{\partial y}$$

Let us compute these derivatives. The definition $R_{\min} \equiv \Delta E - T_0\Delta S + p_0\Delta V$ and the First Principle of thermodynamics lead to $dR_{\min} = dE - T_0 dS + p_0 dV$ and to $dE = TdS - pdV$, respectively. Accordingly,

$$dR_{\min} = (T - T_0)\,dS + (p_0 - p)\,dV$$

Let us select $x = S$. Then $X = \frac{T-T_0}{T_0}$ and the condition $X = 0$ of equilibrium between the body and the environment corresponds to $T = T_0$, i.e. body and environment are at the same temperature. As for Le Châtelier's principle, it corresponds to

$$\left(\frac{\partial T}{\partial S}\right)_y \geq \left(\frac{\partial T}{\partial S}\right)_{LTE} \geq 0 \quad \text{hence} \quad |(\Delta T)_y| \geq |(\Delta T)_{LTE}| \geq 0$$

Attempts to heat (cool) a body at LTE trigger processes which aim at cooling (heating) it.

Let us select $x = V$. Then $X = \frac{p_0-p}{T_0}$ and the condition $X = 0$ of equilibrium between the body and the environment correspond to $p = p_0$, i.e. body and environment are at the same pressure. As for Le Châtelier's principle, it corresponds to

$$\left(\frac{\partial p}{\partial V}\right)_y \leq \left(\frac{\partial p}{\partial V}\right)_{LTE} \leq 0 \quad \text{hence} \quad |(\Delta p)_y| \geq |(\Delta p)_{LTE}| \geq 0$$

Attempts to reduce (increase) the volume of a body at LTE trigger processes which aim at raising (lowering) its pressure.

Let us select $y = V$. Then $\left(\frac{\partial T}{\partial S}\right)_y = \left(\frac{\partial T}{\partial S}\right)_v = \frac{T}{C_v}$ where we have defined the specific heat at constant volume $C_v \equiv T\left(\frac{\partial S}{\partial T}\right)_v$. Moreover, $Y = \frac{1}{T_0}\frac{\partial R_{\min}}{\partial V} = p_0 - p = 0$ for $p = p_0 = $ const., hence $\left(\frac{\partial T}{\partial S}\right)_{LTE} = \left(\frac{\partial T}{\partial S}\right)_{Y=0} = \left(\frac{\partial T}{\partial S}\right)_{p=const.} = \frac{T}{C_p}$ where we have defined the specific heat at constant pressure $C_p \equiv T\left(\frac{\partial S}{\partial T}\right)_p$. Then, the inequality $\left(\frac{\partial T}{\partial S}\right)_y > \left(\frac{\partial T}{\partial S}\right)_{LTE} \geq 0$ reduces to[8]

$$C_p > C_v > 0$$

---

[8] Here is why we postulated in Sect. 3.1 that S has only a global maximum, not local maxima. The First Principle of thermodynamics in the form $dE = TdS - pdV$ gives $C_v = T(\frac{\partial S}{\partial T})_v = (\frac{\partial E}{\partial T})_v$ and $(\frac{dE}{dS})_v = T$, hence $(\frac{dS}{dE})_v = \frac{1}{T}$ and $(\frac{\partial^2 S}{\partial E^2})_v = -\frac{1}{T^2}(\frac{dT}{dE})_v = -\frac{\left[(\frac{dS}{dE})_v\right]^2}{C_v}$, i.e. $C_v = -\frac{\left[(\frac{dS}{dE})_v\right]^2}{(\frac{\partial^2 S}{\partial E^2})_v}$ [6].

If $S = S(E)$ has local maxima, $(\frac{\partial^2 S}{\partial E^2})_v > 0$ and $C_v < 0$ in a neighbourhood of the local minimum of S which is located between two adjacent local maxima of $S(E)$, which in turn correspond to metastable states (see note in Sect. 3.1).

The specific heat at constant pressure is never smaller than the specific heat at constant volume, which in turn is always positive.

Let us select $y = S$. Then $-\left(\frac{\partial p}{\partial V}\right)_y = -\left(\frac{\partial p}{\partial V}\right)_S$ ('adiabatic compressibility').

Moreover, $Y = \frac{1}{T_0}\frac{\partial R_{min}}{\partial S} = T - T_0 = 0$ for $T = T_0 = $ const., hence $-\left(\frac{\partial p}{\partial V}\right)_{LTE} = -\left(\frac{\partial p}{\partial V}\right)_{Y=0} = -\left(\frac{\partial p}{\partial V}\right)_{T=\text{const.}}$ ('isothermal compressibility'). Then, the inequality $\left(\frac{\partial p}{\partial V}\right)_y < \left(\frac{\partial p}{\partial V}\right)_{LTE} \leq 0$ reduces to

$$-\left(\frac{\partial p}{\partial V}\right)_S > -\left(\frac{\partial p}{\partial V}\right)_T \geq 0$$

The adiabatic compressibility is never smaller than the isothermal compressibility, which in turn is always positive.

3.3 Some Consequences of Local Thermodynamic Equilibrium

Some consequences of LTE are listed below.[9]

The First Principle of thermodynamics implies that $C_v = \left(\frac{\partial E}{\partial T}\right)_v$. Together with the definition of G, it implies also that $dG = -SdT + Vdp$. It follows that[10]

$$\left(\frac{\partial G}{\partial p}\right)_T = V \quad ; \quad \left(\frac{\partial G}{\partial T}\right)_p = -S \quad ; \quad \left(\frac{\partial V}{\partial p}\right)_T = -\left(\frac{\partial S}{\partial p}\right)_T$$

The enthalpy is $H = G + TS$. Then

$$\left(\frac{\partial H}{\partial p}\right)_T = V - T\left(\frac{\partial V}{\partial T}\right)_p \quad ; \quad \left(\frac{\partial H}{\partial T}\right)_p = C_p$$

Finally, it is possible to show that $C_p - C_v = -T\left[\left(\frac{\partial V}{\partial T}\right)_p\right]^2 \left[\left(\frac{\partial V}{\partial p}\right)_T\right]^{-1}$. It follows that

$$\left(\frac{\partial H}{\partial T}\right)_p = \left(\frac{\partial E}{\partial T}\right)_v - T\left[\left(\frac{\partial V}{\partial T}\right)_p\right]^2 \left[\left(\frac{\partial V}{\partial p}\right)_T\right]^{-1}$$

If $k = 1, \ldots M$ (chemical, nuclear...) species are present, then the relationships listed above still hold, provided that we replace $()_{T,p} \to ()_{T,p,N_1,\ldots N_M}, dE \to dE +$

---

[9] Here, we refer to Eqs. (4.13), (8.14) (11.3), (11.20) and Problem (11.4) of [7], as well as to [8] and to Chaps. 1 and 2 of [2].

[10] We take into account that $df = \left(\frac{\partial f}{\partial x}\right)_y dx + \left(\frac{\partial f}{\partial y}\right)_x dy$ and that $\frac{\partial^2 f}{\partial x \partial y} = \frac{\partial^2 f}{\partial y \partial x}$ for arbitrary differentiable $f(x, y)$.

$\mu_k dN_k, dH \rightarrow dH + \mu_k dN_k$ and $dG \rightarrow dG + \mu_k dN_k$ everywhere, where N_k and μ_k are the number of particles of the $k-$th species and the chemical potential of the $k-$th species, respectively.

Finally, it is possible to show that $G = N_k \mu_k$. This fact, together with the condition $G = \min$ whenever T and p are kept constant, leads to the inequality:

$$\left(\frac{\partial \mu_k}{\partial N_i}\right)_{T,p} dN_k dN_i \geq 0$$

At LTE we divide H, G, E, etc. by the mass $N_j m_j$ (m_j being the mass of the particle of the j-th species, $j = 1, \ldots M$. We introduce the following quantities per unit mass:

$$h \equiv \frac{H}{N_j m_j} \quad ; \quad u \equiv \frac{E}{N_j m_j} \quad ; \quad g \equiv \frac{G}{N_j m_j} \quad ; \quad s \equiv \frac{S}{N_j m_j} \quad ; \quad v \equiv \frac{V}{N_j m_j}$$

We define also the mass density $\rho \equiv v^{-1}$, the chemical potential per unit mass of the k-th species $\mu_k^0 \equiv \mu_k m_k^{-1}$ and the mass fraction c_k of the k-th species[11]:

$$c_k \equiv \frac{\text{number of particles of k-th species} \cdot \text{mass of 1 particle of k-th species}}{N_j m_j}$$

Since $G = N_k \mu_k$, we write $g = \mu_k^0 c_k$. Analogously, we define the quantities h_k, s_k and v_k such that $h = h_k c_k$, $s = s_k c_k$ and $v = v_k c_k$. We rewrite the thermodynamic relationships just discussed in the following form,[12] which will be useful in Sect. 3.6:

$$h = u + \frac{p}{\rho} \quad ; \quad h_k = \mu_k^0 + T s_k \quad ; \quad c_p \equiv \left(\frac{\partial h}{\partial T}\right)_{v, c_1, \ldots} > c_v \equiv \left(\frac{\partial u}{\partial T}\right)_{v, c_1, \ldots} > 0$$

$$\left(\frac{\partial v}{\partial p}\right)_{T, c_1, \ldots} < 0 \quad ; \quad \left(\frac{\partial \mu_k^0}{\partial c_j}\right)_{T, p, \ldots} dc_j dc_k \geq 0$$

$$\left(\frac{\partial \mu_k^0}{\partial p}\right)_{T, c_1, \ldots} = v_k \quad ; \quad \left(\frac{\partial \mu_k^0}{\partial T}\right)_{p, c_1, \ldots} = -s_k \quad ; \quad \left(\frac{\partial h}{\partial p}\right)_{T, c_1, \ldots} = v - T \left(\frac{\partial v}{\partial T}\right)_{p, c_1, \ldots}$$

$$\left(\frac{\partial h}{\partial T}\right)_{p, c_1, \ldots} = \left(\frac{\partial u}{\partial T}\right)_{v, c_1, \ldots} - \left[\left(\frac{\partial v}{\partial p}\right)_{T, c_1, \ldots}\right]^{-1} T \left[\left(\frac{\partial v}{\partial T}\right)_{p, c_1, \ldots}\right]^2$$

---

[11] Of course, $\sum_k c_k = 1$ identically at all times, hence $\sum_k \frac{dc_k}{dt} = 0$ at all times.

[12] This step is allowed because the number N of particles is $\gg 1$. If N is not $\gg 1$, then different relationships hold [9]. See note in Sect. 1.

The formal analogy between the thermodynamic relationships involving H, G... and the corresponding quantities per unit mass has a physical meaning. It reflects the fact that we may repeat our arguments in Sects. 3.1 and 3.2 concerning a body and its environment to a small part of the body and the rest of the body, respectively, since in equilibrium the temperature and the pressure of all parts of the body are the same. In turn, this requires that the body is somehow homogeneous. We are going to discuss a physically relevant class of strongly inhomogeneous bodies in Sect. 3.4.

3.4 The Role of Gravity

3.4.1 Collapse

What if gravity[13] is present?

Generally speaking, care must be taken when dealing with time-varying gravitational fields. When it comes to a collection of N point-like self-gravitating particles[14] of mass m which has fixed total energy E and which is contained within a spherical region of radius R, it has been shown that in many cases[15] *there is no global maximum to S at any fixed E so that even when maxima exist they are only local maxima* [11]. Correspondingly, the maxima of S describe metastable states,[16] and the stability conditions of Sect. 3.2 based on Le Châtelier's principle may hold for not too large perturbations only. This result is far from surprising: a body whose particles are held together by gravitational forces *will clearly be inhomogeneous, having a higher density towards the centre* [1] and the requirement of homogeneity of Sect. 3.3 is blatantly violated.

For example, compressibility[17] may be < 0 in problems with sufficiently large perturbations. Accordingly, the response of the body to a perturbation which makes it shrink does not necessarily tend to expand the body back to its original volume (by raising the pressure), but can rather facilitate further shrinking (by lowering the pressure). A self-gravitating body may therefore be unstable against 'gravitational collapse', a fundamental process of the utmost importance in astrophysics. For further analysis, let us come back to our system of N self-gravitating particle above, and let us relax the constraint of constant E. For example, a physical process (e.g. radiation towards the external space) may lower E so slowly that the evolution of the gas can be described as a succession of configurations described by the model with $E = $ const. Then, we write $E = K + U = $ const., where $K > 0$ is the total kinetic energy

---

[13] Here, we refer to Sect. 21 of [1,6,10,11].

[14] Here, 'self-gravitating' means that each particle is subject only to the Newtonian gravitational attraction of all other particles. As for the contribution to the entropy balance, e.g. of nuclear fusion in stars, see Ref. [10].

[15] i.e. whenever the particle density at the centre of the sphere is at least 709 times larger than the particle density at the edge of the sphere [11]. For a discussion, see Sect. 3 of Ref. [12].

[16] See note of Sect. 3.1.

[17] Here, we may refer to the adiabatic compressibility with no loss of generality, so that temperature is not yet involved. Its impact will be discussed below.

of all particles and $U \propto -\frac{1}{R}$ is the gravitational potential energy. Let us denote with $\overline{a}(\mathbf{x}, t) = \lim_{t_0 \to \infty} t_0^{-1} \int_t^{t+t_0} a(\mathbf{x}, t') \, dt'$ the time-average of the generic quantity $a = a(\mathbf{x}, t)$. After time-averaging, conservation of energy gives $\overline{E} = \overline{K} + \overline{U}$ with $\overline{K} > 0$ as $K > 0$ (we drop the dependence of a on its arguments for simplicity). Now, the so-called 'virial theorem' of mechanics [13] ensures that gravitational interactions satisfy the condition $2\overline{K} + \overline{U} = 0$. It follows that $\overline{E} = -\overline{K} = +\frac{\overline{U}}{2} \propto -\frac{1}{R}$. Let us apply a perturbation which lowers R. Negative compressibility allows gravitational collapse to occur, and the evolution of our system goes through metastable states (the local maxima of S). The smaller the R, the lower the $\overline{U}$ and $\overline{E}$, the lower the values attained by E over time. No stable equilibrium is ever achieved. Eventually, the gravitational collapse gives birth to a black hole, which cannot be described by our simple model.[18]

Let us drop the assumption of constant E altogether, and focus rather on the impact of temperature. Just like compressibility, it is possible[19] that $C_v < 0$ for the whole body, even if locally $c_v > 0$ for a small part of it: *the specific heat of the body as a whole may be less than zero [...] we may note that this does not contradict the result that the specific heat is positive for every small part of the body, since in these conditions the energy of the whole body is not equal to the energy of its parts; there is also the energy of the gravitational interaction between these parts* [1]. Negative values of C_v have far-reaching consequences. In agreement with physical intuition, the Second Principle of thermodynamics dictates that when two systems with different temperatures interact via a purely thermal connection, heat will flow from the hotter system to the cooler one—see Sect. 4.1.6. Therefore, if such systems have equal temperatures, they are at thermal equilibrium. However, this equilibrium is stable if both systems have positive heat capacities. For such systems, when heat flows from a higher temperature system to a lower temperature one, the temperature of the first decreases and that of the latter increases, so that both approach equilibrium. In contrast, for systems with negative heat capacities, the temperature of the hotter system will further increase as it loses heat, and that of the colder will further decrease, such that they will move farther from equilibrium. Thus, the equilibrium is unstable. Many results follow.[20]

- Two $C_v < 0$ systems in thermal contact do not attain thermal equilibrium: one gets hotter and hotter by losing energy, and the other gets forever colder by gaining energy. Thus, $C_v < 0$ systems cannot be divided into independent parts each with $C_v < 0$. Accordingly, the possibility that different parts of the same system can be at thermodynamic equilibrium with each other even if not with the external world fails, as no such equilibrium can be stable. In turn, this result puts in evidence an often overlooked fact. Unlike the Second Principle of thermodynamics, the fact

---

[18] Realistic models of gravitational collapse include the impact of non-gravitational forces, including the nuclear ones.

[19] See the note in Sect. 3.2.

[20] For a detailed discussion, see Ref. [6].

that the different parts of a body which is not in equilibrium with the environment are in equilibrium with each other is not a fundamental law of Nature; it is rather an assumption. *If* this assumption holds, then the Second Principle leads to consequences like the positiveness of C_v. It is a powerful assumption concerning stability; if it fails, then we cannot take for granted that after compressing a gas in a tank with a piston the gas will expand pushing the piston back and not just shrink further, or that when we put a pan of water on a stove and turn on the stove, the water will end boiling up and not freezing. But it is *not* universal.[21]

- A $C_v < 0$ system cannot achieve thermal equilibrium with a large 'heat bath' (a heat bath is a body with arbitrarily large $C_v > 0$).[22] Any fluctuation that, for example, makes its temporary energy too high will make its temporary temperature too low and the heat flow into it will drive it to ever lower temperatures and higher energies.

- A $C_v < 0$ system can achieve a stable equilibrium in contact with a $C_v > 0$ system provided that their combined heat capacity is negative. To see this, imagine that the $C_v = C_{vA} < 0$ system A is initially a little hotter (higher T) than the $C_v = C_{vB} > 0$ system B. Then heat will flow from A to B. On losing heat A will get hotter (i.e. its temperature increases) but on gaining that heat B will also get hotter. However, because B has a lesser $|C_v|$, its temperature is more responsive to heat gain than A is to heat loss. Thus, B will gain temperature faster than A and a thermal equilibrium will be attained with both A and B hotter than they were to start with. Thus, stable equilibrium is obtained when $C_{vA} + C_{vB} < 0$ [15]. Similar arguments apply if A is initially a little cooler.[23]

---

[21] If we attempt to apply thermodynamics to the entire Universe, regarded as a single isolated system, then [1] *we immediately encounter a glaring contradiction [...] any finite region of it, however large, should have a finite relaxation time and should be in equilibrium. Everyday experience shows us, however, that the properties of Nature bear no resemblance to those of an equilibrium system; and astronomical results show that the same is true throughout the vast region of the Universe accessible to our observation. [...] The reason is that, when large regions of the Universe are considered, the gravitational fields [...] may in a sense be regarded as 'external conditions' to which the bodies are subject.* The statement that an isolated system *must, over a sufficiently long time, reach a state of equilibrium applies of course only to a system in steady external conditions. On the other hand, the general cosmological expansion of the Universe means that [...] the 'external conditions' are by no means steady [...].* Here, it is important that the gravitational field cannot itself be included in an isolated system: the Universe as a whole must be regarded not as an isolated system but as a system in a variable gravitational field. Consequently, *the application of the law of increase of entropy does not prove that [...] equilibrium must necessarily exist.*

[22] In contrast, it has been shown [14] that a self-consistent thermodynamic description exists for a system (a) which is in thermal equilibrium with a thermal bath; (b) which is made of positive and negative electric charges; (c) where every particle interacts with all other particles via electrostatic, Coulombian interactions; (d) and where the net electric charge is zero everywhere. Every particle undergoes both attractive and repulsive interactions on an equal footing, contrary to the self-gravitating particles which undergo attractive interactions only.

[23] The lack of thermal equilibrium between a body A with negative heat capacity and a heat bath B is a particular case, as $C_{vB} \gg |C_{vA}|$ by definition of the heat bath. If the specific heat of B is

Let us apply these results to our self-gravitating particles. Following Ref. [11], *imagine a gravitating*, initially *isothermal gas confined by a sphere [...] and adiabatically expand the sphere. Work is done by the gas so E becomes yet more negative [...] while the sphere's radius expands [...] The inner parts of the gravitating gas are much denser and are held in primarily by gravity so the expansion is mainly taken up by the less dense gas in the outer parts. Thus, the adiabatic fall in temperature of the outer parts due to their expansion will initially be greater than the temperature fall of the inner parts. Thus, there will now be a temperature gradient with the outer parts cooler than the central ones.* As $C_v < 0$, and *as heat flows down the temperature gradient the central parts contract and get hotter while the outer parts held in by the sphere behave like a normal gas* (here, gravity is weaker and particles behave as the particles of a familiar gas with positive specific heat) *so they receive the heat and get hotter too. It is now a race; do the outer parts get hotter faster on gaining the heat than the inner parts do on losing it? Clearly, if the outer have too great a positive heat capacity, they will not respond enough and the inner parts will run away to ever higher temperatures losing more and more heat to the sluggishly responding outer parts.* The inner and the outer parts play the role of systems A and B above, with $C_{vA} < 0$ and $C_{vB} > 0$. Stability is lost whenever $C_v = C_{vA} + C_{vB}$ reaches zero from below; stability requires therefore that C_{vA} takes a value which is quite negative.

Now, $C_{vA} < 0$ implies that the hotter the inner part A of our self-gravitating particles, the lower its energy. But then, this decrease in its energy is precisely what occurs in the gravitational collapse. The final outcome of the gravitational collapse is a black hole; it is obtained as the radius of A shrinks below a critical value (the 'Schwarzschild radius') and is definitely quite stable (as nothing can go out from it, classically at least). Our discussion strongly suggests that the stability of a black hole requires that its specific heat is < 0. If this is true, then the more energy a black hole absorbs, the colder it becomes. But the theory of relativity (which describes black holes) links the energy and the mass of a body with the famous formula $E = mc^2$; then, it follows that the larger the mass which falls into a black hole, the lower its temperature. Now, this falling mass raises the black hole mass; accordingly, the more massive the black hole, the colder. Finally, if a black hole leaks energy slowly enough,[24] so that its evolution can be successfully described as a succession of states at constant energy, then it becomes hotter and hotter until it boils away ('black hole evaporation'). Actually, some black holes at least seem to satisfy the properties listed above: negative specific heat, temperature decreasing with mass and final evaporation [16]. Despite the fact that our discussion is definitely elementary and qualitative, its results are still informative and intriguing as they show how thermodynamics may help physical intuition even when dealing with systems so distant from everyday experience.

---

assumed to be large but finite, then a description of the interaction of A and B is available which is in agreement with Le Châtelier's principle [6].

[24] Through the so-called 'Hawking radiation', a quantum effect

The example provided by gravitational collapse shows that the requirement that entropy has a global maximum—which the arguments of Sect. 3.1 rely upon—needs independent confirmation when it comes to gravitational fields which change in time because of the motion of the particles of the system. Thus, our stability conditions based on Le Châtelier's principle in Sect. 3.2 safely apply only when the gravitational field is constant, like in the case of systems located in the Earth's field. In the following, we shall limit ourselves to constant gravitational fields.

3.4.2 Constant Gravitational Field

The relationships below hold at LTE. We denote with ϕ_g the gravitational potential, which affects chemical potentials as follows[25]

$$\mu_k \to \mu_k + m_k \phi_g \quad \text{hence} \quad \mu_k^0 \to \mu_k^0 + \phi_g$$

Since ϕ_g depends neither on p, nor on T, nor on c_k, it leaves the corresponding derivatives of h and μ_k^0 unaffected. Moreover,[26] the relationship $g = \mu_k^0 c_k$ implies that gravity makes us to replace g as follows: $g = \mu_k^0 c_k \to \mu_k^0 c_k + \phi_g \sum_k c_k = g + \phi_g$. Analogously, $h \to h + \phi_g$ and $u \to u + \phi_g$, in agreement with the theorem of small corrections. In other words, Gibbs' free energy per unit mass, enthalpy per unit mass and internal energy per unit mass transform like chemical potentials per unit mass. In particular, gravity leaves the relationship $h = u + \frac{p}{\rho}$ unaffected, because it adds the same term ϕ_g to h and u. Finally, the relationship $h \to h + \phi_g$ implies $h_k \to h_k + \phi_g$ as no species is privileged. But we have shown that $h_k = \mu_k^0 + T s_k$ and that $\mu_k^0 \to \mu_k^0 + \phi_g$, hence ϕ_g leaves s_k unaffected. In other words, gravity leaves entropy unaffected.

3.5 Continuous Versus Discontinuous Systems

What if the system is in steady state ($\frac{\partial}{\partial t} = 0$) but not at thermodynamic equilibrium?

As we have seen, it is still possible that LTE holds everywhere throughout the system. This is e.g. the case, provided that the system has relaxed to a steady state and that LTE holds in any arbitrary small mass element of the system at all times during the relaxation. Locally, we may define an entropy s per unit mass, so that

$$S = \int_{system} \rho s d\mathbf{x}$$

In this case, both s and the mass density ρ depend on position $\mathbf{x}$ and time t (on $\mathbf{x}$ only in steady state). LTE ensures that all familiar relationships of thermodynamic

---

[25] Here, we refer to Sect. 15 of [1].
[26] Since $\sum_k c_k \equiv 1$, $\sum_k dc_k = d\left(\sum_k c_k\right) \equiv 0$.

equilibrium hold locally. We will refer to such systems as 'continuous systems' below. A particular case of steady state is often met in practical problems, where the system is made of many (say, R) regions where many (say, K) thermodynamic quantities are uniform across each region (even if their values in different regions are different). In this case, entropy S depends on the $N = R \cdot K$ values of the thermodynamic quantities in the various regions, which may be labelled as a finite set $x_1, \ldots x_N$ of values. However, S does not depend on t in a steady state. We retrieve therefore the expression discussed above $S = S(x_1, \ldots x_N)$. We will refer to such systems as 'discontinuous systems' below. In this case, we have shown that $dS = -X_k dx_k$. Derivation on time leads to[27]

$$\frac{dS}{dt} = X_k J_k$$

where we have defined $J_k = -\frac{dx_k}{dt}$, which is usually referred to as the thermodynamic flux 'conjugated' to the thermodynamic force X_k. Analogously, X_k is the thermodynamic force conjugated to J_k. In this form, $\frac{dS}{dt}$ takes into account the entropy production due to all irreversible processes.

3.6   General Evolution Criterion

It is (often implicitly) assumed in many problems that LTE holds at all times within an arbitrary, small mass element, followed along its centre-of-mass motion. Formally

$$da = \frac{da}{dt} \cdot dt \quad ; \quad \frac{da}{dt} = \frac{\partial a}{\partial t} + \mathbf{v} \cdot \nabla a$$

Here and in the following, a in an arbitrary physical quantity and $\mathbf{v}$ is the macroscopic fluid velocity, i.e. the velocity of the centre of mass of the small mass element. Note that $\mathbf{v}$ solves the equation of motion as provided by dynamics, while da is the usual differential of a as provided, for example, by thermodynamic relationships. Accordingly, this assumption is a constraint on the evolution of the small mass element.

---

[27] We stress the following two points. Firstly, in steady state $S = $ constant by definition. Suitable transport processes carry away the amount of entropy produced per unit time at rate $\frac{dS}{dt}$ in order to maintain $S = $ const. throughout our system. Secondly, S is often invoked because of the Second Principle of thermodynamics, and the Second Principle does not involve time in its formulation— see note in Sect. 2.2. It is the assumption of LTE everywhere at all times which allows S to be a differentiable function of t, as S is the volume integral of ρs and both ρ and s may depend on both space and time; in particular, $s = s(T, p, \ldots)$ just like in thermodynamic equilibrium, and p, T, etc. are solutions of the balance equations of energy, momentum, etc. which depend on space and time.

In particular, let us choose T, p, ϕ_g and the c_k's as independent thermodynamic quantities. It follows immediately that $(j, k = 1, \ldots M, v = \frac{1}{\rho})$

$$\frac{d(\rho u)}{dt} = \frac{d(\rho h)}{dt} - \frac{dp}{dt}$$

$$\frac{d(h + \phi_g)}{dt} = \left(\frac{\partial h}{\partial p}\right)_{T, c_1 \ldots} \frac{dp}{dt} + \left(\frac{\partial h}{\partial T}\right)_{p, c_1 \ldots} \frac{dT}{dt} + h_k \frac{dc_k}{dt} + \frac{d\phi_g}{dt}$$

hence $\dfrac{dh}{dt} = \left(\dfrac{\partial h}{\partial p}\right)_{T, c_1 \ldots} \dfrac{dp}{dt} + \left(\dfrac{\partial h}{\partial T}\right)_{p, c_1 \ldots} \dfrac{dT}{dt} + h_k \dfrac{dc_k}{dt}$

and, analogously:

$$\frac{dv}{dt} = \left(\frac{\partial v}{\partial p}\right)_{T, c_1 \ldots} \frac{dp}{dt} + \left(\frac{\partial v}{\partial T}\right)_{p, c_1 \ldots} \frac{dT}{dt} + v_k \frac{dc_k}{dt}$$

$$\frac{d}{dt}\left(\frac{\mu_k^0}{T}\right) = \frac{1}{T}\left(\frac{\partial \mu_k^0}{\partial p}\right)_{T, c_1 \ldots} \frac{dp}{dt} + \left[\frac{\partial}{\partial T}\left(\frac{\mu_k^0}{T}\right)\right]_{p, c_1 \ldots} \frac{dT}{dt} + \frac{1}{T}\left(\frac{\partial \mu_k^0}{\partial c_j}\right)_{T, p, \ldots} \frac{dc_j}{dt}$$

If we start from the assumption of LTE everywhere at all times, then we may apply Le Châtelier's principle to a small mass element at any time, and the inequalities listed in Sect. 3.3 apply therefore to the small mass element all along its trajectory. In particular, it is shown in Sect. A.1 that these relationships lead [8] to the following inequality—to be invoked in Sect. 6.1.3 below

$$\Psi \equiv \frac{d}{dt}\left(\frac{1}{T}\right)\frac{d(\rho u)}{dt} - \rho\frac{d}{dt}\left(\frac{\mu_k^0}{T}\right)\frac{dc_k}{dt} + \frac{\rho}{T}\frac{dp}{dt}\frac{d}{dt}\left(\frac{1}{\rho}\right) - \frac{d}{dt}\left(\frac{1}{T}\right)h\frac{d\rho}{dt} \leq 0$$

This inequality

- involves the time derivatives of thermodynamic quantities;
- is dubbed General Evolution Criterion ('GEC') by some authors; other authors [17] refer rather to a particular inequality as GEC, namely $\frac{d_X P}{dt} \leq 0$—for its meaning, see Sect. 4.3.9;
- is as general as the assumption of LTE at all times.[28] The sign '=' replaces the sign '<' in steady state only.

---

[28] In its original version, GEC includes also an additive, non-positive term $-\frac{\rho}{T}\left[\frac{\partial |v|}{\partial t}\right]^2$. However, the proof of GEC in Sect. A.1 holds regardless of this term. For further applications, see both [8, 18].

References

1. Landau, L.D., Lifshitz, E.: Statistical Physics. Pergamon, Oxford, UK (1960)
2. DeGroot, S.R., Mazur, P.: Non-Equilibrium Thermodynamics. North Holland, Amsterdam (1962)
3. Landau, L.D., Lifshitz, E.: Fluid Mechanics. Pergamon, Oxford, UK (1960)
4. Callen, H.B.: Thermodynamics and an Introduction to Thermostatistics, 2nd edn. Wiley, New York (1985)
5. Fermi, E.: Thermodynamics. Dover Publications (1956)
6. Velazquez, L.: J. Stat. Mech. Theory Exp. **2016**(3), 033105 (2016)
7. Zemansky, M.W.: Heat and Thermodynamics. McGraw Hill, New York (1968)
8. Glansdorff, P., Prigogine, I.: Physica **30**, 351 (1964)
9. Hill, T.L.: Thermodynamics of Small Systems. W.A. Benjamin, Inc. Publisher, New York (1964)
10. Wallace, D.: Br. J. Philos. Sci. **61**(3), 513 (2010)
11. Lynden-Bell, D.: Phys. A **263**(1–4), 293–304 (1998)
12. Padmanabhan, T.: Phys. Rep. **188**(5), 285–362 (1990)
13. Landau, L.D., Lifshitz, E.: Mechanics. Pergamon, Oxford, UK (1960)
14. Lebowitz, J.L., Lieb, E.H.: PRL **22**(13), 631–634 (1969)
15. Thirring, W.: Z. Physik **235**, 339–352 (1970)
16. LoPresto, M.C.: Phys. Teach. **41**, 299–301 (2003)
17. Shimizu, H., Yamaguchi, Y.: Prog. Theor. Phys. **67**, 1 (1982). (Progress Letters)
18. Di Vita, A.: Phys. Rev. E **81**, 041137 (2010)

Linear Non-equilibrium Thermodynamics

4

Abstract

We discuss linear non-equilibrium thermodynamics in both discontinuous and continuous systems. Both Onsager's symmetry, Rayleigh's dissipation function, the minimum entropy production principle and the least dissipation principle are introduced. As examples, we discuss both Kelvin's thermocouple equations, Joule-Thomson expansion, Dufour's, Peltier's, Seebeck's, Soret's and Thomson's effect, as well as Fick's, Knudsen's, Pascal's, Saxen's and Wiedemann-Franz' law. We also discuss the balances of entropy and energy in fluids with viscous heating, Joule heating and many reacting chemical species, as these fluids are examples par excellence of continuous systems.

4.1 Discontinuous Systems

4.1.1 What is the Linear Non-equilibrium Thermodynamics

Here we focus our attention on discontinuous systems.[1] Let $S = S(x_1, \ldots x_N)$ and $x_{i-eq} = 0$ with no loss of generality.[2] We have shown that $\frac{dS}{dt} = X_i J_i$, $J_i = -\frac{dx_i}{dt}$, $X_i = -\frac{\partial S}{\partial x_i} = \beta_{ik} x_k$, $S - S_{eq} = -\frac{1}{2}\beta_{ik} x_i x_k$ and $\beta_{ik} = \beta_{ki}$. We introduce the probability $w(x_1, \ldots x_N) dx_1 \ldots dx_N$ that the 1st quantity takes a value between x_1 and $x_1 + dx_1$ and that the 2nd quantity takes a value between x_2 and $x_2 + dx_2$... and that the Nth quantity takes a value between x_N and $x_N + dx_N$.[3] With the help of statistical mechanics, it is possible to show that (k_B Boltzmann's constant):

---

[1] We refer to Sects. 120 , 121 of [1] and to [2]. For a concise but crystal-clear review of the whole topic of linear non-equilibrium thermodynamics, see [3], Chap. 14.

[2] Here and in the following $i, j, k, l, p, q = 1, \ldots N$, N number of relevant quantities.

[3] The function $w = w(x_1, \ldots x_N)$ is usually referred to in the literature as the 'distribution function'

© The Author(s), under exclusive license to Springer Nature Switzerland AG 2022
A. Di Vita, *Non-equilibrium Thermodynamics*, Lecture Notes in Physics 1007,
https://doi.org/10.1007/978-3-031-12221-7_4

$$w \propto \exp\left[\frac{\left(S - S_{eq}\right)}{k_B}\right]$$

This relationship is referred to as 'Einstein's formula' in the literature. We define also the average $< a > \equiv \int a w dx_1, \dots dx_N$ of the generic quantity a and the correlation function $< ab >$ of the generic quantities a and b. We recollect some relevant results below.

Firstly, the symmetry $X_i \leftrightarrow x_i$ (Sect. 2.5) ensures that:

$$< X_i x_k > = < X_k x_i >$$

Secondly, with the help of statistical mechanics it is possible to show that:

$$< X_i x_k > = k_B \delta_{ik}$$

where $\delta_{ik} = 1$ if $i = k$ and $\delta_{ik} = 0$ if $i \neq k$ (Kronecker' delta).

Thirdly, time invariance of the laws of physics implies that:

$$< x_i(t) x_k(t - \tau) > = < x_i(t + \tau) x_k(t) >$$

for arbitrary times t and τ. Physically, this means that physics does not depend on the choice of the origin of time.

Fourthly, microscopic reversibility of the laws of physics ensures that:

$$x_i(t) = \pm x_i(-t)$$

for arbitrary t. This means that physics makes no distinction between future and past.[4] As a consequence, we write:

$$< x_i(t) x_k(t - \tau) > = \pm < x_i(t) x_k(t + \tau) >$$

for arbitrary times t and τ. The sign is $+$ if either $x_i(t) = +x_i(-t)$ and $x_k(t) = +x_k(-t)$ or $x_i(t) = -x_i(-t)$ and $x_k(t) = -x_k(-t)$; the sign is $-$ otherwise.

---

[4] Even (odds) functions of $\mathbf{v}$ have the sign $+$ (-), as we reverse the sign of t but not of $\mathbf{x}$ in $\mathbf{v} = \frac{d\mathbf{x}}{dt}$. Thus, mass, electric charge, acceleration, force, energy, temperature, entropy, and position have the sign $+$. Vorticity and magnetic field have the sign -; in fact, vorticity is linear in $\mathbf{v}$ by definition, and Maxwell's equations of electromagnetism make the magnetic field to be linear in the current density, which in turn scales linearly with the velocity of charge carriers. According to the same equations, the electric field scales as a cross product of velocity and magnetic field and has therefore the sign $+$. Neither simultaneous change of sign of t and magnetic field nor simultaneous change of sign of t and vorticity affect x_i.

Here we focus our attention on the case with the sign +. The case with the sign $-$ is briefly discussed in the following.[5]

For problems with $x_i \neq x_{i-eq}$ ('far from equilibrium' or 'non-equilibrium') two further assumptions are customary:

$$-\frac{dx_i}{dt} = J_i = \lambda_{ik}x_k \quad \text{with} \quad Im(\lambda_{ik}) = 0 \quad \text{and} \quad < x_i\,(t+\tau) - x_i\,(t) >= \tau \cdot \frac{dx_i}{dt}$$

i.e. relaxation on a time-scale τ occurs and follows the same linear laws for both the physical quantities and their averages.[6] Hence the name Linear Non-Equilibrium Thermodynamics ('LNET') which appears in the literature.

Intuitively, LNET applies to states which are not too far from thermodynamic equilibrium so that linearity of the relationships between the J_i's and the x_k's is justified. However, this is no rigorous argument, as we miss a definition of 'not too far': validity of LNET is to be assessed for each problem separately.

The relationship $X_i = \beta_{ik}x_k$ gives $x_k = \left(\beta^{-1}\right)_{kj} X_j$, hence:

$$J_i = L_{ij}X_j \quad ; \quad L_{ij} \equiv \lambda_{ik}\left(\beta^{-1}\right)_{kj} \quad \text{Onsager's coefficient matrix}$$

All elements of Onsager's coefficient matrix are real and depend on no x_k. Symmetry $x_k \leftrightarrow X_k$ implies therefore that they depend also on no X_k.[7] Moreover, all elements of Onsager's coefficient matrix are constant: $\frac{d\lambda_{ik}}{dt} = 0$ and $\frac{d\beta_{ik}}{dt} = 0$ imply $\frac{dL_{ik}}{dt} = 0$.

4.1.2 Onsager's Symmetry

The following, fundamental result of LNET is due to Onsager. We multiply both sides of $< x_i\,(t+\tau) - x_i\,(t) >= \tau \cdot \frac{dx_i}{dt}$ by $x_l\,(t)$, take the average, invoke the results of Sect. 4.1.1 and obtain:

$$< x_l\,(t) < x_i\,(t+\tau) - x_i\,(t) >>=< x_l\,(t) \cdot \tau \cdot \frac{dx_i}{dt} >=$$

$$= - < x_l\,(t) > \cdot \tau \cdot J_i >= -\tau \cdot L_{ik} < x_l X_k >=$$

$$= -\tau \cdot k_B \cdot L_{ik} \cdot \delta_{lk} = -\tau \cdot k_B \cdot L_{il}$$

---

[5] It is experimentally observed in the Brownian motion of the needle in a galvanometer, provided that x_i and x_k are the deflection and the angular velocity of the needle respectively.

[6] Of course, relaxation of $x_i\,(t)$ starts from a given initial condition $x_i\,(t = 0)$. Fluctuations affect $x_i\,(t)$ during the relaxation; when averaging at a given time t during a relaxation process, we must keep $x_i\,(t = 0)$ fixed.

[7] For instance, if $\mathbf{q} = -\chi\nabla T$ is the heat transport equation and the thermodynamic forces and fluxes are $\nabla\left(T^{-1}\right)$ and the heat flux $\mathbf{q}$ respectively, then the heat transport equation requires $\mathbf{q} = \chi T^2\nabla\left(T^{-1}\right)$ and LNET requires $\chi T^2 = \text{const}$.

Now, there is nothing special about the indices i and l. We may swap them and obtain:

$$< x_i (t) < x_l (t + \tau) - x_l (t) >>= -\tau \cdot k_B \cdot L_{li}$$

Term-by-term subtraction leads to:

$$L_{li} - L_{il} = (\tau k_B)^{-1} \left[< x_l (t) < x_i (t + \tau) - x_i (t) >> - < x_i (t) < x_l (t + \tau) - x_l (t) >> \right]$$

Time invariance and microscopic reversibility imply that the R.H.S. vanishes,[8] hence:

$$L_{li} = L_{il} \quad ; \quad \text{Onsager's symmetry}$$

If $< x_i (t) x_k (t - \tau) >= -x_i (t) x_k (t + \tau)$ then term-by term sum of the relationships $< x_l (t) < x_i (t + \tau) - x_i (t) >>= -\tau \cdot k_B \cdot L_{il}$ and $< x_i (t) < x_l (t + \tau) - x_l (t) >>= -\tau \cdot k_B \cdot L_{li}$ leads to $L_{il} = -L_{li}$. However, no ambiguity ever arises: since entropy is even in $\mathbf{v}$, if no magnetic field and no vorticity occur then $S - S_{eq}$ is the sum of two quadratic forms, the former and the latter containing only quantities even and odd in $\mathbf{v}$ respectively, and $L_{il} = L_{li}$. Moreover, if L_{il} depends on magnetic field (vorticity) then nothing changes provided that the sign of the magnetic field (vorticity) is changed as we swap i and l.

Finally, we stress the point that Onsager's symmetry is not equivalent to linearity of relaxation-ruling laws and to lack of self-organisation; see, e.g. the counterexamples in [4,5] respectively.

4.1.3   Rayleigh's Dissipation Function

We define Rayleigh's dissipation function [1]:

$$f \equiv \frac{1}{2} L_{ik} X_i X_k$$

This quantity had originally been introduced by Rayleigh [6] in his treatment of damped mechanical vibrations, in order to describe the impact of those forces *which vary in direct proportion to the component velocities of the parts of the system*[9] on the equations of motion. In that context, the dissipation function is

---

[8] Trivial algebra leads to: $< x_l (t) < x_i (t + \tau) - x_i (t) >>=<< x_l (t) x_i (t + \tau) - x_l (t) x_i (t) >>=<< x_l (t) x_i (t + \tau) >> - << x_l (t) x_i (t) >>$ where $<<>>$ denotes double averaging (we compute the first and the second average for fixed $x_l (t = 0)$ and $x_i (t = 0)$ respectively). Microscopic reversibility and time invariance imply $<< x_l (t) x_i (t + \tau) >>=<< x_l (t) x_i (t - \tau) >>$ and $<< x_l (t) x_i (t - \tau) >>=<< x_l (t + \tau) x_i (t) >>$ respectively. The looked-for conclusion follows, as $< x_l (t) < x_i (t + \tau) - x_i (t) >>=<< x_i (t) x_l (t + \tau) >> - << x_i (t) x_l (t) >>=< x_i (t) < x_l (t + \tau) - x_l (t) >>$.

[9] Typically, friction forces.

a necessarily positive quadratic function of the coordinates and represents the rate at which energy is dissipated. In recent times, f has been routinely utilized in non-equilibrium thermodynamics (see, e.g. [7,8]). Since $\frac{dS}{dt} = X_i J_i$ and $J_i = L_{ij} X_j$, we obtain $\frac{dS}{dt} = 2f$ so that $\frac{dS}{dt} \geq 0$ corresponds to $f \geq 0$.[10] We recall that if our body is in thermal contact with an environment at temperature T_0 and pressure p_0, then $R_{\min} = \Delta E - T_0 \Delta S + p_0 \Delta V$. But $X_i = \frac{1}{T_0} \frac{\partial R_{\min}}{\partial x_i}$ and $J_i = -\frac{dx_i}{dt}$, then $2f = X_i J_i = -\frac{1}{T_0} \frac{\partial R_{\min}}{\partial x_i} \frac{dx_i}{dt} = -\frac{1}{T_0} \frac{dR_{\min}}{dt}$ and $-2fT_0 = \frac{dR_{\min}}{dt} = \frac{dE}{dt} - T_0 \frac{dS}{dt} + p_0 \frac{dV}{dt}$. We take advantage of this result in two distinct cases below.

Firstly, the definition of Helmholtz' free energy $F = E - TS$ allows us to write:

$$\frac{dF}{dt} = \frac{dE}{dt} - T\frac{dS}{dt} - S\frac{dT}{dt} = -S\frac{dT}{dt} - 2fT_0 + (T_0 - T)\frac{dS}{dt} - p_0 \frac{dV}{dt}$$

It follows that:

$$\frac{dF}{dt} = -2fT_0 \quad \text{if} \quad V = \text{const. and} \quad T = T_0 = \text{const.}$$

Secondly, the definition of Gibbs' free energy $G = F + pV$ allows us to write:

$$\frac{dG}{dt} = \frac{dF}{dt} + p\frac{dV}{dt} + V\frac{dp}{dt} = V\frac{dp}{dt} - S\frac{dT}{dt} - 2fT_0 + (T_0 - T)\frac{dS}{dt} + (p - p_0)\frac{dV}{dt}$$

It follows that:

$$\frac{dG}{dt} = -2fT_0 \quad \text{if} \quad p = p_0 = \text{const. and} \quad T = T_0 = \text{const.}$$

Positiveness of f (i.e. irreversibility, $\frac{dS}{dt} > 0$) agrees with the fact that F and G are decreasing functions of time for relaxation processes at constant T, V and constant T, p respectively; F and G relax to a minimum (and $f \to 0$) when the system achieves thermodynamic equilibrium. In the following, positiveness of f is ensured provided that we take Onsager's symmetry in the usual form $L_{il} = L_{li}$, i.e. if we neglect terms in $S - S_{eq}$ which are bilinear in quantities proportional to odd and even powers of $\mathbf{v}$; as a matter of principle, such terms could arise if either the magnetic field or the vorticity are non-zero.[11] In the general case, we may replace L_{il} with its symmetrized version $L_{il}^{(s)} \equiv \frac{1}{2}(L_{il} + L_{li})$ in the definition of f, and our results concerning F and G remain unaffected. The contribution of the antisymmetric matrix $L_{il}^{(a)} \equiv L_{il} - L_{il}^{(s)} = \frac{1}{2}(L_{il} - L_{li})$ to f, indeed, vanishes identically as $X_i X_l = X_l X_i$.

---

[10] If Onsager's symmetry holds and $f > 0$ then L_{ik} is both positive-definite and invertible.

[11] Mathematically, Onsager's symmetry implies that f is a positive-definite form, i.e. that $f = 0$ if and only if all thermodynamic forces vanish.

4.1.4   Minimum Entropy Production in Discontinuous Systems

We have shown that $f = 0$ at thermodynamic equilibrium. In this case, the definition of f implies $X_i = 0$: all thermodynamic forces vanish at thermodynamic equilibrium. A system at thermodynamic equilibrium is therefore just a particular case of a system with $X_i = $ const. We are going to proof the following theorem[12]:

If Onsager's symmetry holds and if an integer N' exists such that $1 \leq N' \leq N$, that $X_i = $ const. for $1 \leq i \leq N'$ and that $J_i = 0$ for $N' + 1 \leq i \leq N$ then $\frac{dS}{dt} = $ min with respect to perturbations of both thermodynamic forces and fluxes.

This result is called 'minimum entropy production principle' and is the most renowned example of principle of minimization of entropy production ('MinEP'); further examples of MinEP will be discussed in the following. In a system where a MinEP holds, a relaxed state far from thermodynamic equilibrium corresponds to a minimum of the amount of entropy produced per unit time. Since $\frac{dS}{dt} = X_i J_i$, indeed, we obtain:

$$\frac{d^2 S}{dt^2} = X_i \frac{dJ_i}{dt} + J_i \frac{dX_i}{dt} \quad \text{hence} \quad \frac{d^3 S}{dt^3} = X_i \frac{d^2 J_i}{dt^2} + J_i \frac{d^2 X_i}{dt^2} + 2 \frac{dX_i}{dt} \frac{dJ_i}{dt}$$

Substitution of $L_{ij} = L_{ji}$, $\frac{dL_{ij}}{dt} = 0$ and $J_i = L_{ij} X_i$ gives:

$$\frac{d^2 S}{dt^2} = 2 J_i \frac{dX_i}{dt} \quad ; \quad \frac{d^3 S}{dt^3} = 2 J_i \frac{d^2 X_i}{dt^2} + 2 L_{ij} \frac{dX_i}{dt} \frac{dX_i}{dt}$$

Note that $L_{ij} = L_{ji}$ implies $L_{ij} \frac{dX_i}{dt} \frac{dX_i}{dt} \geq 0$. Moreover, $X_i = $ const. for $1 \leq i \leq N'$ and that $J_i = 0$ for $N' + 1 \leq i \leq N$ imply $J_i \frac{dX_i}{dt} = 0$ and $J_i \frac{d^2 X_i}{dt^2} = 0$ for all i's. Then, we conclude that:

$$\frac{d^2 S}{dt^2} = \frac{d}{dt} \left(\frac{dS}{dt} \right) = 0 \quad \text{and} \quad \frac{d^3 S}{dt^3} = \frac{d}{dt} \left(\frac{d^2 S}{dt^2} \right) \geq 0 \quad \text{i.e.} \quad \frac{dS}{dt} = \text{min}$$

This is a minimum property which describes a state which is both far from equilibrium ($f > 0$) and steady (constant thermodynamic forces drive the system far from equilibrium); furthermore, it is open to exchanges with the external world. This minimum property involves the entropy produced per unit time by *all* irreversible processes. Its proof requires validity of Onsager's symmetry, hence of all the assumptions underlying LNET. Unfortunately, preliminary analysis of validity of such assumption for a given problem is often overlooked in the literature.

---

[12] See Chaps. 5 and 6 of [8].

4.1.5 The Least Dissipation Principle

What if $J_i = $ const. - i.e. if there is no arbitrarily chosen N' and only the X_i's may change[13] ?

Let us compute the perturbation $d\left(\frac{dS}{dt} - f\right)$ of $\frac{dS}{dt} - f$ due to perturbations of thermodynamic forces. To this purpose, we invoke the relationships $\frac{dS}{dt} = X_i J_i$, $J_i = L_{ij} X_j$, $dL_{ik} = dt \cdot \frac{dL_{ik}}{dt} = 0$ and $f = \frac{1}{2} L_{ik} X_i X_k$, as well as Onsager's symmetry and the assumption $dJ_i = 0$. We write:

$$d\left(\frac{dS}{dt} - f\right) = d\left(X_i J_i - \frac{1}{2} L_{ik} X_i X_k\right) = (J_i - L_{ik} X_k) \, dX_i = 0$$

$$d^2\left(\frac{dS}{dt} - f\right) = d\left[(J_i - L_{ik} X_k) \, dX_i\right] = (J_i - L_{ik} X_k) \, d^2 X_i + dJ_i \, dX_i - L_{ik} \, dX_k \, dX_i =$$

$$= -L_{ik} \, dX_k \, dX_i \leq 0$$

Accordingly, $\frac{dS}{dt} - f$ is a maximum with respect to perturbations of thermodynamic forces only.

The symmetry $x_k \leftrightarrow X_k$ allows us to generalize this result. We start with $f = \frac{1}{2} L_{ij} X_i X_j$, $J_i = L_{ij} X_j$, $J_i = \lambda_{ik} x_k$, $L_{ij} = \lambda_{ik} \left(\beta^{-1}\right)_{kj}$, $\beta_{kj} = \beta_{jk}$ and Onsager's symmetry. The latter ensures that the Onsager coefficient matrix is not singular. Then:

$$f = \frac{1}{2} L_{ik} X_i X_k = \frac{1}{2} L_{ik} \left(L^{-1}\right)_{ip} J_p \left(L^{-1}\right)_{kq} J_q = \frac{1}{2} L_{ki} \left(L^{-1}\right)_{ip} J_p \left(L^{-1}\right)_{kq} J_q =$$

$$= \frac{1}{2} \delta_{kp} J_p \left(L^{-1}\right)_{kq} J_q = \frac{1}{2} \left(L^{-1}\right)_{kq} J_k J_q = \frac{1}{2} \beta_{pj} \left(\lambda^{-1}\right)_{jq} J_k J_q =$$

$$= \frac{1}{2} \beta_{pj} \left(\lambda^{-1}\right)_{jq} \lambda_{pk} x_k \lambda_{qi} x_i = \frac{1}{2} \delta_{ij} \beta_{pj} \lambda_{pk} x_i x_k = \frac{1}{2} \beta_{ip} \lambda_{pk} x_i x_k$$

so that f depends on the x_i's only. Now, the symmetry $x_i \leftrightarrow X_i$ implies $dS = -x_i \, dX_i$, hence $\frac{dS}{dt} = -x_i \frac{dX_i}{dt}$. Moreover, $X_i = \beta_{ik} x_k$ and $-\frac{dx_i}{dt} = J_i = \lambda_{ik} x_k$ imply: $\frac{dX_i}{dt} = \beta_{ik} \frac{dx_k}{dt} = -\beta_{ik} \lambda_{ik} x_k$. As expected, the three last results lead back to $\frac{dS}{dt} = 2f$.

Now, we are able to discuss the problem where only the J_i's may change and all X_i's are kept fixed ($dX_i = 0$). The same arguments discussed above show[14] that $\frac{dS}{dt} - f$ is a maximum with respect to perturbations of thermodynamic fluxes only, with $f = \frac{1}{2} \left(L^{-1}\right)_{pq} J_p J_q$.

---

[13] Here we refer to [8] and [9].
[14] We recall that if a matrix is symmetric then its inverse is symmetric.

What if both $dX_i \neq 0$ and $dJ_i \neq 0$?

Physically, we may consider an evolution with both $dX_i \neq 0$ and $dJ_i \neq 0$ as a succession made of many processes with $dX_i = 0$, $dJ_i \neq 0$ (hence $2f = \left(L^{-1}\right)_{pq} J_p J_q$) and $dX_i \neq 0$, $dJ_i = 0$ (hence $2f = L_{ik} X_i X_k$ piecewise. For each process separately $\frac{dS}{dt} - f$ is a maximum. After summing on all processes, we obtain the same result for the evolution as a whole (the contributions of $L_{ik} X_i X_k$ and $\left(L^{-1}\right)_{pq} J_p J_q$ vanish for the processes with $dX_i = 0$ and $dJ_i = 0$ respectively). It follows that:

$\frac{dS}{dt} - f$ *is a maximum either with respect to perturbations of thermodynamic fluxes at fixed thermodynamic forces* $(dX_p = 0,\ f = \frac{1}{2}\left(L^{-1}\right)_{pq} J_p J_q)$ *or with respect to perturbations of thermodynamic forces at fixed thermodynamic fluxes* $(dJ_i = 0,\ f = \frac{1}{2} L_{ij} X_i X_j)$.

This result is usually referred to as 'least dissipation principle' or 'Onsager-Machlup's principle'. It is massively utilized in analyzing solutions of the kinetic treatment of charge carriers in solids.

In the following we investigate some applications of Onsager's symmetry to particular discontinuous systems.

4.1.6   The Balance of Entropy in a Copper Wire

We focus our attention upon a copper wire at LTE between two heat reservoirs 1, 2 at temperatures T_1, T_2 respectively[15] –see Fig. 4.1.

We denote with Q the amount of heat which flows spontaneously per unit time across the wire from 1 towards 2. This transport of heat leaves the copper wire unaffected in steady state, hence the wire entropy S_{wire} does not change:

$$\frac{dS_{wire}}{dt} = 0$$

Of course:

$$\frac{dS_1}{dt} = -\frac{Q}{T_1} \quad ; \quad \frac{dS_1}{dt} = +\frac{Q}{T_2}$$

At all times, the total entropy S_{TOT} is:

$$S_{TOT} = S_1 + S_2 + S_{wire}$$

The Second Principle of thermodynamics requires that $\frac{dS_{TOT}}{dt} > 0$ for our spontaneous process, i.e.:

$$0 < \frac{dS_{TOT}}{dt} = \frac{dS_1}{dt} + \frac{dS_2}{dt} + \frac{dS_{wire}}{dt} = Q\left(\frac{1}{T_2} - \frac{1}{T_1}\right)$$

---

[15] Here we refer to Sects. 9 and 13 of [10].

Fig. 4.1 A copper wire between two reservoirs

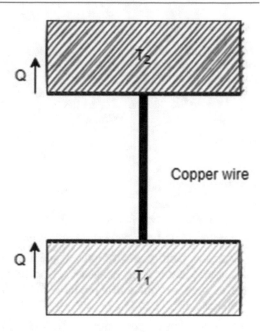

which for $Q > 0$ (heat flows from 1 to 2) leads to the expected result:

$$T_1 > T_2$$

i.e. heat flows spontaneously from the hotter body towards the cooler body. Now, let us focus our attention on the wire. We may split $\frac{dS_{wire}}{dt}$ as follows:

$$0 = \frac{dS_{wire}}{dt} = \left(\frac{dS}{dt}\right)_{\text{produced in the wire}} + \left(\frac{dS}{dt}\right)_{\text{coming into the wire from the external world}}$$

$$\left(\frac{dS}{dt}\right)_{\text{coming into the wire from the external world}} = \left(\frac{dS}{dt}\right)_{\text{coming from 1}} + \left(\frac{dS}{dt}\right)_{\text{coming from 2}}$$

$$\left(\frac{dS}{dt}\right)_{\text{coming from 1}} = -\frac{dS_1}{dt} \quad ; \quad \left(\frac{dS}{dt}\right)_{\text{coming from 2}} = -\frac{dS_2}{dt}$$

The relationships above give at once:

$$\left(\frac{dS}{dt}\right)_{\text{produced in the wire}} = \frac{dS_{TOT}}{dt} = Q\left(\frac{1}{T_2} - \frac{1}{T_1}\right) > 0$$

Thus, a 'steady state' corresponds to $\frac{dS_{wire}}{dt} = 0$. If a steady state is *also* at thermodynamic equilibrium, then $S_{TOT} = \max$ and $\frac{dS_{TOT}}{dt} = 0$. Far from equilibrium, in contrast, a steady state has $0 < \frac{dS_{TOT}}{dt} = \left(\frac{dS}{dt}\right)_{\text{produced in the wire}}$; the system (our wire) is kept far from equilibrium by suitable boundary conditions ($T_1 \neq T_2$).

Fig. 4.2 A copper wire
carries an electric current
and is in thermal contact
with a heat reservoir all
along the length of the wire

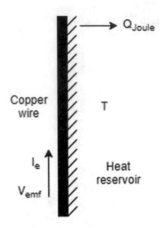

A crucial (and often overlooked) point is that all irreversible processes raise entropy, not just heating processes. Here the wire undergoes no heating. All the same, entropy is continuously produced in it (in this case, through heat conduction).

What about LNET? Let $T_2 = T_1 - \Delta T$, $\Delta T \ll T_1 (\equiv T)$. Then $\frac{1}{T_2} - \frac{1}{T_1} \approx \frac{\Delta T}{T^2}$. Let us define the 'entropy current' $I_S \equiv \frac{Q}{T}$. It is the time derivative of a physical quantity $\left(I_S = -\frac{dS_1}{dt} \right)$. We may therefore select I_S as a thermodynamic flux. The corresponding thermodynamic force is $I_S^{-1} \cdot \left(\frac{dS}{dt} \right)_{\text{produced in the wire}} = \frac{\Delta T}{T}$. If the thermodynamic force vanishes then we retrieve thermodynamic equilibrium.

Now, we focus our attention on a copper wire at LTE which lies upon a heat reservoir at temperature T with entropy $S_{reservoir}$ and is in thermal contact with this reservoir all along the length of the wire. Let an electric current I_e flow across the wire under the effect of an e.m.f. V_{emf} (see Fig. 4.2).

The Joule heating power is $Q_{Joule} = V_{emf} I_e$. We write in steady state:

$$\frac{dS_{TOT}}{dt} = \frac{dS_{wire}}{dt} + \frac{dS_{reservoir}}{dt} \quad ; \quad \frac{dS_{wire}}{dt} = 0 \quad ; \quad \frac{dS_{reservoir}}{dt} = +\frac{Q_{Joule}}{T}$$

$$\frac{dS_{wire}}{dt} = \left(\frac{dS}{dt} \right)_{\text{produced in the wire}} + \left(\frac{dS}{dt} \right)_{\text{coming into the wire from the reservoir}}$$

$$\left(\frac{dS}{dt} \right)_{\text{coming into the wire from the reservoir}} = -\frac{Q_{Joule}}{T}$$

where of course the amount of entropy coming into the wire from the reservoir per unit time has the same amplitude and opposite sign of the amount of entropy coming from the wire into the reservoir, which in turn is equal to the growth of $S_{reservoir}$ per unit time.

The relationships above lead to:

$$\left(\frac{dS}{dt}\right)_{\text{produced in the wire}} = +\frac{Q_{Joule}}{T}$$

The current I_e is the time derivative of a physical quantity—namely, the telectric charge—and is therefore a viable thermodynamic flux. The corresponding thermodynamic force is $I_e^{-1} \cdot \left(\frac{dS}{dt}\right)_{\text{produced in the wire}} = \frac{V_{emf}}{T}$. The wire is kept far from thermodynamic equilibrium by the applied e.m.f. If the thermodynamic force vanishes then we retrieve thermodynamic equilibrium.

4.1.7 Wiedemann-Franz' Law

What if both I_S, I_e, ΔT and V_{emf} are $\neq 0$? Two distinct, irreversible processes occur simultaneously: heat conduction and Joule heating. We denote the length and the cross section of the wire with l_{wire} and A_{wire} respectively. Both I_S and I_e are time derivatives of physical quantities, hence both are eligible as thermodynamic currents. According to the notation above, we write $J_1 = I_S$ and $J_2 = I_e$. The corresponding thermodynamic forces are $X_1 = \frac{\Delta T}{T}$ and $X_2 = \frac{V}{T}$; both vanish at thermodynamic equilibrium. Our model of heat transport is linearized, as $\Delta T \ll T$. If I_e then we define the thermal conductivity χ such that $Q = \chi \frac{A_{wire}}{l_{wire}} \Delta T$ [16]; thus, $I_s = L_{11} \frac{\Delta T}{T}$. Since $\left(\frac{dS}{dt}\right)_{\text{produced in the wire}} > 0$, $\chi > 0$.

In copper, the relationship between V_{emf} and I_e is linear. If $I_S = 0$ then we define the electrical conductivity σ_Ω such that $V_{emf} = R_\Omega I_e$ where the electrical resistance R_Ω is equal to $R_\Omega = \frac{l_{wire}}{A_{wire}} \sigma_\Omega$. Thus, $I_e = L_{22} \frac{V_{emf}}{T}$ where $L_{22} = T \sigma_\Omega \frac{A_{wire}}{l_{wire}}$. Since $\left(\frac{dS}{dt}\right)_{\text{produced in the wire}} > 0$, $\sigma_\Omega > 0$.

In the general case, $J_i = L_{ij} X_j$ $(i, j = 1, 2)$ with $L_{12} = T \left(\frac{\partial I_S}{\partial V_{emf}}\right)_{\Delta T \to 0}$ and $L_{21} = T \left(\frac{\partial I_e}{\partial \Delta T}\right)_{V_{emf} \to 0}$. In order to understand if $L_{12} = +L_{21}$ or $L_{12} = -L_{21}$, we investigate the behaviour of the quantities which I_S and I_e are time derivatives of under the transformation $\mathbf{v} \to -\mathbf{v}$.[17]

As for I_e, it is the time derivative of the electric charge, which does not depend on $\mathbf{v}$ and therefore does not change sign. As for I_S, it is equal to $-\frac{dS_1}{dt}$, where S_1 is the entropy of the reservoir at temperature $T_1 = T$. When applied to this heat reservoir,[18] the First Principle of thermodynamics reads $dE_1 = T dS_1$, and since $T = $ const. we write $I_S = $ const. $\cdot \frac{dE_1}{dt}$. Energy is quadratic in $\mathbf{v}$; hence E_1 does not change sign. Since

---

[16] We recall that $Q > 0$ for $\Delta T > 0$.

[17] i.e. $t \to -t$, as $\mathbf{v} = \frac{d\mathbf{r}}{dt}$ and $\mathbf{r}$ does not change.

[18] Which we suppose to have constant volume V_1 with no loss of generality, as V_1 affects neither heat conduction nor Joule heating.

the electric charge and E_1 behave the same way under reversal of $\mathbf{v}$, we conclude that Onsager's symmetry $L_{12} = L_{21}$ holds and $\left(\frac{\partial I_S}{\partial V_{emf}}\right)_{\Delta T \to 0} = \left(\frac{\partial I_e}{\partial \Delta T}\right)_{V_{emf} \to 0}$.

It is customary to define the 'Seebeck coefficient' $\varepsilon \equiv \left(\frac{I_S}{I_e}\right)_{\Delta T \to 0}$. Since $L_{12} = L_{21}$, the definitions of X_1 and X_2 imply $\varepsilon = \frac{L_{12}}{L_{22}} = \frac{L_{21}}{L_{22}} = -\left(\frac{V_{emf}}{\Delta T}\right)_{I_e \to 0}$. Accordingly, applied e.m.f. induce differences of temperature, and applied differences of temperature induce differences of e.m.f.. In the following, we take advantage of the fact that $dL_{ij} = 0$, divide the relationships concerning I_S and I_e by the same constant factor L_{22}, and introduce the new thermodynamic fluxes $I'_S \equiv \frac{I_S}{L_{22}}$ and $I'_e \equiv \frac{I_e}{L_{22}}$. In contrast, we leave the thermodynamic forces unchanged. Onsager's symmetry and the definition of ε give[19] :

$$I'_S = \left(\frac{\chi}{T\sigma_\Omega}\right)\left(\frac{\Delta T}{T}\right) + \varepsilon \frac{V_{emf}}{T} \quad ; \quad I'_e = \varepsilon \frac{\Delta T}{T} + \frac{V_{emf}}{T}$$

Now, with this new, perfectly allowable set of thermodynamic fluxes the Onsager coefficient L_{11} is equal to $L_{11} = \frac{\chi}{T\sigma_\Omega}$. Since L_{11} is constant, we obtain Wiedemann-Franz' law:

$$\chi \propto T \cdot \sigma_\Omega$$

4.1.8 Seebeck Effect

What if more wires are present[20] ? Many effects are possible. Here we consider the cases with zero magnetic field. An example is the so-called Seebeck effect: if two wires made of different metals (say, A and B) are in contact at two points I and II at different temperatures T_I and T_{II} and no net electric current is allowed to flow in steady state, then there is a non-zero e.m.f. between I and II (see Fig. 4.3).

Let us denote with ε_A and ε_B the Seebeck coefficient (Sect. 4.1.7) of metal A and metal B respectively. For the moment we assume that both ε_A and ε_B are $\ll 1$; we drop this assumption below. Since no electric current flows across A, the e.m.f. $V_{I-II(A)}$ between I and II across A is:

$$V_{I-II(A)} = -\varepsilon_A (T_I - T_{II})$$

Since no electric current flows across B, the e.m.f. $V_{I-II(A)}$ between I and II across B is:

$$V_{I-II(B)} = -\varepsilon_B (T_I - T_{II})$$

---

[19] $\varepsilon = $ const. as both L_{21} and L_{22} are constant.
[20] Here we refer to Sects. 13.8, 13.9 of [10].

Fig. 4.3 Seebeck effect

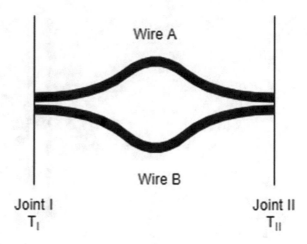

Wire A

Wire B

Joint I
T_I

Joint II
T_{II}

Term-by-term subtraction provides us with the net e.m.f.:

$$V_{AB} = V_{I-II(A)} - V_{I-II(B)} = (\varepsilon_A - \varepsilon_B)(T_I - T_{II})$$

For finite temperature jump, Seebeck's law follows:

$$V_{AB} = \int_{T_I}^{T_{II}} dT\,(\varepsilon_A - \varepsilon_B)$$

4.1.9 Peltier Effect

We discuss Peltier's effect[21] : if two wires made of different metals (say, A and B) are in contact at one junction point with no constraint on the electric current I_e flowing across this junction point between the two wires, then it is possible to keep both wires at the same temperature T in steady state provided that we subtract from the junction the Joule heating power Q_{Joule} plus an amount $Q_{Peltier} = \pi_{AB} \cdot I_e$ of heat, π_{AB} being a coefficient of proportionality which is usually referred to as Peltier's coefficient (see Fig. 4.4). The net amount $Q_{junction} \equiv Q_{Joule} + Q_{Peltier}$ is $\approx Q_{Peltier}$ for small I_e, as $Q_{Joule} \propto I_e^2$. It is precisely the approximation of small I_e which ensures the linearity required for LNET to be valid.

The energy balance at the junction reads $Q_{junction} = Q_A - Q_B$, where Q_A and Q_B are the amounts of heat flowing per unit time across wires A and B respectively. Since T is uniform, $I_S = \frac{Q}{T}$. Accordingly, we write $Q_A = T \cdot I_{S(A)}$ and $Q_B = T \cdot I_{S(B)}$, where $I_{S(A)} = \varepsilon_A \cdot I_e$ and $I_{S(B)} = \varepsilon_B \cdot I_e$. The relationships above lead to Peltier's law:

$$\pi_{AB} = (\varepsilon_A - \varepsilon_B) \cdot T$$

---

[21] And refer to Sects. 13.8 and 13.9 of [10]

Fig. 4.4 Peltier effect

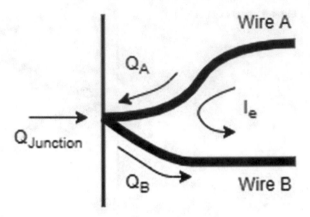

Fig. 4.5 Thomson effect

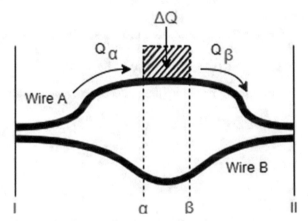

4.1.10 Thomson Effect

We discuss Thomson's effect (and refer to Sects. 13.8 and 13.9 of [10]): If two wires made of different metals (say, A and B) are in contact at two points I and II at different temperatures T_I and T_{II} and no constraint is put on the net electric current I_e , then the heat ΔQ exchanged in steady state between a small segment of one of the wires (say, between two near points α and β of wire A at temperatures T and $T + \Delta T$ respectively, see Fig. 4.5) and a heat reservoir at temperature T is the Joule heating power Q_{Joule} minus an amount $Q_{Thomson} = \sigma_A \cdot I_e \cdot \Delta T$, σ_A being a proportionality coefficient usually referred to as Thomson's coefficient.

The heat flow balance of wire A between α and β reads $\Delta Q = Q_{Joule} - Q_{Thomson}$. As usual, $Q_{Joule} = I_e V_{\alpha\beta}$, where $V_{\alpha\beta} = \varepsilon_A \Delta T$ is the e.m.f. between points α and β on wire A with Seebeck's coefficient ε_A.[22] Accordingly, we write:

$$\Delta Q = I_e \Delta T \, (\varepsilon_A - \sigma_A)$$

---

[22] Following the literature, here we defined ΔT with a sign opposite to the sign utilised before.

But $\Delta Q = Q_\beta - Q_\alpha$, and $Q = T \cdot I_S$, as $\Delta T \ll T$; this is the familiar linearization assumption, which, by the way, allows us to write $I_S = \varepsilon_A I_e$. Accordingly, we obtain $Q = T \varepsilon_A I_e$. We invoke this result in order to compute ΔQ.

Formally, $\Delta Q = \Delta (T \varepsilon_A I_e) = \varepsilon_A I_e \Delta T + T I_e \Delta \varepsilon_A + T \varepsilon_A \Delta I_e$. But conservation of electric charge requires $\Delta I_e = 0$, then:

$$\Delta Q = \varepsilon_A I_e \Delta T + T I_e \Delta \varepsilon_A$$

Comparison of the two expressions above for ΔQ leads to $\sigma_A = -T \frac{\Delta \varepsilon_A}{\Delta T}$, which in turn reduces to $\sigma_A = -T \frac{d\varepsilon_A}{dT}$ provided that the distance between α and β is so small that we can replace Δa with da for an arbitrary quantity a. There is nothing special about metal A. As for B, we obtain $\sigma_B = -T \frac{d\varepsilon_B}{dT}$. Term-by-term subtraction leads to Thomson's law:

$$\sigma_A - \sigma_B = -T \frac{d (\varepsilon_A - \varepsilon_B)}{dT}$$

4.1.11 Kelvin's Thermocouple Equations

Straightforward algebra [10] shows that Seebeck's law, Peltier's law and Thomson's law lead to the two fundamental equations ruling thermocouples, i.e. the first Kelvin's equation of thermocouples:

$$\pi_{AB} = T \frac{dV_{AB}}{dT}$$

and the second Kelvin's equation of thermocouples:

$$\sigma_A - \sigma_B = -T^2 \frac{d^2 V_{AB}}{dT^2}$$

4.1.12 Knudsen Versus Pascal

We have been discussed systems of metallic conductors in the last Sections. This is not to say that LNET provides us with information on such systems only. For example, let us consider two regions of space, say 1 and 2, which are separated by a wall with a hole [2]–see Fig. 4.6.

Regions 1 and 2 are filled with the same monoatomic perfect gas with mole number n, temperature T, volume V, energy $E = nC_v T$ and entropy $S = nC_v \ln T + R \ln V - nR \ln n$, with $C_v = C_p - R$, C_p and R molar specific heat at constant volume, molar specific heat at constant pressure and constant of perfect gases respectively. Statistical mechanics shows that $C_v = \frac{3R}{2}$. Moreover, the pressure p satisfies the perfect gas equation of state $pV = nRT$. We denote with a_1, a_2 and with δa

Fig. 4.6 Two regions are separated by a wall with a hole

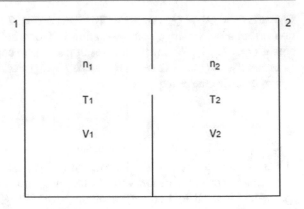

the value of the physical quantity a in 1, its value in 2 and the perturbation of a respectively. We take $V_1 = V_2$ respectively. At equilibrium:

$$n_1 = n_2 = n \quad ; \quad T_1 = T_2 = T$$

Let us perturb this equilibrium. Total mass is conserved during the perturbation, hence:

$$\delta n_1 + \delta n_2 = 0$$

Total energy is conserved during the perturbation, hence $\delta\,(n_1 T_1) + \delta\,(n_2 T_2) = 0$, i.e.:

$$n_1\,(\delta T_1 + \delta T_2) + (T_2 - T_1)\,\delta n_2 = 0$$

The total entropy is $S_{TOT} = S_1 + S_2$. The relationships above lead to:

$$dS_{TOT} = -\frac{R}{n}\,(\delta n_2)^2 - \frac{nC_v}{T^2}\,(\delta T_2)^2$$

Now, we may choose $x_1 = \delta n_2$ and $x_2 = \delta E_2$. Both x_1 and x_2 are unchanged as $\mathbf{v} \to -\mathbf{v}$, hence Onsager's symmetry $L_{ik} = L_{ki}$ holds, with $i, k = 1, 2$. After straightforward algebra, the relationship $X_i = \beta_{ik} x_k$ leads to:

$$\beta_{11} = \frac{2\,(R + C_v)}{n} \quad ; \quad \beta_{22} = \frac{2}{nT^2 C_v} \quad ; \quad \beta_{12} = \beta_{21} = -\frac{2}{nT}$$

$$X_1 = \frac{2R}{n}\delta n_2 - 2C_v\frac{\delta T_2}{T} \quad ; \quad X_2 = \frac{2\delta T_2}{T^2}$$

Physically, fluctuations in mole number and temperature drive particle flows:

$$\frac{d\,(\delta n_2)}{dt} = A\delta n_2 + B\frac{\delta T_2}{T^2}$$

where A and B are constant coefficients.[23] Moreover, fluctuations in particle flow and temperature drive energy flow:

$$\frac{d\,(\delta E_2)}{dt} = \Gamma\delta\frac{d\,(\delta n_2)}{dt} + W\frac{\delta T_2}{T^2}$$

where Γ and W are constant coefficients. Together with the definitions of the X_i's, these two relationships lead to the following system of equations:

$$\frac{d\,(\delta n_2)}{dt} = \frac{An}{2R}X_1 + \frac{1}{2}\left(\frac{AnC_vT}{R} + B\right)X_2$$

$$\frac{d\,(\delta E_2)}{dt} = \frac{An\Gamma}{2R}X_1 + \frac{1}{2}\left(\frac{AnC_vT\Gamma}{R} + B\Gamma + W\right)X_2$$

This system of equations is precisely of the form $J_i = L_{ik}X_k$, with $J_i = -\frac{dx_i}{dt}$. Onsager's symmetry $L_{12} = L_{21}$ leads to:

$$\frac{B}{A} = \frac{n\,(\Gamma - C_vT)}{R}$$

Now, suppose that a difference in T is artificially maintained, but that we wait until a steady state is reached (we say that the system undergoes 'relaxation' to this final state). In such state $\frac{d\delta n_2}{dt} = 0$.[24] This means:

$$A\delta n_2 + B\frac{\delta T_2}{T^2} = 0$$

The last two relationships lead to:

$$\frac{\delta n_2}{n} = \frac{\delta T_2}{TR}\left(C_v - \frac{\Gamma}{T}\right)$$

We add $\frac{\delta T_2}{T}$ to both sides and obtain:

$$\frac{\delta n_2}{n} + \frac{\delta T_2}{T} = \psi\frac{\delta T_2}{T} \quad ; \quad \psi \equiv \frac{1}{R}\left(C_v - \frac{\Gamma}{T}\right) + 1$$

The equation of state for perfect gases applied to 2 reads $p_2V_2 = n_2RT_2$. Since $V_2 = \text{const.}$, after taking the logarithm of both sides and differentiating

---

[23] When writing down the contribution of the perturbation of temperature we have to take $n_2 = \text{const.}$; conservation of total energy gives therefore: $\delta T_1 + \delta T_2 = 0$, hence $T_2 - T_1 = 2\delta T_2$.

[24] The condition $\frac{d\delta E_2}{dt} = 0$ provides us with the value for W, which will of no interest in the following.

we obtain $\delta \ln p_2 = \delta \ln n_2 + \delta \ln T_2$. Accordingly, the relationship above gives $\delta \ln p_2 = \psi \delta \ln T_2$. There is nothing special about region 2. The same result holds therefore whenever a relaxation at constant density occurs, and generally speaking we write $\delta \ln p = \psi \delta \ln T$. Then, integration from 1 to 2 gives:

$$\frac{p_2}{p_1} = \left(\frac{T_2}{T_1} \right)^{\psi}$$

This is a link between pressures and temperatures of regions 1 and 2 of equal volume, filled with the same perfect gas and separated by a wall with a hole, provided that fluctuations of density have relaxed. The only information which cannot be deduced from thermodynamics only is the quantity $\Gamma = \left[\frac{d(n_2)}{dt} \right]_{\delta T_2 = 0}^{-1} \left[\frac{d(E_2)}{dt} \right]_{\delta T_2 = 0}$ inside the exponent ψ. Physically, Γ is the ratio of the energy flux and the flux of particles hitting on the hole in the wall when the temperature is fixed. We discuss two particular cases in the following.

Firstly, if the diameter of the hole is $\ll$ the mean free path then kinetic theory applies and shows that $\Gamma = 2RT$. Accordingly, $\psi = 0.5$ and we get Knudsen's law:

$$\frac{p_2}{p_1} = \sqrt{\frac{T_2}{T_1}}$$

Secondly, if the diameter of the hole is $\gg$ the mean free path then fluid dynamics applies and shows that $\Gamma = C_p T$.[25] Accordingly, $\psi = 0$ and we get Pascal's law:

$$\frac{p_2}{p_1} = 1$$

4.2   Entropy Balance in Fluids

4.2.1   Dissipationless Fluids

Before discussing continuous systems, it is useful to review the various forms of both energy and entropy balance in the continuous system *par excellence*, i.e. a fluid. Fluids are routinely supposed to be at LTE [11]; validity of LNET in fluids, however, is far form trivial, and detailed discussion of their balances of energy and entropy is therefore helpful when it comes to grasp some difficulties in the application of LNET to continuous systems even beyond fluid dynamics. Accordingly, no LNET is assumed to hold in the fluids described in Sects. 4.2.1–4.2.9. To start with, we discuss

---

[25] See also Sect. 4.2.3.

dissipationless fluids.[26] We assume that there is no net source or pit of mass. Then, the balance of mass is:

$$\frac{\partial \rho}{\partial t} + \nabla \cdot (\rho \mathbf{v}) = 0$$

The balance of momentum (Euler's equation) is:

$$\rho \frac{\partial \mathbf{v}}{\partial t} + \rho \mathbf{v} \cdot \nabla \mathbf{v} + \nabla p = 0$$

We assume LTE at all times in an arbitrary small mass element of the fluid.[27] The First Principle of thermodynamics applied to a small mass element reads $du = T ds - p d \left(\rho^{-1}\right)$ and may be rewritten as $dh = T ds + \rho^{-1} dp$. In turn, these two formulas lead to:

$$\rho \frac{\partial u}{\partial t} = \rho T \frac{\partial s}{\partial t} + \frac{p}{\rho} \frac{\partial \rho}{\partial t} \quad ; \quad \nabla h = T \nabla s + \frac{1}{\rho} \nabla p$$

as $da = dt \cdot \frac{da}{dt} = dt \cdot \left(\frac{\partial a}{\partial t} + \mathbf{v} \cdot \nabla a\right)$ for arbitrary $a\,(\mathbf{x}, t)$. The relationships above allow us to write[28] the energy balance as follows:

$$\frac{\partial}{\partial t}\left(\rho \frac{|\mathbf{v}|^2}{2} + \rho u\right) = \frac{|\mathbf{v}|^2}{2} \frac{\partial \rho}{\partial t} + \rho \mathbf{v} \cdot \frac{\partial \mathbf{v}}{\partial t} + \rho \frac{\partial u}{\partial t} + u \frac{\partial \rho}{\partial t} =$$

$$= -\frac{|\mathbf{v}|^2}{2} \nabla \cdot (\rho \mathbf{v}) - \rho \mathbf{v} \cdot (\mathbf{v} \cdot \nabla) \mathbf{v} - \mathbf{v} \cdot \nabla p + \rho T \frac{\partial s}{\partial t} + \frac{p}{\rho} \frac{\partial \rho}{\partial t} - \left(h - \frac{p}{\rho}\right) \nabla \cdot (\rho \mathbf{v}) =$$

$$= -\frac{|\mathbf{v}|^2}{2} \nabla \cdot (\rho \mathbf{v}) - \rho \mathbf{v} \cdot \nabla \left(\frac{|\mathbf{v}|^2}{2}\right) - \mathbf{v} \cdot \nabla p + \rho T \frac{\partial s}{\partial t} - h \nabla \cdot (\rho \mathbf{v}) =$$

$$= -\nabla \cdot \left[\rho \mathbf{v}\left(\frac{|\mathbf{v}|^2}{2} + h\right)\right] + \rho T \left(\frac{\partial s}{\partial t} + \mathbf{v} \cdot \nabla s\right)$$

In this energy balance, the L.H.S. $\frac{\partial}{\partial t}\left(\rho \frac{|\mathbf{v}|^2}{2} + \rho u\right)$ is the time derivative of the sum of kinetic $(\rho \frac{|\mathbf{v}|^2}{2})$ and internal (ρu) energy per unit volume. In the R.H.S., the quantity $\nabla \cdot \left[\rho \mathbf{v}\left(\frac{|\mathbf{v}|^2}{2} + h\right)\right]$ is the sum of the flux $\nabla \cdot \left[\rho \mathbf{v}\left(\frac{|\mathbf{v}|^2}{2} + u\right)\right]$ of (kinetic + internal) energy due to the fluid motion and of $\nabla \cdot (p\mathbf{v})$, where $p\mathbf{v}$ is the mechanical work done by pressure per unit volume and time.

---

[26] We refer to Sects. 1, 15, 16, 26, 49, 57 of [11].

[27] A macroscopic description of a fluid which does not satisfy LTE everywhere at all times is available, namely the so called 'extended thermodynamics'–not to be confused with the EIT of Sect. 6.1.6. For a review see Ref. [12] and Refs. therein.

[28] Here we invoke the identities $(\mathbf{a} \cdot \nabla)\,\mathbf{a} = \nabla \left(\frac{|\mathbf{a}|^2}{2}\right) - \mathbf{a} \wedge \nabla \wedge \mathbf{a}\,; \mathbf{a} \cdot \mathbf{a} \wedge \mathbf{b} = 0$.

No dissipation occurs. Then, the energy may either decrease (increase) in doing (receiving) mechanical work upon (from) the external world or be exported (imported) by the fluid moving outwards (inwards). In other words, the quantities listed above in the energy balance compensate each other exactly.[29] Thus, the energy balance leads to:

$$\frac{ds}{dt} = \frac{\partial s}{\partial t} + \mathbf{v} \cdot \nabla s = 0$$

i.e., entropy is constant. This is obvious, given the fact that we discarded all irreversible process. We conclude that for dissipationless fluids the mass balance, the momentum balance and the approximation of LTE at all times give the following balances of energy and entropy:

$$\frac{\partial}{\partial t} \left(\rho \frac{|\mathbf{v}|^2}{2} + \rho u \right) + \nabla \cdot \left[\rho \mathbf{v} \left(\frac{|\mathbf{v}|^2}{2} + h \right) \right] = 0$$

$$\rho \frac{ds}{dt} = 0$$

4.2.2   Viscous Fluids

What if we include viscosity [11]?

A new term appears in the momentum balance (Navier-Stokes equation; $i, k = 1, 2, 3$):

$$\rho \frac{\partial \mathbf{v}}{\partial t} + \rho \mathbf{v} \cdot \nabla \mathbf{v} + \nabla p = \frac{\partial \sigma'_{ik}}{\partial x_k}$$

where σ'_{ik} is the ik−th component of the viscous stress tensor and x_k is the k-th spatial coordinate. After taking into account the identity $\frac{\partial (v_i \sigma'_{ik})}{\partial x_k} = v_i \frac{\partial \sigma'_{ik}}{\partial x_k} + \sigma'_{ik} \frac{\partial v_i}{\partial x_k}$ and writing $\frac{\partial (v_i \sigma'_{ik})}{\partial x_k} = \nabla \cdot (\mathbf{v} \cdot \sigma')$, the same arguments discussed above lead to the energy balance:

$$\frac{\partial}{\partial t} \left(\rho \frac{|\mathbf{v}|^2}{2} + \rho u \right) + \nabla \cdot \left[\rho \mathbf{v} \left(\frac{|\mathbf{v}|^2}{2} + h \right) - \mathbf{v} \cdot \sigma' \right] = \rho T \left(\frac{\partial s}{\partial t} + \mathbf{v} \cdot \nabla s \right) - \sigma'_{ik} \frac{\partial v_i}{\partial x_k}$$

Now, $\int_{fluid} d\mathbf{x} \nabla \cdot (\mathbf{v} \cdot \sigma')$ is the amount of energy spent per unit time because of viscosity at the boundary of the fluid. Gauss' theorem of divergence transforms the volume integral into a surface integral; the former and the latter are computed on the

---

[29] If $\frac{\partial}{\partial t} = 0$ then mass and momentum balance lead to Bernoulli's law $\mathbf{v} \cdot \left(\frac{|\mathbf{v}|^2}{2} + h \right) = 0$. If the kinetic energy term is negligible then we retrieve Joule-Thomson expansion.

volume and the boundary respectively. Again, then, the L.H.S. vanishes. In this case
the R.H.S. gives:

$$\rho T \left(\frac{\partial s}{\partial t} + \mathbf{v} \cdot \nabla s \right) = \sigma'_{ik} \frac{\partial v_i}{\partial x_k}$$

The L.H.S. is just equal to $T \cdot \rho \frac{ds}{dt}$, i.e. the non-exact differential of heat supplied
by viscosity per unit volume and time. (We wrote 'non-exact' as the the R.H.S. shows
that value of this quantity depends on the values of velocity, and not just on the initial
and final state, of the small mass element). Since viscous heating is an irreversible
process, $\frac{ds}{dt} > 0$. Correspondingly, $\sigma'_{ik} \frac{\partial v_i}{\partial x_k}$ is the viscous power density.

We conclude that for viscous fluids the mass balance, the momentum balance and
the approximation of LTE at all times give the following balances of energy and
entropy:

$$\frac{\partial}{\partial t} \left(\rho \frac{|\mathbf{v}|^2}{2} + \rho u \right) + \nabla \cdot \left[\rho \mathbf{v} \left(\frac{|\mathbf{v}|^2}{2} + h \right) - \mathbf{v} \cdot \sigma' \right] = 0$$

$$\rho \frac{ds}{dt} = \frac{\sigma'_{ik}}{T} \frac{\partial v_i}{\partial x_k}$$

Viscosity affects both the balance of energy and the balance of entropy.

4.2.3 Joule-Thomson Throttled Expansion

Let us discuss in more detail the result of fluid mechanics $\Gamma = \left[\frac{d(n_2)}{dt} \right]^{-1}_{\partial T_2 = 0}$
$\left[\frac{d(E_2)}{dt} \right]_{\partial T_2 = 0} = C_p T$ quoted in Sect. 4.1.12 [2]. In a perfect gas, $C_p T$ is just equal to
the molar enthalpy $m_m h$, where m_m is the molar mass and $h \equiv u + \frac{p}{\rho}$ is the enthalpy
per unit mass, u, $\rho = m_m n_m$ being the internal energy per unit mass, the mass density
and the molar density respectively. We write the mass balance (Sect. 4.2.1):

$$\frac{\partial \rho}{\partial t} + \nabla \cdot (\rho \mathbf{v}) = 0$$

and the energy balance (Sect. 4.2.8, no electromagnetic field, no gravity, no heat
flux):

$$\frac{\partial}{\partial t} \left(\rho \frac{|\mathbf{v}|^2}{2} + \rho u \right) + \nabla \cdot \left[\rho \mathbf{v} \left(\frac{|\mathbf{v}|^2}{2} + h \right) - \mathbf{v} \cdot \sigma' \right] = 0$$

where the viscous stress tensor σ' is linear in the components of the velocity $\mathbf{v}$.

If $|\mathbf{v}|$ is so small[30] that terms $\propto O\left(|\mathbf{v}|^2 \right)$ are negligible, then the energy balance
reduces to $\frac{\partial}{\partial t} (\rho u) + \nabla \cdot (\rho \mathbf{v} h) = 0$ and this result, together with the balance of mass,

---

[30] This is where we assume linearization in this discussion.

leads to[31] :

$$\frac{\partial}{\partial t}(\rho u) + \rho \mathbf{v} \cdot \nabla h - h\frac{\partial \rho}{\partial t} = 0$$

Moreover, multiplication of $\Gamma = C_p T = h m_m$ by $\frac{\partial n_m}{\partial t}$ gives $\Gamma \frac{\partial n_m}{\partial t} = h\frac{\partial \rho}{\partial t}$ since m_m is obviously constant. For the same reason we may also write $\rho \mathbf{v} \cdot \nabla h = n_m \mathbf{v} \cdot \nabla \Gamma$. The energy balance reduces further to:

$$\frac{\partial}{\partial t}(\rho u) = \Gamma \frac{\partial n_m}{\partial t} - n_m \mathbf{v} \cdot \nabla \Gamma$$

Now, we rewrite Γ as $\Gamma = \left[\frac{d(n_m)}{dt}\right]_{T_2=\text{const.}}^{-1}\left[\frac{d(\rho u)}{dt}\right]_{T_2=\text{const.}}$, which simplifies further to[32]: $\Gamma = \left[\frac{\partial(n_m)}{\partial t}\right]_{T_2=\text{const.}}^{-1}\left[\frac{\partial(\rho u)}{\partial t}\right]_{T_2=\text{const.}}$ as $\mathbf{v}$ is small. Furthermore, since T_2 is constant (a difference in T is artificially maintained by the external world) then we may drop the subscript, and write: $\Gamma = \left(\frac{\partial n_m}{\partial t}\right)^{-1}\frac{\partial(\rho u)}{\partial t}$. Thus, the energy balance takes the simple form $n_m \mathbf{v} \cdot \nabla \Gamma = 0$. Integration of both sides of this equation on a volume enclosing the hole connecting regions 1 and 2 leads to[33] (this volume is displayed with dotted boundaries in Fig. 4.7):

$$\oint d\mathbf{a} \cdot \mathbf{v} n_m \Gamma = 0$$

i.e. the net enthalpy flux vanishes.

Accordingly, total enthalpy is conserved across the hole. This is the well- known Joule-Thomson throttled expansion. Implicitly, however, we have extended our result to region 1 as well when integrating: we neglected both $|\mathbf{v}|^2$ and ∇T.

4.2.4   Fluids with Electromagnetic Fields

Let us include electromagnetic fields [13]. Here we assume that there is no net electric charge[34] and refer to the electric field, the electric current density, the velocity of light in vacuum, the vacuum magnetic permittivity and the vacuum dielectric constant as

---

[31] We invoke the identity $\nabla \cdot (a\mathbf{b}) = a\nabla \cdot \mathbf{b} + \mathbf{b} \cdot \nabla a$ for arbitrary scalar a and vector $\mathbf{b}$.

[32] By definition, $\frac{d}{dt} = \frac{\partial}{\partial t} + \mathbf{v} \cdot \nabla$

[33] Here we invoke both the mass balance in steady state $\left(\frac{\partial}{\partial t} = 0\right)$, the relationship $\rho = n_m m_m$, Gauss' theorem of divergence and the identity $\nabla \cdot (w\mathbf{b}) = w\nabla \cdot \mathbf{b} + \mathbf{b} \cdot \nabla w$ for arbitrary scalar w and vector $\mathbf{b}$.

[34] Locally, if the net electric charge $\neq 0$ then strong electrostatiuc repulsion occurs and prevents the system from attaining LTE at all times.

Fig. 4.7 The region of interest in the discussion on Joule-Thomson throttled expansion

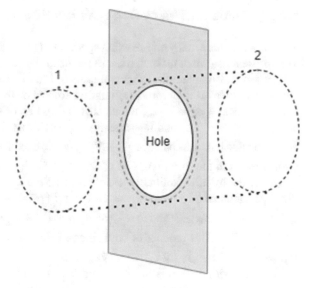

to $\mathbf{E}, \mathbf{j}_{el}, c = 3 \cdot 10^8$ m·s^{-1}, $\mu_0 = 4 \cdot \pi \cdot 10^{-7}$ T·A^{-1}·m and $\varepsilon_0 = (\mu_0 c^2)^{-1}$ respectively. Lorenz force density $\mathbf{j}_{el} \wedge \mathbf{B}$ adds to the R.H.S. of Navier-Stokes' equation. Then, we replace[35] :

$$\mathbf{v} \cdot \rho \frac{\partial \mathbf{v}}{\partial t} \to \mathbf{v} \cdot \rho \frac{\partial \mathbf{v}}{\partial t} + \mathbf{v} \cdot \mathbf{j}_{el} \wedge \mathbf{B} \quad ; \quad -\mathbf{j}_{el} \wedge \mathbf{B} \to -\mathbf{j}_{el} \cdot (\mathbf{E} + \mathbf{v} \wedge \mathbf{B}) + \mathbf{j}_{el} \cdot \mathbf{E}$$

where 2 of the 4 Maxwell's equations of electromagnetism, namely $\nabla \wedge \mathbf{E} + \frac{\partial \mathbf{B}}{\partial t} = 0$ and $\nabla \wedge \mathbf{B} - \frac{1}{c^2} \frac{\partial \mathbf{E}}{\partial t} - \mu_0 \mathbf{j}_{el} = 0$, lead to[36] $\mathbf{j}_{el} \cdot \mathbf{E} = -\frac{1}{\mu_0} \nabla \cdot (\mathbf{E} \wedge \mathbf{B}) - \frac{\partial}{\partial t} \left(\frac{1}{2\mu_0} |\mathbf{B}|^2 + \frac{1}{2} \varepsilon_0 |\mathbf{E}|^2 \right)$. Together with these substitutions, our discussion leads to the conclusion that for viscous fluids with electromagnetic fields the mass balance, the momentum balance and the approximation of LTE at all times give the following balances of energy and entropy:

$$\frac{\partial}{\partial t} \left(\rho \frac{|\mathbf{v}|^2}{2} + \rho u + \frac{1}{2\mu_0} |\mathbf{B}|^2 + \frac{1}{2} \varepsilon_0 |\mathbf{E}|^2 \right) + \nabla \cdot \left[\rho \mathbf{v} \left(\frac{|\mathbf{v}|^2}{2} + h \right) - \mathbf{v} \cdot \sigma' + \frac{\mathbf{E} \wedge \mathbf{B}}{\mu_0} \right] = 0$$

$$\rho \frac{ds}{dt} = \frac{\sigma'_{ik}}{T} \frac{\partial v_i}{\partial x_k} + \frac{\mathbf{j}_{el} \cdot (\mathbf{E} + \mathbf{v} \wedge \mathbf{B})}{T}$$

Joule heating is irreversible for $\frac{\mathbf{j}_{el} \cdot (\mathbf{E}+\mathbf{v}\wedge\mathbf{B})}{T} > 0$, with $\mathbf{j}_{el} \cdot (\mathbf{E} + \mathbf{v} \wedge \mathbf{B})$ Joule power density. Both the balance of energy and the balance of entropy are affected.

---

[35] We invoke the identity $\mathbf{a} \cdot \mathbf{b} \wedge \mathbf{c} + \mathbf{b} \cdot \mathbf{a} \wedge \mathbf{c} = 0$ for arbitrary $\mathbf{a}$, $\mathbf{b}$ and $\mathbf{c}$.
[36] We invoke the identity $\nabla \cdot (\mathbf{a} \wedge \mathbf{b}) = \mathbf{b} \cdot \nabla \wedge \mathbf{a} - \mathbf{a} \cdot \nabla \wedge \mathbf{b}$ for arbitrary $\mathbf{a}$ and $\mathbf{b}$.

4.2.5   Fluids with Many Non-reacting Species

What if we include many non-reacting species [11]? We assume $k = 1, \ldots N$ species to be present in a small mass element of the fluid. Diffusion is irreversible, but leaves both mass and momentum balance unaffected, as it occurs inside the small mass element and leaves therefore $\mathbf{v}(t)$ unaffected.[37] For the k-th species, let us define the mass density $\rho_k \equiv \rho \cdot c_k$, the electric charge e_k of one particle,[38] the particle density $n_k \equiv m_k^{-1} \cdot \rho_k$ and the mass flow $\mathbf{j}_k$ such that $\nabla \cdot \mathbf{j}_k + \rho \frac{dc_k}{dt} = 0$; the latter definition and the mass balance $\frac{\partial \rho}{\partial t} + \nabla \cdot (\rho \mathbf{v}) = 0$ lead to the balance of mass of the k-the species in the form: $\frac{\partial \rho_k}{\partial t} + \nabla \cdot (\rho_k \mathbf{v}) = -\nabla \cdot \mathbf{j}_k$.

Under the by now familiar assumption of LTE everywhere at all times, the First Principle of thermodynamics $du = T ds - pd\left(\rho^{-1}\right) + \mu_k^0 dc_k$ leads both to $\rho \frac{\partial u}{\partial t} = \rho T \frac{\partial s}{\partial t} + \frac{p}{\rho} \frac{\partial \rho}{\partial t} + \rho \mu_k^0 \frac{\partial c_k}{\partial t}$ and to $dh = T ds + \frac{1}{\rho} dp + \mu_k^0 dc_k$, hence to $\nabla h = T \nabla s + \frac{1}{\rho} \nabla p + \mu_k^0 \nabla c_k$. Dot product of both sides of the last relationship by $\mathbf{v}$ gives: $-\mathbf{v} \cdot \nabla p = \rho \mathbf{v} T \cdot \nabla s - \rho \mathbf{v} \cdot \nabla h + \rho \mathbf{v} \mu_k^0 \nabla c_k$.

Together with these relationships, our discussion leads to the conclusion that for viscous fluids with electromagnetic fields and many non-reacting species the mass balance, the momentum balance and the approximation of LTE at all times give the following balances of energy and entropy:

$$\frac{\partial}{\partial t}\left(\rho \frac{|\mathbf{v}|^2}{2} + \rho u + \frac{1}{2\mu_0}|\mathbf{B}|^2 + \frac{1}{2}\varepsilon_0|\mathbf{E}|^2\right) + \nabla \cdot \left[\rho \mathbf{v}\left(\frac{|\mathbf{v}|^2}{2} + h\right) - \mathbf{v} \cdot \sigma' + \frac{\mathbf{E} \wedge \mathbf{B}}{\mu_0}\right] = 0$$

$$\rho \frac{ds}{dt} = \frac{\sigma'_{ik}}{T}\frac{\partial v_i}{\partial x_k} + \frac{\mathbf{j}_{el} \cdot (\mathbf{E} + \mathbf{v} \wedge \mathbf{B})}{T} + \frac{\mu_k^0 \nabla \cdot \mathbf{j}_k}{T}$$

Entropy balance only is affected.

4.2.6   Fluids with Many Species Reacting with Each Other

What if reactions among species occur [11]? We take into account $r = 1, \ldots R$ (chemical, nuclear...) reactions and define the following quantities: the stoichiometric coefficient ν_{kr} of the k-th species in the r-th reaction, the reaction rate Υ_r of the r-th reaction, and the affinity $A_r^0 \equiv -\mu_k^0 \nu_{kr}$ of the r-th reaction. The mass balance of the k-th species reads $\nabla \cdot \mathbf{j}_k + \rho \frac{dc_k}{dt} = \nu_{kr}\Upsilon_r$. Accordingly, the same results of the case

---

[37] $\mathbf{v}$ is the macroscopic velocity of the center-of-mass of the small mass element.

[38] LTE at all times requires vanishing electrostatic repulsion inside the fluid, and lack of electrostatic repulsion requires vanishing net electric charge $n_k e_k$ inside the small mass element. Accordingly, even if the electrostatic potential ϕ_{el} does not vanish it leaves the mass density unaffected (i.e. we have $d\rho = 0$) but it leads to the replacement $\mu_k \to \mu_k + e_k \phi_{el}$. It follows that $\mu_k^0 c_k \to \mu_k^0 c_k + \rho^{-1}\phi_{el}e_k n_k = \mu_k^0 c_k$, so that the electrostatic potential leaves the entropy balance unaffected.

with no reactions hold, provided that $-\nabla \cdot \mathbf{j}_k \rightarrow -\nabla \cdot \mathbf{j}_k + \nu_{kr} \Upsilon_r$. Accordingly, the balance of mass of the k-th species reads : $\frac{\partial \rho_k}{\partial t} + \nabla \cdot (\rho_k \mathbf{v}) = -\nabla \cdot \mathbf{j}_k + \nu_{kr} \Upsilon_r$.

Physically, reactions are due to inter-particle interactions inside the small mass element, and leave therefore $\mathbf{v}(t)$ unaffected. Thus, our discussion leads to the conclusion that for viscous fluids with electromagnetic fields and many reacting species the mass balance, the momentum balance and the approximation of LTE at all times give the following balances of energy and entropy:

$$\frac{\partial}{\partial t}\left(\rho \frac{|\mathbf{v}|^2}{2} + \rho u + \frac{1}{2\mu_0}|\mathbf{B}|^2 + \frac{1}{2}\varepsilon_0|\mathbf{E}|^2\right) + \nabla \cdot \left[\rho \mathbf{v}\left(\frac{|\mathbf{v}|^2}{2} + h\right) - \mathbf{v} \cdot \sigma' + \frac{\mathbf{E} \wedge \mathbf{B}}{\mu_0}\right] = 0$$

$$\rho \frac{ds}{dt} = \frac{\sigma'_{ik}}{T}\frac{\partial v_i}{\partial x_k} + \frac{\mathbf{j}_{el} \cdot (\mathbf{E} + \mathbf{v} \wedge \mathbf{B})}{T} + \frac{\mu_k^0 \nabla \cdot \mathbf{j}_k}{T} + \frac{A_r^0 \Upsilon_r}{T}$$

Formally, entropy balance only is affected. However, there is something new in the balance of energy. The definition of Gibbs' free energy $G = H - TS$ gives $\mu_k^0 = h_k - T s_k$. In turn, this implies that A_r^0 contains terms $\propto h_k$, whose sum is the reaction heat. Reactions may affect the energy balance through the reaction heat, i.e. through the actual value of h.

4.2.7 Fluids with Gravity

What if gravity is present? Here we refer to Ref. [1] and assume constant gravitational field (Sect. 3.4.2), the case opposite to gravitational collapse (Sect. 3.4.1). We have shown that the gravitational potential ϕ_g affects chemical potentials but leaves entropy unaffected. Moreover, both enthalpy and internal energy per unit mass transform like chemical potentials per unit mass. Our discussion shows therefore that for viscous, gravitating fluids with electromagnetic fields and many reacting species the mass balance, the momentum balance and the approximation of LTE at all times give the following balances of energy and entropy:

$$\frac{\partial}{\partial t}\left(\rho \frac{|\mathbf{v}|^2}{2} + \rho u + \frac{1}{2\mu_0}|\mathbf{B}|^2 + \frac{1}{2}\varepsilon_0|\mathbf{E}|^2 + \rho\phi_g\right) + \nabla \cdot \left[\rho \mathbf{v}\left(\frac{|\mathbf{v}|^2}{2} + h\right) - \mathbf{v} \cdot \sigma' + \frac{\mathbf{E} \wedge \mathbf{B}}{\mu_0}\right] = 0$$

$$\rho \frac{ds}{dt} = \frac{\sigma'_{ik}}{T}\frac{\partial v_i}{\partial x_k} + \frac{\mathbf{j}_{el} \cdot (\mathbf{E} + \mathbf{v} \wedge \mathbf{B})}{T} + \frac{\mu_k^0 \nabla \cdot \mathbf{j}_k}{T} + \frac{A_r^0 \Upsilon_r}{T}$$

As a particular case, we retrieve Bernoulli equation $\mathbf{v} \cdot \nabla\left(\frac{|\mathbf{v}|^2}{2} + h + \phi_g\right) = 0$ in the steady state of a dissipation-free system immersed in a static gravitational field and with no electromagnetic field. Entropy balance is not affected.

4.2.8   Local form of the Entropy Balance

Any further scalar quantity (with the proper dimensions) which adds to the energy balance adds also to the balance of entropy. Let such scalar be the divergence of a vector, say $\mathbf{q}$. Then, the most general form of both energy and entropy balance in a steady gravitational field in a small mass element at LTE at all times is [11]:

$$\frac{\partial}{\partial t}\left(\rho\frac{|\mathbf{v}|^2}{2} + \rho u + \frac{1}{2\mu_0}|\mathbf{B}|^2 + \frac{1}{2}\varepsilon_0|\mathbf{E}|^2 + \rho\phi_g\right) +$$

$$+\nabla\cdot\left[\rho\mathbf{v}\left(\frac{|\mathbf{v}|^2}{2} + h\right) - \mathbf{v}\cdot\sigma' + \frac{\mathbf{E}\wedge\mathbf{B}}{\mu_0} + \mathbf{q}\right] = 0$$

$$\rho\frac{ds}{dt} = \frac{\sigma'_{ik}}{T}\frac{\partial v_i}{\partial x_k} + \frac{\mathbf{j}_{el}\cdot(\mathbf{E}+\mathbf{v}\wedge\mathbf{B})}{T} + \frac{\mu_k^0\nabla\cdot\mathbf{j}_k}{T} + \frac{A_r^0\Upsilon_r}{T} - \frac{\nabla\cdot\mathbf{q}}{T}$$

The first equation *defines* the so-called 'heat flux' $\mathbf{q}$. Generally speaking, $\mathbf{q}$ is therefore the time derivative of *no* physical quantity. We will refer to this fact in Sect. 5.2.1.

As for the second equation, we rewrite it in the form below,[39] which we are going to refer to in the following as to the 'local form of the entropy balance':

$$\rho\frac{ds}{dt} = \sigma - \nabla\cdot\mathbf{j}_s = \frac{P_h}{T} - \frac{\nabla\cdot\mathbf{q}}{T}$$

where we have defined both the entropy production density

$$\sigma \equiv \frac{\sigma'_{ik}}{T}\frac{\partial v_i}{\partial x_k} + \frac{\mathbf{j}_{el}\cdot(\mathbf{E}+\mathbf{v}\wedge\mathbf{B})}{T} - \mathbf{j}_k\cdot\nabla\frac{\mu_k^0}{T} + \frac{A_r^0\Upsilon_r}{T} + \mathbf{q}\cdot\nabla\frac{1}{T}$$

the entropy flow

$$\mathbf{j}_s \equiv \frac{\mathbf{q}}{T} - \frac{\mu_k^0\mathbf{j}_k}{T}$$

and the heating power density

$$P_h \equiv \nabla\cdot\mathbf{q} + \rho T\frac{ds}{dt} = \sigma'_{ik}\frac{\partial v_i}{\partial x_k} + \mathbf{j}_{el}\cdot(\mathbf{E}+\mathbf{v}\wedge\mathbf{B}) + \mu_k^0\nabla\cdot\mathbf{j}_k + A_r^0\Upsilon_r$$

---

[39] We invoke the identity $\nabla\cdot(\mathbf{a}w) = \mathbf{a}\cdot w + w\nabla\cdot\mathbf{a}$ for arbitrary scalar w and vector $\mathbf{a}$.

An alternative form of this entropy balance is[40] (see Sect. 4.1.2):

$$\frac{\partial (\rho s)}{\partial t} = \sigma - \nabla \cdot \mathbf{j}'_s \;\; ; \;\; \mathbf{j}'_s \equiv \mathbf{j}_s + \rho s \mathbf{v}$$

4.2.9 Global form of the Entropy Balance: Back to the Copper Wire

We are going to perform a volume integration of both sides of the balance of mass and of the entropy balance in local form (Sect. 4.2.8) on a volume V of fluid[41] with the help of Gauss' theorem of divergence. As for the mass balance $\frac{\partial \rho}{\partial t} + \nabla \cdot (\rho \mathbf{v}) = 0$, for an arbitrary quantity a it leads to[42] :

$$\frac{d}{dt} \int \rho a d\mathbf{x} = \int \rho \frac{da}{dt} d\mathbf{x}$$

Accordingly, volume integration of both sides of the entropy balance in local form leads to[43] the entropy balance in global form:

$$\frac{dS}{dt} = \frac{d_i S}{dt} + \frac{d_e S}{dt}$$

where $\frac{dS}{dt} = \int_{fluid} \rho \frac{ds}{dt} d\mathbf{x}$, $\frac{d_i S}{dt} = \int_{fluid} \sigma d\mathbf{x}$ and $\frac{d_e S}{dt} = -\int_{boundary} \mathbf{j}_s \cdot d\mathbf{a}$ are the total time derivative of the entropy of the fluid, the amount of entropy produced per unit time within the fluid and the amount of entropy coming into the fluid per unit time respectively. Clearly, the entropy balance of the fluid in global form is just like the entropy balance of our copper wire. Just as in the copper wire, in case of steady state ($\frac{dS_{fluid}}{dt} = 0$) the amount of entropy produced within the fluid per unit time is carried across the boundary and goes to the external world. Comparison with the steady state of discontinuous systems shows therefore that in continuous systems it is $\frac{d_i S}{dt}$ which plays the role played by $\frac{dS}{dt} = X_i J_i$ in discontinuous systems. The fundamental reason of this strict analogy between a copper wire and a fluid—in spite of the possible occurrence of complex physical phenomena like viscosity, Joule heating, electromagnetism and gravitation—is the fact that our descriptions of both

---

[40] The assumption of constant and uniform (and possibly vanishing) gravitational field is crucial here. In the general case, gravity is described by General Relativity. In this theory, an entropy flow which is linear in s and $\frac{q}{T}$ leads to paradoxical results, like e.g. a failure to ensure maximum entropy at equilibrium. Self-consistency is restored if nonlinear term are added, and information concerning the corresponding phenomenological coefficients is removed by assuming Onsager's symmetry [14].

[41] As usual, $d\mathbf{x}$ and $d\mathbf{a}$ are the volume element and the surface element on the boundary.

[42] Here we invoke the identity $\frac{d}{dt} \int_V d\mathbf{x} a = \int_V d\mathbf{x} \frac{\partial a}{\partial t} + \int_V d\mathbf{x} \nabla \cdot (a\mathbf{u})$, where $\mathbf{u}$ is the speed of the generic point on the boundary surface of the integration volume V. If V is the fluid volume then $\mathbf{u} = \mathbf{v}$ on the boundary of the fluid.

[43] The notation $P \equiv \frac{d_i S}{dt}$ is often utilized in the literature.

the wire and the fluid share a common, fundamental assumption: the validity of LTE at all times everywhere throughout the system. In particular, the introduction of $\mathbf{q}$ shows that the range of validity of the entropy balance in global form in a physical system is broader than the range of validity of Navier-Stokes' equation, as it coincides with the range of validity of the assumption of LTE at all times everywhere throughout the system. In contrast, LNET is required nowhere.

4.3 Continuous Systems

4.3.1 A Slight Abuse of Notation

When it comes to fluid/continuous systems, the question arises [15,16] whether variational principles like the minimum entropy production principle and the least dissipation principle describe steady states, in analogy to the corresponding results regarding discontinuous systems. We have seen that the assumption of LTE at all times and everywhere throughout the system leads to general results like Le Châtelier's principle, the general evolution criterion and the entropy balance in its global form. We have also shown that further, additional assumptions—like e.g. the linearity of the underlying phenomenological laws–lead to LNET in discontinuous systems at least. In particular, we have discussed examples where the heat flux is the time derivative of an energy and is coupled to electric conduction with negligible Joule heating—see our discussion of thermocouples. In contrast, LNET never holds when it comes to throttled expansion, as the latter phenomenon involves both viscosity and turbulence: the former and the latter phenomenon, indeed, are represented in Navier-Stokes' equation by terms like $\mathbf{v} \cdot \nabla \mathbf{v}$ and $\mathbf{v} \cdot \boldsymbol{\sigma}'$, both of which are $\propto O\left(|\mathbf{v}|^2\right)$ and are therefore not described by any model based on linearity. Now, throttled expansion occurs in fluids, i.e. in continuous systems. If we are able to decide if and when LNET holds for continuous systems, we are also able to apply the related variational principles provided that some kind of relaxation occurs in such systems. For simplicity, and with a slight abuse of notation, we refer to the collection of all results of LNET such as Onsager's symmetry, the least dissipation principle etc. discussed so far once more as to LNET, here and in the following.

4.3.2 Thermodynamic Forces and Fluxes in Continuous Systems

We have seen in Sect. 4.2.9 that in continuous systems the quantity which corresponds to the quantity $\frac{dS}{dt} = X_i J_i$ in discontinuous systems is $\frac{d_i S}{dt}$. The definition of the latter quantity and the fact that σ is a bilinear form suggest that:

$$\sigma = X_i J_i \quad ; \quad J_i = L_{ik} X_k \quad ; \quad i, k = 1, \ldots N \quad \text{thermodynamic quantities}$$

in continuous systems.[44] LNET still means $L_{ik} = L_{ki}$. In discontinuous systems where LNET holds the thermodynamic flux J_i is the time derivative of something, i.e. it represents the rate of change of some physical quantity.

In continuous systems J_i is often a vector; we denote its components as J_{ip} ($p = 1, 2, 3$). In this case, J_i plays a role which is similar to the role played in discontinuous systems provided that $\frac{\partial J_{ip}}{\partial x_p} + \frac{\partial b_i}{\partial t} = 0$ for some quantity b_i.[45]

This equation reduces to a conservation law—and may therefore be compared with available physical models—of the form $\frac{\partial}{\partial x_p}\left(D_{ikpq}\frac{\partial a_k}{\partial x_q}\right) + \frac{\partial b_i}{\partial t} = 0$ $(i, k = 1, \ldots N;$ $q = 1, 2, 3)$ provided that $J_{ip} = D_{ikpq}\frac{\partial a_k}{\partial x_q}$ (where both D_{ikpq} and a_k are unknown so far).

Linearity of the latter relationship suggests that the k-th thermodynamic force X_k conjugated to our thermodynamic flow is a vector whose q-th component is $X_{kq} = \frac{\partial a_k}{\partial x_q}$, so that a) X_k is the gradient of a physical quantity; b) $J_{ip} = D_{ikpq}X_{kq}$; c) LNET corresponds to $D_{ikpq} = D_{kipq}$.

Remarkably, the relationships $D_{ikpq} = D_{kipq}$, $X_{kq} = \frac{\partial a_k}{\partial x_q}$, $J_{ip} = D_{ikpq}\frac{\partial a_k}{\partial x_q}$ and $\frac{\partial J_{ip}}{\partial x_p} + \frac{\partial b_i}{\partial t} = 0$ listed above are consistent with each other in steady state $(\frac{\partial}{\partial t} = 0)$ at least. In fact, LNET is related to the least dissipation variational principle

$$\delta \int d\mathbf{x}\left(X_i J_i - \frac{1}{2}L_{ik}X_i X_k\right) = 0 \quad \text{for} \quad X_i \to X_i + \delta X_i \quad ; \quad \delta J_i = 0$$

We show that the Euler-Lagrange equations[46] of this variational principle reduce to the conservation laws ruling the system in steady state.[47] To this purpose, we rewrite the variational principle as[48]

$$\delta \int L d\mathbf{x} = 0 \quad ; \quad L \equiv J_{ip}\frac{\partial a_i}{\partial x_p} - \frac{1}{2}D_{ikpq}\frac{\partial a_i}{\partial x_p}\frac{\partial a_k}{\partial x_q} \quad ; \quad a_i \to a_i + \delta a_i$$

---

[44] Here we refer to Sect. 6, Chap. VI of [8].

[45] If this is true, indeed, then according to Gauss' theorem of divergence applied to a volume V with fixed boundary Ω the integral quantity $\oint_\Omega J_{ip}da_p$ is $= \frac{d}{dt}\int_V b_i d\mathbf{x}$, i.e. is the time derivative of a volume integral, and we may apply to it the arguments concerning LNET in discontinuous systems.

[46] See Sect. A.2

[47] Here we cannot help referring to the fascinating generalization of this result due to Sienutycz et al. [17], where a duly generalized version of the least dissipation principle leads to Euler-Lagrange equations which are the conservation laws ruling the system in a time-dependent state. This is basically a generalization of LNET to the general, time-dependent case. As usual with LNET, all phenomenological coefficients remain constant.

[48] Being a variational principle with a constraint, in a rigorous treatment here we should utilize the technique of Lagrange multipliers λ_{ip}–see Sect. A.3. In contrast, we have just chosen to keep the value of J_i constant in the following of this Section, for simplicity. Had we followed the rigorous treatment, we would have obtained the same results, plus a relationship $\lambda_{ip} = \frac{\partial a_i}{\partial x_p}$ invoked no more below.

Fig. 4.8 The region of interest in the discussion on entropy production due to diffusion

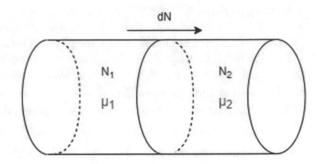

When writing down Euler-Lagrange equations $\frac{\partial L}{\partial a_i} - \frac{\partial}{\partial x_p} \frac{\partial L}{\partial \left(\frac{\partial a_i}{\partial x_p} \right)} = 0$ we observe

that $\frac{\partial L}{\partial a_i} = 0$. Moreover, $D_{ikpq} = D_{kipq}$ implies that $\frac{\partial L}{\partial \left(\frac{\partial a_i}{\partial x_p} \right)} = J_{ip} - D_{ikpq} \frac{\partial a_k}{\partial x_q}$, and,

as a consequence, $0 = \frac{\partial}{\partial x_p} \frac{\partial L}{\partial \left(\frac{\partial a_i}{\partial x_p} \right)} = \frac{\partial J_{ip}}{\partial x_p} - \frac{\partial}{\partial x_p} \left(D_{ikpq} \frac{\partial a_k}{\partial x_q} \right)$ just as expected from

relationships $J_{ip} = D_{ikpq} \frac{\partial a_k}{\partial x_q}$ and $\frac{\partial J_{ip}}{\partial x_p} + \frac{\partial b_i}{\partial t} = 0$ in steady state. Thus, validity of
LNET in a continuous system may require not only linearity of phenomenological
laws, constant coefficients etc. but also that the thermodynamic forces are gradients
of physical quantities and that the divergence of each thermodynamic flux is equal
to the partial time derivative of a physical quantity.

4.3.3   Entropy Production Due to Diffusion

We describe the diffusion[49] from region 1 to region 2 located at different lengths in
a region of cross section Σ as a 'reaction' $1 \rightarrow 2$ where one particle disappears from
region 1 (with length l and volume $l \cdot \Sigma$) and appears in region 2–see Fig. 4.8.
    Our formalism gives then $-\nu_1 = \nu_2 = 1, -dn_1 = dn_2 = d\xi', T_1 = T_2 = T$ and
$A = -\nu_1 \mu_1 - \nu_2 \mu_2 = \mu_1 - \mu_2$. If diffusion is one-dimensional and occurs only
in the x direction perpendicular to the cross section with j_{px} x-th component
of the particle flow $\mathbf{j}_p$, then volume integration on region 1 of the the balance
of mass leads to: $\frac{dn_1}{dt} \cdot l \cdot \Sigma + j_{px} \cdot \Sigma = 0$, hence $\frac{dn_1}{dt} = -\frac{j_{px}}{l}$. It follows that
$T\sigma = \Upsilon' A = \frac{d\xi'}{dt} A = -\frac{dn_1}{dt} A = \frac{j_{px} A}{l} = -\frac{j_{px}(\mu_2 - \mu_1)}{l}$. Straightforward generalization to three dimensions for arbitrarily small l leads to $\sigma = -\mathbf{j}_p \cdot \nabla \left(\frac{\mu}{T} \right)$. Moreover,
if many species are present and the particle flow of the k-th species is $\mathbf{j}_{pk} \equiv \frac{\mathbf{j}_k}{m_k}$,
further generalization gives:

$$\sigma = -\mathbf{j}_{pk} \cdot \nabla \left(\frac{\mu_k}{T} \right) = -\mathbf{j}_k \cdot \nabla \left(\frac{\mu_k^0}{T} \right)$$

---

[49] See Sect. 4.3 of [18]

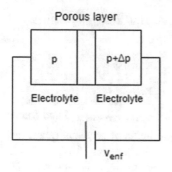

Porous layer

p   p+Δp

Electrolyte   Electrolyte

V_{enf}

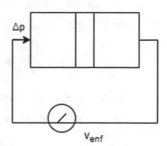

Δp

V_{enf}

Fig. 4.9 The system of interest in the discussion on Saxen's law. Two electrolytes are separated by a porous layer. If a voltage is applied, then a pressure jump is observed (left). If a pressure jump is applied, then a voltage is observed (right)

What about LNET? The quantity $\nabla\left(\frac{\mu_k}{T}\right)$ is a gradient, and the divergence of $\mathbf{j}_{pk}$ is proportional to a time derivative. Moreover, Fick's law[50] is a linear relationship between these quantities. Finally, τ is related to the diffusion time-scale. We conclude that LNET holds. It is therefore far from surprising that Prigogine has proposed a MinEP for this problem [19]–see Sect. 4.3.10.

4.3.4 Saxen's Laws

This is a generalisation of the problem above.[51] There are $k = 1, \ldots N$ species; the particle of the k-th species has electrical charge e_k. If a voltage $V_{emf} = \phi_{el2} - \phi_{el1}$ (ϕ_{el} electrostatic potential) is applied between electrolytes 1 and 2, then a pressure jump $\Delta p = p_2 - p_1$ is observed (see Fig. 4.9). All the way around, if a pressure jump Δp is applied, then a voltage V_{emf} is observed. The quantities Δp and V_{emf} are referred to as 'electroosmotic pressure' and 'streaming potential' in the literature. Here we neglect Joule heating.

Should $\phi_{el} = 0$ everywhere, we would write $-\nu_{k1} = +\nu_{k2} = +1$ and $-dn_{k1} = +dn_{k2} = d\xi'_k$ for the particles of k-th species crossing the boundary between region 1 and 2. Moreover, the relationship $T_1 = T_2 = T$ implies that $\frac{\partial \mu_k}{\partial p} = \left(\frac{\partial \mu_k}{\partial p}\right)_T = v_k$, so that $\mu_{k1} - \mu_{k2} = -v_k \Delta p$. Furthermore, $A_k = -\nu_{k1}\mu_{k1} - \nu_{k2}\mu_{k2} = \mu_{k1} - \mu_{k2} = -v_k \Delta p$. Finally, $\Upsilon'_1 = \frac{d\xi'_1}{dt}$ and $\sigma = \frac{\Upsilon'_k A_k}{T}$. If $\phi_{el} \neq 0$ then we replace $\mu_{k1} \rightarrow \mu_{k1} + e_k\phi_{el}$ and $\mu_{k2} \rightarrow \mu_{k2} + e_k\phi_{el}$, so that $A_k = -v_k\Delta p - e_k V_{emf}$ and

$$\sigma = \frac{\Upsilon'_k A_k}{T} = -\frac{V_{emf}}{T}e_k\frac{dn_{k2}}{dt} - \frac{\Delta p}{T}v_k\frac{dn_{k2}}{dt}$$

---

[50] For $\frac{\mu}{T} \propto \ln n$ at least–as e.g. in perfect gases–so that $\nabla\left(\frac{\mu}{T}\right) \propto \nabla \ln n$. See Sect. 4.3.5.
[51] We refer to Sect. 16.7 of [18].

In a linearised treatment, we define the 'streaming current' $I_{streaming} \equiv -e_k \frac{dn_{k2}}{dt}$ and the other flow $J_{streaming} \equiv -v_k \frac{dn_{k2}}{dt}$ and write:

$$I_{streaming} = L_{11}\frac{V_{emf}}{T} + L_{12}\frac{\Delta p}{T} \quad ; \quad J_{streaming} = L_{21}\frac{V_{emf}}{T} + L_{22}\frac{\Delta p}{T}$$

We stress the point that thermodynamic fluxes ($I_{streaming}$, $J_{streaming}$) and forces ($\frac{V_{emf}}{T}$, $\frac{\Delta p}{T}$) are proportional to time derivatives and gradients of physical quantities respectively.

Experimenters work with the so called 'electroosmotic coefficient' $\left(\frac{J_{streaming}}{I_{streaming}}\right)_{\Delta p=0} = \frac{L_{21}}{L_{11}}$ as well as with other coefficients, namely

$$\left(\frac{I_{streaming}}{J_{streaming}}\right)_{V_{emf}=0} = \frac{L_{12}}{L_{22}}$$

$$\left(\frac{V_{emf}}{\Delta p}\right)_{I_{streaming}=0} = -\frac{L_{12}}{L_{11}}$$

$$\left(\frac{\Delta p}{V_{emf}}\right)_{J_{streaming}=0} = -\frac{L_{21}}{L_{22}}$$

Now, LNET $L_{12} = L_{21}$ corresponds to the experimentally verified Saxen's laws:

$$\left(\frac{V_{emf}}{\Delta p}\right)_{I_{streaming}=0} = \left(\frac{J_{streaming}}{I_{streaming}}\right)_{\Delta p=0}$$

$$\left(\frac{\Delta p}{V_{emf}}\right)_{J_{streaming}=0} = -\left(\frac{I_{streaming}}{J_{streaming}}\right)_{V_{emf}=0}$$

4.3.5 Fick's Law

The thermodynamic fluxes and forces in the diffusion-related entropy production density $\sigma = -\mathbf{j}_{pk} \cdot \nabla\left(\frac{\mu_k}{T}\right)$ are $\mathbf{j}_{pk}$ and $-\nabla\left(\frac{\mu_k}{T}\right)$ respectively. Accordingly, if $\nabla T = 0$ then the linear phenomenological laws[52] are of the type $\mathbf{j}_{pk} = -\frac{L_{pk}}{T}\nabla\mu_k$. In a linear treatment of a system with N species it is usually assumed that:

$$\mathbf{j}_{pk} = -D_{ik}\nabla n_k \quad ; \quad i, k = 1 \ldots N$$

---

[52] We refer to Chap. 16 of [18]

('Fick's law'). If $N = 1$ then $\nabla \mu = \frac{\partial \mu}{\partial n} \nabla n$ and Fick's law implies $\mathbf{j}_p = -D\nabla n$, hence $L_{11} = \frac{TD}{\frac{\partial \mu}{\partial n}}$. If $N = 2$ then it is customary to deal with the case $\nabla T = 0$, $\nabla p = 0$ for simplicity. As usual in such cases, we start with the condition of extremum of Gibbs' free energy G; $dG = 0$ implies $\mu_k dn_k = 0$. But $G = \mu_k n_k$, then $\mu_k dn_k = 0$ implies $n_k d\mu_k = 0$. Moreover, LTE at all times implies $da = \frac{da}{dt} \cdot dt = \left(\frac{\partial a}{\partial t} + \mathbf{v} \cdot \nabla a\right) \cdot dt = \left(\frac{\partial a}{\partial t} + \frac{d\mathbf{r}}{dt} \cdot \nabla a\right) \cdot dt$ for the generic quantity a. In steady state $\frac{\partial}{\partial t} = 0$, then $da = d\mathbf{r} \cdot \nabla a$. Let $a = \mu_k$. Accordingly, $n_k d\mu_k = 0$ implies $(n_k \nabla \mu_k) \cdot d\mathbf{r} = 0$ for arbitrary $d\mathbf{r}$, hence:

$$0 = n_k \nabla \mu_k = n_1 \nabla \mu_1 + n_2 \nabla \mu_2$$

Correspondingly:

$$\sigma = -\mathbf{j}_{pk} \cdot \nabla \left(\frac{\mu_k}{T}\right) = -\frac{\mathbf{j}_{pk}}{T} \cdot \nabla \mu_k = -\frac{\mathbf{j}_{p1}}{T} \cdot \nabla \mu_1 - \frac{\mathbf{j}_{p2}}{T} \cdot \nabla \mu_2 = -\frac{1}{T}\left(\mathbf{j}_{p1} - \frac{n_1}{n_2}\mathbf{j}_{p2}\right) \cdot \nabla \mu_1$$

Further simplification follows from the fact that $d\left(\frac{1}{\rho}\right) = v_k dc_k$ for $\nabla T = 0$ and $\nabla p = 0$, as the definition of the mass flow $\mathbf{j}_k$ for the k-th species implies $\rho \frac{d}{dt}\left(\frac{1}{\rho}\right) = v_k \rho \frac{dc_k}{dt} = v_k \nabla \cdot \mathbf{j}_k$. Usually, diffusion leaves ρ unaffected, as it is a slow, incompressible process. Then $v_k \nabla \cdot \mathbf{j}_k = 0$. Moreover, $v_k \nabla \cdot \mathbf{j}_k = \nabla \cdot (v_k \mathbf{j}_k) - \mathbf{j}_k \cdot \nabla v_k = \nabla \cdot (v_k \mathbf{j}_k) - \mathbf{j}_k \cdot \nabla \left(\frac{\partial \mu_k^0}{\partial p}\right)_T$ and $\mathbf{j}_k \cdot \nabla \left(\frac{\partial \mu_k^0}{\partial p}\right)_T = \mathbf{j}_{pk} \cdot \nabla \left(\frac{\partial \mu_k}{\partial p}\right)_T = \left[\frac{\partial (\mathbf{j}_{pk} \cdot \nabla \mu_k)}{\partial p}\right]_T \propto \left[\frac{\partial (\mathbf{j}_{pk} \cdot \nabla \mu_k)}{\partial p}\right]_T = O\left(n_k \nabla \mu_k\right) = 0$ because of $n_k \nabla \mu_k = 0$. Consequently, incompressibility implies $\nabla \cdot (v_k \mathbf{j}_k) = 0$. Together with the definitions of $\mathbf{j}_{pk} \equiv \frac{\mathbf{j}_k}{m_k}$ and of $v_k' \equiv m_k v_k$, this gives[53] $\mathbf{j}_{p1} v_1' + \mathbf{j}_{p2} v_2' = 0$. Then

$$\sigma = -\frac{1}{T}\left(1 + \frac{n_1 v_1'}{n_2 v_2'}\right)\mathbf{j}_{p1} \cdot \nabla \mu_1 = -\frac{1}{T}\left(1 + \frac{n_1 v_1'}{n_2 v_2'}\right)\left(\frac{\partial \mu_1}{\partial n_1}\right)_{p,T} \mathbf{j}_{p1} \cdot \nabla n_1$$

Comparison of this formula with our discussion of the $N = 1$ problem suggests that Fick's law for the species 1 takes the form $\mathbf{j}_{p1} = -D_1 \nabla n_1$, i.e. the same form of the $N = 1$ case but with a modified form for the coefficient. The formula for σ shows that this suggestion is valid provided that

$$D_1 = \frac{L_{11}}{T}\left(1 + \frac{n_1 v_1'}{n_2 v_2'}\right)\left(\frac{\partial \mu_1}{\partial n_1}\right)_{p,T}$$

---

[53] The argument $v_k \mathbf{j}_k$ of the divergence is a harmonic function of spatial coordinates because of Fick's law, because $\nabla \cdot (v_k \mathbf{j}_k) = 0$ vanishes and because we neglect the contributions of the gradients of v_k to $\nabla \cdot (v_k \mathbf{j}_k)$ as higher-order terms. Now, a harmonic function which vanishes on the boundary—as in case of far-away boundaries—vanishes everywhere.

Here we limit ourselves to write down the results of the (cumbersome but straightforward) algebra for $N = 3$.

$$n_1 \nabla \mu_1 + n_2 \nabla \mu_2 + n_3 \nabla \mu_3 = 0 \quad ; \quad \mathbf{j}_{p1} v_1' + \mathbf{j}_{p2} v_2' + \mathbf{j}_{p3} v_3' = 0$$

$$\mathbf{j}_{p1} = -D_{11} \nabla n_1 - D_{12} \nabla n_2 \quad ; \quad \mathbf{j}_{p2} = -D_{21} \nabla n_1 - D_{22} \nabla n_2$$

$$\sigma = -\frac{\mathbf{j}_{p1}}{T} \cdot \nabla \mu_1 - \frac{\mathbf{j}_{p2}}{T} \cdot \nabla \mu_2 - \frac{\mathbf{j}_{p3}}{T} \cdot \nabla \mu_3 = \mathbf{F}_1 \cdot \mathbf{j}_{p1} + \mathbf{F}_2 \cdot \mathbf{j}_{p2}$$

$$\mathbf{F}_1 = -\frac{1}{T} \left(\nabla \mu_1 + \frac{n_1 v_1'}{n_3 v_3'} \nabla \mu_1 + \frac{n_2 v_1'}{n_3 v_3'} \nabla \mu_2 \right)$$

$$\mathbf{F}_2 = -\frac{1}{T} \left(\nabla \mu_2 + \frac{n_2 v_2'}{n_3 v_3'} \nabla \mu_2 + \frac{n_1 v_2'}{n_3 v_3'} \nabla \mu_1 \right)$$

$$\nabla \mu_1 = \left(\frac{\partial \mu_1}{\partial n_1} \right)_{p,T} \nabla n_1 + \left(\frac{\partial \mu_1}{\partial n_2} \right)_{p,T} \nabla n_2 \quad ; \quad \nabla \mu_2 = \left(\frac{\partial \mu_2}{\partial n_1} \right)_{p,T} \nabla n_1 + \left(\frac{\partial \mu_2}{\partial n_2} \right)_{p,T} \nabla n_2$$

$$L_{11} = \frac{D_{11} d - D_{12} b}{ad - bc} \quad ; \quad L_{12} = \frac{D_{12} a - D_{11} c}{ad - bc}$$

$$L_{21} = \frac{D_{21} d - D_{22} b}{ad - bc} \quad ; \quad L_{22} = \frac{D_{22} a - D_{21} c}{ad - bc}$$

$$a \equiv \left(1 + \frac{n_1 v_1'}{n_3 v_3'} \right) \left(\frac{\partial \mu_1}{\partial n_1} \right)_{p,T} + \left(\frac{n_2 v_1'}{n_3 v_3'} \right) \left(\frac{\partial \mu_2}{\partial n_1} \right)_{p,T}$$

$$b \equiv \left(1 + \frac{n_2 v_2'}{n_3 v_3'} \right) \left(\frac{\partial \mu_2}{\partial n_1} \right)_{p,T} + \left(\frac{n_2 v_2'}{n_3 v_3'} \right) \left(\frac{\partial \mu_1}{\partial n_1} \right)_{p,T}$$

$$c \equiv \left(1 + \frac{n_1 v_1'}{n_3 v_3'} \right) \left(\frac{\partial \mu_1}{\partial n_2} \right)_{p,T} + \left(\frac{n_2 v_1'}{n_3 v_3'} \right) \left(\frac{\partial \mu_2}{\partial n_2} \right)_{p,T}$$

$$d \equiv \left(1 + \frac{n_2 v_2'}{n_3 v_3'} \right) \left(\frac{\partial \mu_2}{\partial n_2} \right)_{p,T} + \left(\frac{n_2 v_2'}{n_3 v_3'} \right) \left(\frac{\partial \mu_1}{\partial n_2} \right)_{p,T}$$

According to LNET $L_{12} = L_{21}$; this corresponds to:

$$D_{12} a + D_{22} b = D_{21} d + D_{11} c$$

a result which has been experimentally confirmed with accuracy $\approx O\left(10^{-2} \right)$.

4.3.6 Soret Effect and Dufour Effect

When discussing diffusion, we have assumed $\nabla T = 0$ insofar. Generally speaking, however, particle diffusion drives heat transport in steady state (Dufour effect). Correspondingly, heat transport drives particle diffusion (Soret effect). The entropy production density when both particle diffusion and heat transport occur is:

$$\sigma = -\mathbf{j}_k \cdot \nabla \left(\frac{\mu_k^0}{T} \right) + \mathbf{q} \cdot \nabla \left(\frac{1}{T} \right) = (\mathbf{q} - \mu_k^0 \mathbf{j}_k) \cdot \nabla \left(\frac{1}{T} \right) - \frac{\mathbf{j}_k}{T} \cdot \nabla \mu_k^0$$

Let us rewrite σ with the help of the following relationships[54] concerning μ_k^0: $d\mu_k^0 = \left(d\mu_k^0 \right)_{p,T} + \left(\frac{\partial \mu_k^0}{\partial T} \right)_p dT + \left(\frac{\partial \mu_k^0}{\partial p} \right)_T dp$, $\left(\frac{\partial \mu_k^0}{\partial T} \right)_p - s_k$, $\left(\frac{\partial \mu_k^0}{\partial p} \right)_T = v_k$ and $h_k = \mu_k^0 + T s_k$, hence $\nabla \mu_k^0 = \left(\nabla \mu_k^0 \right)_{p,T} - s_k \nabla T + v_k \nabla p = \left(\nabla \mu_k^0 \right)_{p,T} + T^2 s_k \nabla \left(\frac{1}{T} \right) + v_k \nabla p$. If $\nabla p = 0$ then these relationships lead to:

$$\sigma = \mathbf{q}_{reduced} \cdot \nabla \left(\frac{1}{T} \right) - \frac{\mathbf{j}_k}{T} \cdot \left(\nabla \mu_k^0 \right)_{p,T} \quad ; \quad \mathbf{q}_{reduced} \equiv \mathbf{q} - h_k \mathbf{j}_k$$

Remarkably, the quantity $\mathbf{q}$ includes the contribution $h_k \mathbf{j}_k$ of particle diffusion; it may differ form zero even if no heat conduction occurs.

Now, we take $N = 2$ in the $\nabla T = 0$, $\nabla p = 0$ case, invoke the results of our discussion of Fick's law and write $n_1 \left(\nabla \mu_1 \right)_{p,T} + n_2 \left(\nabla \mu_2 \right)_{p,T} = 0$ and $\mathbf{j}_{p1} v_1' + \mathbf{j}_{p2} v_2' = 0$. Moreover, $\nabla \mu_1 = \left(\frac{\partial \mu_1}{\partial n_1} \right)_{p,T} \nabla n_1$. It follows that:

$$\sigma = \mathbf{q}_{reduced} \cdot \nabla \left(\frac{1}{T} \right) - \frac{1}{T} \left(1 + \frac{n_1 v_1'}{n_2 v_2'} \right) \left(\frac{\partial \mu_1}{\partial n_1} \right)_{p,T} \mathbf{j}_{p1} \cdot \nabla n_1$$

The thermodynamic forces are $\nabla \left(\frac{1}{T} \right)$ and $-\frac{1}{T} \left(1 + \frac{n_1 v_1'}{n_2 v_2'} \right) \left(\frac{\partial \mu_1}{\partial n_1} \right)_{p,T} \nabla n_1$ respectively. The corresponding conjugated fluxes are $\mathbf{q}_{reduced}$ and $\mathbf{j}_{p1}$ respectively. The linear phenomenological relationships are

$$\mathbf{q}_{reduced} = L_{qq} \nabla \left(\frac{1}{T} \right) + L_{qp} (-1) \frac{1}{T} \left(1 + \frac{n_1 v_1'}{n_2 v_2'} \right) \left(\frac{\partial \mu_1}{\partial n_1} \right)_{p,T} \nabla n_1$$

$$\mathbf{j}_{p1} = L_{pq} \nabla \left(\frac{1}{T} \right) + L_{pp} (-1) \frac{1}{T} \left(1 + \frac{n_1 v_1'}{n_2 v_2'} \right) \left(\frac{\partial \mu_1}{\partial n_1} \right)_{p,T} \nabla n_1$$

Not surprisingly, we retrieve the diffusion coefficient D_1 of species 1 in a $N = 2$ problem with $\nabla T = 0$ in the last addendum on the R.H.S. of the second equation,

---

[54] We refer to Chap. 16 of [18]

provided that we identify L_{pp} with L_{11}. It is customary to define the thermodiffusion coefficient $D_T \equiv \frac{L_{pq}}{n_1 T^2}$, the Dufour coefficient $D_D \equiv L_{qp} \frac{1}{n_1 T} \left(1 + \frac{n_1 v_1'}{n_2 v_2'}\right) \left(\frac{\partial \mu_1}{\partial n_1}\right)_{p,T}$ and the Soret coefficient $S_T \equiv \frac{D_T}{D_1}$. Onsager's symmetry $L_{qp} = L_{pq}$ leads to the following, experimentally confirmed relationship which does not involve L_{qq}, i.e. heat conduction.

$$\frac{D_D}{S_T D_1} = T \left(1 + \frac{n_1 v_1'}{n_2 v_2'}\right) \left(\frac{\partial \mu_1}{\partial n_1}\right)_{p,T}$$

4.3.7 Entropy Production Due to Reactions Among Species

Let us consider $r = 1, \ldots R$ reactions among $k = 1, \ldots N$ chemical (nuclear) species. Here we neglect all other processes,[55] so that both the energy u per unit mass and the mass density ρ are constant and $\sigma \equiv \rho \frac{ds}{dt}$. The First Principle of thermodynamics $du = T ds - p d \left(\frac{1}{\rho}\right) + \mu_k^0 dc_k$ leads to $\frac{ds}{dt} = -\frac{\mu_k^0}{T} \frac{dc_k}{dt}$, a quantity which is strictly positive because of the Second Principle of thermodynamics. Then, irreversibility leads to $\sigma = \frac{A_r^0 \Upsilon_r}{T} > 0$ (Sect. 4.2.6). We stress the point that it is the sum on r which is positive, not each addendum separately.

As for the r-th reaction, we define the 'reaction degree' $\xi_r(t)$ as the solution of the differential equation $\Upsilon_r = \frac{d\xi_r}{dt}$ with the initial condition $\xi_r(t = 0) = 0$. The balance of mass of the k-th species leads therefore to $\rho_k(t) = \rho_k(t = 0) + v_{kr} \xi_r(t)$. The reaction rate is the total time derivative of the reaction degree. However, here we say nothing about $\frac{A_r^0}{T}$.

In order to check whether LNET holds, let us discuss the problem of 3 species (A, B and C) undergoing 3 reactions described by the following intertwined equations:

$$\frac{dn_A}{dt} = k_{-1} n_B - k_{+1} n_A \quad ; \quad \frac{dn_B}{dt} = k_{-2} n_C - k_{+2} n_B \quad ; \quad \frac{dn_C}{dt} = k_{-3} n_A - k_{+3} n_C$$

Here $n_A = m_A^{-1} \rho_A, n_B = m_B^{-1} \rho_B, n_C = m_C^{-1} \rho_C, \mu_A^0 = m_A^{-1} \mu_A, \mu_B^0 = m_B^{-1} \mu_B, \mu_C^0 = m_C^{-1} \mu_C, \rho_A = \rho \cdot c_A, \rho_B = \rho \cdot c_B, \rho_C = \rho \cdot c_C, v_{A1} = v_{B2} = v_{C3} = -1, v_{B1} = v_{C2} = v_{A3} = +1, A_1^0 = -v_{A1} \mu_A^0 - v_{B1} \mu_B^0 = \mu_A^0 - \mu_B^0, A_2^0 = \ldots = \mu_B^0 - \mu_C^0, A_3^0 = \ldots = \mu_C^0 - \mu_A^0, d\xi_1 = v_{A1}^{-1} d\rho_A = v_{B1}^{-1} d\rho_B, d\xi_2 = v_{B2}^{-1} d\rho_B = v_{C2}^{-1} d\rho_C, d\xi_3 = v_{C3}^{-1} d\rho_C = v_{A3}^{-1} d\rho_A, \Upsilon_1 = \frac{d\xi_1}{dt}, \Upsilon_2 = \frac{d\xi_2}{dt}, \Upsilon_3 = \frac{d\xi_3}{dt}$ and

$$T\sigma = \Upsilon_1 A_1^0 + \Upsilon_2 A_2^0 + \Upsilon_3 A_3^0$$

We rewrite σ in order to check the validity of LNET. We write $\Upsilon_1 A_1^0 = -v_{A1} \mu_A^0 \frac{d\xi_1}{dt} - v_{B1} \mu_B^0 \frac{d\xi_1}{dt} = -v_{A1} \mu_A^0 \frac{d\rho_A}{v_{A1} dt} - v_{B1} \mu_B^0 \frac{d\rho_B}{v_{B1} dt} = -\mu_A^0 \frac{d\rho_A}{dt} -$

---

[55] Here we refer to Sects. 2.5 and 4.1 of [18]

$\mu_B^0 \frac{d\rho_B}{dt} = -\mu_A \frac{dn_A}{dt} - \mu_B \frac{dn_B}{dt} = \Upsilon_1' A_1$ where $\Upsilon_1' = \frac{d\xi_1'}{dt}$, $d\xi_1' = \frac{dn_A}{\nu_{A1}} = \frac{dn_B}{\nu_{B1}}$ and $A_1 \equiv -\nu_{A1}\mu_A - \nu_{B1}\mu_B$, hence $A_1 = \mu_A - \mu_B$. Similar results holds for reactions 2 and 3. It follows that:

$$A_1 + A_2 + A_3 = 0$$

$$\Upsilon_1' = k_{+1}n_A - k_{-1}n_B \quad ; \quad \Upsilon_2' = k_{+2}n_B - k_{-2}n_C \quad ; \quad \Upsilon_3' = k_{+3}n_C - k_{-3}n_A$$

$$T\sigma = \Upsilon_1' A_1 + \Upsilon_2' A_2 + \Upsilon_3' A_3 = \left(\Upsilon_1' - \Upsilon_3'\right) A_1 + \left(\Upsilon_2' - \Upsilon_3'\right) A_2$$

For the sake of simplicity, we assume that both A, B and C are perfect gases. Accordingly, we write $\mu_A = -k_B T \ln (n_A)$, $\mu_B = -k_B T \ln (n_B)$ and $\mu_C = -k_B T \ln (n_C)$. It follows that $A_1 = -k_B T \ln \left(n_A^{-1} n_B\right)$, $A_2 = -k_B T \ln \left(n_B^{-1} n_C\right)$ and $A_3 = -k_B T \ln \left(n_C^{-1} n_A\right)$.

Finally, let we assume linearization, i.e.

$$a = a_0 + \delta a \quad ; \quad |\delta a| \propto O\left(\epsilon\right) \quad ; \quad \epsilon \ll 1$$

for the generic quantity a. We discuss below the range of validity of this assumption. In particular, we write $\left(\Upsilon_1'\right)_0 = \left(\Upsilon_2'\right)_0 = \left(\Upsilon_3'\right)_0 = 0$, hence $\frac{n_{B0}}{n_{A0}} = \frac{k_{+1}}{k_{-1}}$, $\frac{n_{C0}}{n_{B0}} = \frac{k_{+2}}{k_{-2}}$ and $\frac{n_{A0}}{n_{C0}} = \frac{k_{+3}}{k_{-3}}$. It follows that[56]:

$$A_1 = -k_B T \ln \left(\frac{n_B}{n_A}\right) = -k_B T \ln \left(\frac{n_{B0} + \partial n_B}{n_{A0} + \partial n_A}\right) \approx -k_B T \ln \left(\frac{n_{B0}}{n_{A0}} + \frac{\delta n_B}{n_{A0}} - \frac{n_{B0}\delta n_A}{n_{A0}^2}\right) =$$

$$= -k_B T \ln \left[\left(\frac{n_{B0}}{n_{A0}}\right)\left(1 + \frac{\delta n_B}{n_{B0}} - \frac{\delta n_A}{n_{A0}}\right)\right] =$$

$$= -k_B T \ln \left(\frac{n_{B0}}{n_{A0}}\right) - k_B T \ln \left(1 + \frac{\delta n_B}{n_{B0}} - \frac{\delta n_A}{n_{A0}}\right) \approx$$

$$\approx -k_B T \ln \left(\frac{n_{B0}}{n_{A0}}\right) + k_B T \left(\frac{\delta n_A}{n_{A0}} - \frac{\delta n_B}{n_{B0}}\right) = A_{10} + \delta A_1$$

where $\quad A_{10} = -k_B T \ln \left(\frac{n_{B0}}{n_{A0}}\right) \quad$ and $\quad \delta A_1 = k_B T \left(\frac{\delta n_A}{n_{A0}} - \frac{\delta n_B}{n_{B0}}\right) = \frac{k_B T}{n_{A0}} \left(\delta n_A - \frac{k_{-1}\delta n_B}{k_{+1}}\right). \quad$ Moreover $\quad \Upsilon_1' = \left(\Upsilon_1'\right)_0 + \delta \Upsilon_1' = \delta \Upsilon_1' = k_{+1}\partial n_A -$

---

[56] If $|x| \ll 1$ then $(1 + x)^{-1} \approx 1 - x$ and $\ln (1 + x) \approx x$.

$k_{-1} \partial n_B = \frac{k_{+1} n_{A0}}{k_B T} \partial A_1$. The same line of reasoning leads to similar results for B and C. We obtain:

$$\delta \Upsilon_1' = \frac{k_{+1} n_{A0}}{k_B T} \partial A_1 \;\; ; \;\; \delta \Upsilon_2' = \frac{k_{+2} n_{B0}}{k_B T} \partial A_2 \;\; ; \;\; \delta \Upsilon_3' = \frac{k_{+2} n_{C0}}{k_B T} \partial A_3$$

The relationship $A_1 + A_2 + A_3 = 0$ implies $\delta A_1 + \delta A_2 + \delta A_3 = 0$, hence $\delta \Upsilon_3' = -\frac{k_{+2} n_{C0}}{k_B T} (\partial A_1 + \partial A_2)$ and we may conclude writing down our results in Onsager-symmetric form:

$$\delta \left(\Upsilon_1' - \Upsilon_3' \right) = L_{11} \delta A_1 + L_{12} \delta A_2 \;\; ; \;\; \delta \left(\Upsilon_2' - \Upsilon_3' \right) = L_{21} \delta A_1 + L_{22} \delta A_2$$

$$L_{11} = \frac{k_{+1} n_{A0} + k_{+3} n_{C0}}{k_B T} \;\; ; \;\; L_{22} = \frac{k_{+2} n_{B0} + k_{+3} n_{C0}}{k_B T} \;\; ; \;\; L_{12} = L_{21} = \frac{k_{+3} n_{C0}}{k_B T}$$

We are left with a relevant physical question: when is our linearization procedure valid? Let us come back to our estimate for A_1. At thermodynamic equilibrium, $A_1 = 0$ (this is just the familiar 'rule of phases'). If our steady state differs from equilibrium just a bit, then we expect $|A_1|$ to be somehow small. Now, the relationships $A_{10} = -k_B T \ln \left(\frac{n_{B0}}{n_{A0}} \right)$ and $\frac{n_{B0}}{n_{A0}} = \frac{k_{+1}}{k_{-1}}$ lead to a well-known formula for the reaction velocity $V_1 \equiv k_{-1} - k_{+1}$:

$$V_1 = k_{-1} \left[1 - \exp \left(-\frac{A_{10}}{k_B T} \right) \right]$$

If $|A_1| \ll k_B T$ then $V_1 \ll 1$, $k_{+1} \approx k_{-1}$ and $n_{A0} \approx n_{B0}$ in steady state, and linearization makes sense. The inequality $|A_1| \ll k_B T$ is a condition of validity for LNET to be valid for reaction 1. Similar considerations hold for reactions 2 and 3 too.[57]

4.3.8   Coupling of Diffusion and Reactions

We have seen in Sects. 4.3.3 and 4.3.7 that LNET applies to particle diffusion and that, if $|A_r| \ll k_B T$ for the r-th reaction ($r = 1, \ldots R$) among particles of $k = 1, \ldots N$ species, then LNET applies also to reactions. Here we discuss the case of diffusion and reactions occurring together.[58] The relevant relationships are:

$$\sigma = -\mathbf{j}_k \cdot \nabla \left(\frac{\mu_k^0}{T} \right) + \frac{A_r^0 \Upsilon_r}{T} \;\; ; \;\; \rho \frac{dc_k}{dt} + \nabla \cdot \mathbf{j}_k = \nu_{kr} \Upsilon_r \;\; ; \;\; A_r^0 = -\mu_k^0 \nu_{kr}$$

---

[57] Biophysicists are routinely involved with problems with a large number of interacting chemical species. For a review of LNET in biophysics, Ref. [20] is an excellent review.

[58] We refer to Chap. V, Sect. 4.b of [8] as well as to [18].

Thermodynamic forces and fluxes[59] are X_i and Y_i, where $i = 1, \ldots R + N$, $\mathbf{X}_i = -\nabla\left(\frac{\mu_k^0}{T}\right)$, $\mathbf{Y}_i = \mathbf{j}_i$ for $i = 1, \ldots N$ and $X_i = \frac{A_i^0}{T}$, $Y_i = \Upsilon_i$ for $i = N + 1, \ldots R + N$. Some authors define $\Gamma_k = -\frac{\mu_k^0}{T}$, a compact notation we are going to make use of in the following. Accordingly, we write $\frac{A_r^0}{T} = v_{kr}\Gamma_k$ and $\sigma = \mathbf{j}_k \cdot \nabla\Gamma_k + v_{kr}\Gamma_k\Upsilon_r = X_m Y_m$, $m = 1, \ldots N + R$. The total amount of entropy produced inside the system is

$$P \equiv \frac{d_i S}{dt} = \int \sigma\, d\mathbf{x} = \int X_m Y_m d\mathbf{x} = \int \left(\mathbf{j}_k \cdot \nabla\Gamma_k + v_{kr}\Gamma_k\Upsilon_r\right) d\mathbf{x}$$

The notation $\frac{d_X P}{dt} \equiv \int Y_m \frac{dX_m}{dt} d\mathbf{x}$; $\frac{d_Y P}{dt} \equiv \int X_m \frac{dY_m}{dt} d\mathbf{x}$ is also commonly adopted in the literature. Note that $\frac{dP}{dt} = \frac{d_X P}{dt} + \frac{d_Y P}{dt}$. LNET ($L_{mn} = L_{nm}$ where $m, n = 1, \ldots R + N$ and $Y_m = L_{mn}X_n$) implies $\frac{d_X P}{dt} = \frac{d_Y P}{dt}$ and $\frac{dP}{dt} = 2\frac{d_X P}{dt}$.

Let us write the Euler-Lagrange equations of the minimum entropy production principle,[60] which reads[61]:

$$\min = P = \int \left(\mathbf{j}_k \cdot \nabla\Gamma_k + v_{kr}\Gamma_k\Upsilon_r\right) d\mathbf{x} = \int \left(L_{ki}\nabla\Gamma_i\nabla\Gamma_k + v_{kr}\Gamma_k L_{rp}\frac{A_p^0}{T}\right) d\mathbf{x} =$$

$$= \int \left(L_{ki}\nabla\Gamma_i\nabla\Gamma_k + v_{kr}\Gamma_k L_{rp}\Gamma_i v_{ip}\right) d\mathbf{x} = \int L d\mathbf{x} \; ; \quad L \equiv L_{ki}\nabla\Gamma_i\nabla\Gamma_k + L_{rp}v_{kr}v_{ip}\Gamma_k\Gamma_i$$

Taking the Γ_i's as the Lagrangian coordinates[62] and invoking LNET, we have

$$\frac{\partial L}{\partial \Gamma_q} = 2L_{rp}v_{kr}v_{ip}\Gamma_k\delta_{iq} = 2L_{rp}v_{kr}v_{qp}\Gamma_k = 2v_{qp}L_{pr}\frac{A_r^0}{T} = 2v_{qp}\Upsilon_p$$

Moreover[63]

$$\nabla \cdot \left[\frac{\partial L}{\partial\left(\nabla\Gamma_q\right)}\right] = 2\nabla \cdot \left(L_{ki}\delta_{kq}\nabla\Gamma_i\right) = 2\nabla \cdot \left(L_{qi}\nabla\Gamma_i\right) = 2\nabla \cdot \mathbf{j}_q$$

and the Euler-Lagrange equations $\frac{\partial L}{\partial\Gamma_q} - \nabla \cdot \left[\frac{\partial L}{\partial(\nabla\Gamma_q)}\right] = 0$ give at once

$$\nabla \cdot \mathbf{j}_q = v_{qp}\Upsilon_p$$

---

[59] As usual, both thermodynamic forces and fluxes may be scalars, component of vectors...

[60] Remember that the Υ_r's are related to the time derivative of the local values of the particle densities of reacting particles; as such, they depend on the affinities.

[61] For simplicity, we skip dummy indices running on spatial coordinates unless necessary.

[62] See Sect. A.2

[63] Here $\nabla \cdot \left[\frac{\partial L}{\partial(\nabla\Gamma_q)}\right]$ is a shortcut for $\frac{\partial}{\partial x_l}\left[\frac{\partial L}{\partial\left(\frac{\partial\Gamma_q}{\partial x_l}\right)}\right]$ where $l = 1, 2, 3$ is a dummy index running on spatial coordinates.

i.e. the correct particle balance equations for $\frac{dc_k}{dt} = 0$. The case $\frac{dc_k}{dt} \neq 0$ is discussed in Sect. 4.3.9.

4.3.9   Stability Versus the Coupling of Diffusion and Reactions

Let us compute the sign of $\frac{d_X P}{dt}$ with the help of Gauss' theorem of divergence, while invoking no LNET. Regardless of the actual value of $\frac{dc_k}{dt}$, $\frac{d_X P}{dt} = \int \left(\mathbf{j}_k \cdot \nabla \frac{d\Gamma_k}{dt} + \Upsilon_r v_{kr} \frac{d\Gamma_k}{dt} \right) d\mathbf{x} = \oint \frac{d\Gamma_k}{dt} \mathbf{j}_k \cdot d\mathbf{a} + \int \left(-\frac{d\Gamma_k}{dt} \nabla \cdot \mathbf{j}_k + \Upsilon_r v_{kr} \frac{d\Gamma_k}{dt} \right) d\mathbf{x}$.
If either fluxes $\mathbf{j}_k$ vanish or thermodynamic forces Γ_k are constant at the boundary, then the surface integral vanishes and $\frac{d_X P}{dt} = \int \left(-\nabla \cdot \mathbf{j}_k + \Upsilon_r v_{kr} \right) \frac{d\Gamma_k}{dt} d\mathbf{x} =$
$\int \rho \frac{dc_k}{dt} \frac{d\Gamma_k}{dt} d\mathbf{x} = -\int \rho \frac{dc_k}{dt} \frac{d}{dt} \left(\frac{\mu_k^0}{T} \right) d\mathbf{x} = \int (\Psi - \Psi_{N=1}) \, d\mathbf{x}$. Here Ψ is the quantity involved in the GEC of Sect. 3.6, and is always ≤ 0 because LTE holds everywhere at all times. The quantity $\Psi_{N=1}$ is just the value of Ψ whenever $N = 1$, i.e. if there is only 1 species and no reaction. In particular, the fact that $\Psi \leq 0$ regardless of the numbers of species and reactions implies that $\Psi_{N=1} \leq 0$. But then, as we raise these numbers, $\Psi = \Psi_{N=1} + (\Psi - \Psi_{N=1})$ remains ≤ 0 everywhere at all times regardless of N and the number of reactions only if $\Psi - \Psi_{N=1}$ too remains ≤ 0 everywhere at all times regardless of N and the number of reactions.[64] Accordingly: $\int (\Psi - \Psi_{N=1}) \, d\mathbf{x} \leq 0$, hence the result which some authors [21] refer to as 'general evolution criterion' (Sect. 3.6; the sign '=' replaces the sign '<' in steady state only):

$$\frac{d_X P}{dt} \leq 0$$

4.3.10   Minimum Entropy Production in Continuous Systems

If LNET holds in the fluid with diffusion and reactions of Sect. 4.3.9, then $\frac{d_X P}{dt} = \frac{d_Y P}{dt}$ and therefore $\frac{dP}{dt} = 2 \frac{d_X P}{dt}$. The inequality $\frac{d_X P}{dt} \leq 0$ gives:

$$\frac{dP}{dt} \leq 0$$

The operator '=' replaces the sign '<' in steady state only. At thermodynamic equilibrium no entropy whatsoever is produced at any time and $P = 0$ at all times,

---

[64] As we add more and more species and more and more reactions among them, the contribution to Ψ of every new species and every new reaction separately must be ≤ 0 regardless of the detailed description of each species and of each reaction, as the sign of Ψ depends only on the assumption of LTE.

hence $\frac{dP}{dt} = 0$. Far from thermodynamic equilibrium, both particle diffusion and reactions among particles may raise entropy, hence $P > 0$. In particular, $P \equiv P_{st} > 0$ and $\frac{dP}{dt} = 0$ in steady states far from thermodynamic equilibrium. In our system, validity of LNET is a 'sufficient' condition, not a 'necessary' condition[65] for $\frac{dP}{dt} < 0$ during relaxation towards the steady state.[66] The relevance of the inequality $\frac{dP}{dt} \leq 0$ is made clear by the following argument: together, the conditions $\frac{d(P - P_{st})}{dt} = \frac{dP}{dt} \leq 0$ (the operator '=' replacing the sign '<' in steady state only), $P - P_{st} = 0$ in steady state and $P - P_{st} \geq 0$ (i.e. P achieves its minimum value in steady state) are sufficient conditions for the eventual relaxation of P to P_{st} according to Lyapunov's theory of stability (and $P - P_{st}$ is called 'Lyapunov function'). Correspondingly, a sufficient condition for the stability of a steady state is that the latter satisfies the MinEP

$$P = \min$$

This extremum condition is universally referred to in the literature as to 'Prigogine's minimum entropy production principle'. Just as for discontinuous systems, the proof of this MinEP requires[67] both the assumption of LTE everywhere at all times and the LNET symmetry in the form $L_{mn} = L_{nm}$. Without this symmetry, LTE alone leads to $\frac{d_X P}{dt} \leq 0$ only, through GEC. Note also that the boundary condition of vanishing fluxes and/or constant thermodynamic forces resembles the condition $J_i = 0$ and/or $X_i =$ constant of MinEP in discontinuous systems (Sect. 4.1.4). However, here the entropy production due to diffusion and reactions among particles only is involved, in contrast with the case of discontinuous systems.

This MinEP has a simple interpretation in terms of Le Châtelier's principle at LTE [24]. Let us suppose that a physical system involves just one thermodynamic force, a vector field $\mathbf{X} = -\nabla \phi$ where ϕ plays the role of a potential. Correspondingly, there is one thermodynamic flux $\mathbf{J}$, which must be also a vector because $\mathbf{J} \cdot \mathbf{X}$ is a scalar by definition. For simplicity, let us assume $\mathbf{J} = L\mathbf{X}$ where $L = L(\phi)$ too is a scalar. In steady state the conservation equation is just $\nabla \cdot \mathbf{J} = 0$, i.e. $\nabla \cdot (L\nabla\phi) = 0$. According to our MinEP, the steady state satisfies $\delta P = 0$ where $P = \int (\mathbf{J} \cdot \mathbf{X}) \, d\mathbf{x} = \int L|\nabla\phi|^2 d\mathbf{x}$ and the volume integral is performed on the volume of the system. Volume integration of the conservation equation leads to the constraint of zero net flux across the boundary, i.e. $\oint \mathbf{J} \cdot d\mathbf{a} = 0$ (all surface integrals being performed

---

[65] We recall that in the statement 'A implies B' A is the 'sufficient' condition of B and B is the 'necessary' condition of A. If 'nonA' is true whenever 'A' is false, then 'A implies B' if and only if 'nonB implies nonA'.

[66] There is at least one example of a system–namely, glucose catabolism in epithelial breast cells where entropy is produced by both chemical reactions, heat conduction and particle transport and the results of numerical simulations show that $\frac{dP}{dt} < 0$ holds even if the validity of assumptions which LNET relies upon is not ensured [22].

[67] As explicitly stated by Glansdorff and Prigogine in their seminal paper, Ref. [23].

on the system boundary). Accordingly[68] we write $\delta\left(P + \lambda \oint \mathbf{J} \cdot d\mathbf{a}\right) = 0$ with λ Lagrange multiplier and:

$$\delta P = \delta \int L|\nabla\phi|^2 d\mathbf{x} = \delta \int [\nabla (L\phi) \cdot \nabla\phi - \phi\nabla L \cdot \nabla\phi] \, d\mathbf{x} =$$

$$= \delta \int [\nabla (L\phi) \cdot \nabla\phi - \phi (\nabla \cdot (L\nabla\phi) - L\Delta\phi)] \, d\mathbf{x} = \delta \int [\nabla (L\phi) \cdot \nabla\phi + L\phi\Delta\phi] \, d\mathbf{x} =$$

$$= \delta \int \nabla \cdot (L\phi\nabla\phi) \, d\mathbf{x} = \delta \oint L\phi\nabla\phi \cdot d\mathbf{a} = \oint [(L\nabla\phi)\,\delta\phi + \phi\delta\,(L\nabla\phi)] \cdot d\mathbf{a} = -\oint (\phi\delta\mathbf{J} + \mathbf{J}\delta\phi) \cdot d\mathbf{a}$$

$$\delta\left(\lambda \oint \mathbf{J} \cdot d\mathbf{a}\right) = (\delta\lambda) \oint \mathbf{J} \cdot d\mathbf{a} + \lambda\left(\oint \delta\mathbf{J} \cdot d\mathbf{a}\right) = \lambda\left(\oint \delta\mathbf{J} \cdot d\mathbf{a}\right)$$

Thus, the solution of $\delta\left(P + \lambda \oint \mathbf{J} \cdot d\mathbf{a}\right) = 0$ is[69] $\oint [(\phi - \lambda)\,\delta\mathbf{J} + \mathbf{J}\delta\phi] \cdot d\mathbf{a} = 0$. The boundary surface is arbitrary; our result holds for arbitrary boundary surface provided that

$$\delta\mathbf{J} = -\frac{\mathbf{J}\delta\phi}{\lambda - \phi}$$

In other words, a surface element of a physical system satisfying our MinEP initially faces an external environment characterized by the value ϕ of the state variable and has a steady flow $\mathbf{J}$ through itself. These two quantities, of course, can be functions of the space variables. If the external state variable changes locally by $\delta\phi$ then the response of the system is to change the flow by $\delta\mathbf{J}$, in agreement with le Châtelier's principle at LTE. Such a change ensures minimum entropy production in the new altered environment.

References

1. Landau, L.D., Lifshitz, E.: Statistical Physics. Pergamon, Oxford, UK (1960)
2. Casimir, H.B.G.: Rev. Mod. Phys. **17**, 343 (1945)
3. Callen, H.B.: Thermodynamics and an Introduction to Thermostatistics, 2nd edn. Wiley, New York (1985)
4. Molvig, K.: Phys. Fluids **27**(12), 2847 (1984)
5. Helbing, D., Vilcsek, T.: New J. Phys. **1**(13), 1–13 (1999)
6. Strutt, J.W.: (Lord Rayleigh) Proc. Lond. Math. Soc. **s1-4**(1), 357–368 (1871)
7. Lavenda, B.: Thermodynamics of Irreversible Processes. McMillan, London (1979)

---

[68] After integration by parts and invoking both Gauss' theorem of divergence, the conservation equation, the constraint and the results of Sect. A.3. We stress the point that we are not requiring that the perturbation vanishes on the boundary.

[69] We take $\lambda = $ const. as $\delta\lambda$ has disappeared altogether from the computation. Thus, we may take λ inside the integral.

8. Gyarmati, I.: Non-equilibrium Thermodynamics. Springer, Berlin (1970)
9. Ichiyanagi, M.: Phys. Rep. **243**, 125–182 (1994)
10. Zemansky, M.W.: Heat and Thermodynamics. McGraw Hill, New York (1968)
11. Landau, L.D., Lifshitz, E.: Fluid Mechanics. Pergamon, Oxford, UK (1960)
12. Arima, T., Taniguchi, S., Ruggeri, T., Sugiyama, M.: Contin. Mech. Thermodyn. **24**(4), 271–292 (2012)
13. Landau, L.D., Lifshitz, E.: Electrodynamics of Continuous Media. Pergamon, Oxford, UK (1960)
14. Andersson, N., Comer, G.L.: Living Rev. Relat. **24**, 3 (2021)
15. Di Vita, A.: Phys. Rev. E **81**, 041137 (2010)
16. Rebhan, E.: Phys. Rev. A **42**, 781 (1990)
17. Sienutycz, S., Berry, R.S.: Phys. Rev. A **46**(10), 6359–6370 (1992)
18. Kondraputi, D., Prigogine, I.: Modern Thermodynamics. Wiley, New York (1998)
19. Prigogine, I.: Prigogine. I. Étude thermodynamique des Phenoménes Irreversibles Desoer, Liége, Belgium (1947). (in French)
20. Demirel, Y., Sandler, S.I.: Biophys. Chem. **97**, 87–111 (2002)
21. Shimizu, H., Yamaguchi, Y.: Prog. Theor. Phys. **67**, 1 (1982). (Progress Letters)
22. Zivieri, R., Pacini, N., Finocchio, G., Carpentieri, M.: Nat. Sci. Rep. **7**, 9134 (2017)
23. Glansdorff, P., Prigogine, I.: Physica **20**, 773–780 (1954). (in French)
24. Richardson, I.W.: Biophys. J. **9**(2), 265–267 (1969)

Beyond Linear Non-equilibrium Thermodynamics

5

Abstract

The non-existence theorem of Gage et al. shows that no general-purpose variational principle of non-equilibrium thermodynamics exists. As an example, we discuss stability in problems of heat conduction described by Fourier's law. We discuss the minimization of entropy production in problems with convection at moderate Rayleigh's number, Joule and viscous heating in fluid dynamics, astrophysics, physiology, hydrology, geology and porous media. We introduce both Busse's, Chandrasekhar's, Kirchoff's, Korteweg–Helmholtz' and 'maximum economy' principles, Malkus' conjecture as well as the principle of minimum energy expenditure rate per unit volume. We discuss maximization of entropy production in both Sawada's thought experiment, Paltridge's model of Earth's ocean and atmosphere, and problems concerning heat conduction, convection at large Rayleigh's number, crystallization processes, detonation waves, dunes, solidification processes, shock waves and tokamaks. Rules of selection between steady and oscillating stable configurations in different problems are highlighted and include Rayleigh's criterion and Rauschenbach's hypothesis in thermoacoustics, Welander's model on thermohaline circulation and Eddington's model of Cepheid stars. Outside physics, we discuss Bejan's constructal law, Lotka and Odum's maximum power principle, Zipf's principle of least effort, Zipf-Mandelbrot's law, Pareto's distribution, as well as the gravity model and the entropy model of urban planning.

5.1   Gage et al.'s Theorem

In LNET all coefficients L_{ij} are constant and uniform quantities. What if they are not? To start with, we limit ourselves to a monodimensional system with coordinate $a \leq y \leq b$ (generalization to 3 dimensions is straightforward). Moreover, let us assume that the L_{ij}'s still link thermodynamic forces X_i and fluxes J_i the usual way ($J_i = L_{ij}X_j$ where $X_i = \frac{\partial \Gamma_i}{\partial y}$, Γ_i are the thermodynamic quantities of interest, L_{ij} is

© The Author(s), under exclusive license to Springer Nature Switzerland AG 2022   73
A. Di Vita, *Non-equilibrium Thermodynamics*, Lecture Notes in Physics 1007,
https://doi.org/10.1007/978-3-031-12221-7_5

invertible—see Sect. 4.1.3—and, as usual, $i, j, k \ldots = 1, \ldots N$. The novelty is that we allow the L_{ij}'s to depend on the Γ_i's and write formally: $L_{jk} = L_{jk} (\Gamma_i)$. We may wonder whether a variational principle—in the rigorous meaning described in Sect. A.2—exists which describes the steady state of the system, where the Γ_i's depend on y (and where y plays the role of the variable t in the Appendix).

Let us take advantage of the fact that we know the Euler–Lagrange equations. Indeed, they are the conservation equations in steady state, which read:

$$\frac{\partial J_i}{\partial y} = 0$$

We may rewrite our question in other words: does a function $\Phi = \Phi \left(\Gamma_i, L_{jk}, X_i \right)$ exist such that the unconstrained[1] variational principle:

$$\Theta \equiv \int_a^b \Phi dy = \text{extremum}$$

leads to the Euler-Lagrange equations $\frac{\partial J_i}{\partial y} = 0$ above? We know the answer at thermodynamic equilibrium: $\Theta = S$ and Φ is an entropy density. What is the solution outside of thermodynamic equilibrium?

To fix the ideas, here we discuss the simple physical system of Sect. 4.3.10 [1] and assume MinEP: the steady state satisfies $\delta P = 0$ where $P = \int (\mathbf{J} \cdot \mathbf{X}) \, d\mathbf{x} = \int L|\nabla\phi|^2 d\mathbf{x}$ and the volume integral is performed on the volume of the system. Then[2]:

$$0 = \delta P = \int \left[\frac{\partial L}{\partial \phi} |\nabla\phi|^2 \delta\phi + 2L\nabla\phi \cdot \delta\nabla\phi \right] d\mathbf{x} = \int \left[\frac{\partial L}{\partial \phi} |\nabla\phi|^2 \delta\phi + 2L\nabla\phi \cdot \nabla\delta\phi \right] d\mathbf{x} =$$

$$= \int (\delta\phi) \left[\frac{\partial L}{\partial \phi} |\nabla\phi|^2 - \nabla \cdot (2L\nabla\phi) \right] d\mathbf{x} + \oint (\delta\phi) \, 2L\nabla\phi \cdot d\mathbf{a} =$$

$$= \int (\delta\phi) \left[\frac{\partial L}{\partial \phi} |\nabla\phi|^2 - \nabla \cdot (2L\nabla\phi) \right] d\mathbf{x} = -\int (\delta\phi) \left[\frac{\partial L}{\partial \phi} |\nabla\phi|^2 + 2L\Delta\phi \right] d\mathbf{x}$$

$$= -\int (\delta\phi) \left(2\sqrt{L} \right) \nabla \cdot \left(\sqrt{L}\nabla\phi \right) d\mathbf{x}$$

where we have taken into account that $\delta\phi = 0$ everywhere on the boundary surface of the volume the surface integral is computed upon and that $\nabla L = \frac{\partial L}{\partial \phi}\nabla\phi$. Our result

---

[1] Here the word *unconstrained* is a shortcut for the following words: *where no constraint like e.g.* $\partial J_i = 0$ *applies (unlike e.g. the minimization in Sect. 4.1.4), so that no quantity other than the* Γ_i's, *the* L_{jk}'s *and the* X_i's *acts as Lagrange multiplier (see Sect. A.3) in the variational principle, in contrast e.g. with the minimization in Sect. 4.3.2.*

[2] after integration by parts and invoking Gauss' theorem of divergence.

holds for arbitrary $\delta\phi$, hence we obtain $\nabla \cdot \left(\sqrt{L}\nabla\phi\right) = 0$ rather than the correct conservation equation $\nabla \cdot (L\nabla\phi) = 0$.[3]

We may ask if this disappointing result is just an unfortunate case, and if there is some other solution (different e.g. from MinEP) of our problem outside of thermodynamic equilibrium. Surprising as it may seem, the answer is that there is *no* solution. Literally, no unconstrained variational principle holds with the steady-state balance equations as Euler-Lagrange equations in non-equilibrium thermodynamics—see Sect. A.4 for the proof of this statement, due to Gage et al. [2].

This result seems surprising[4] because the stability of systems kept at steady state far from thermodynamic equilibrium is an everyday experience: think e.g of a resistor where the electric current remains constant in spite of unavoidable fluctuations whenever a given voltage is applied.[5] It is scarcely credible that this stability is only due to the linearity of phenomenological relationship which LNET and its treatment of stability relies upon, because in many cases phenomenological relationships are nonlinear.[6]

In order to solve this conundrum, we recall that the steady state of a physical system is described by a solution of the equations of motion (i.e., of the conservation equations of mass, momentum, energy, electric charge...). But steady solutions may be either stable or unstable with respect to perturbations. In some particular case, it is possible to obtain information concerning stability from the equations of motion themselves. For example, the steady states of a particle subject to an external potential force correspond to extrema of the potential; minima (maxima) are stable (unstable) steady states. Dynamical systems where a Lyapunov function [5] exist provide a further example.

Accordingly, even if it is impossible to write the conservation equations as the Euler-Lagrange equations of a variational principle in the general case, it is still conceivable to write the necessary condition of stability for a particular class of

---

[3] Unless $L \equiv 1$, i.e. unless the thermodynamic force identically coincides with the thermodynamic flux. The conservation equation differs from the Euler-Lagrange equation by having L instead of $\sqrt{L}$; correspondingly, the variational principle whose Euler-Lagrange equation coincides with the conservation equation is $\delta \int L^2|\nabla\phi|^2 d\mathbf{x} = \delta \int |\mathbf{J}|^2 d\mathbf{x} = 0$. If we assume here that no constraint affects $\mathbf{J}$ (in contrast with Sect. 4.3.10), the only extremum (zero) corresponds to $\mathbf{J} \equiv 0$, hence $\mathbf{X} \equiv 0$ and the system is at thermodynamic equilibrium (see text below). Of course, a constraint actually follows from the volume integration of the conservation equation, namely the vanishing net flux at the boundary; but here we are discussing an unconstrained variational principle. In the unconstrained (constrained) variational principle, physical information concerning the conservation equation is a consequence of the principle itself (is to be provided independently, from scratch). See Sect. 5.3.4 for a solution of the variational problem $\int |\mathbf{J}|^2 d\mathbf{x} = \min$ with the constraint $\nabla \cdot \mathbf{J} = 0$.

[4] And makes it wrong to speak of 'variational principle' outside thermodynamic equilibrium, rigorously speaking [3].

[5] At thermodynamic equilibrium, its precisely the condition of maximum S which ensures that large deviations from this equilibrium are unlikely, because of Einstein's formula—see Sect. 4.1.1.

[6] Things get even worse when unsteady gravitational fields are taken into account. In the framework of General Relativity, indeed, it is possible to write general-purpose variational principle for dissipative fluids for particular, dedicated models of matter only: *problems can be fixed by introducing additional dynamical fields*, however *we should not expect to be able to use observations to single out a preferred theoretical description* [4].

systems in steady state as the requirement that such state corresponds to an extremum of some quantity.[7] In this case, a physically allowable steady state is a solution of the equations of motion; furthermore, if it corresponds also to an extremum of this hypothetical quantity, then it is not only steady, but also stable.

Non-equilibrium thermodynamics is therefore vindicated. Its new goal shall not be the formulation of a (non-existent) general-purpose variational principle, but to provide rules for finding the quantity whose extremization is the necessary condition of stability of steady states in a given physical system. Being of thermodynamic nature, such rules are likely to depend on the detailed dynamics of the system only weakly, even if the actual quantity to be minimized (or maximized) is system-dependent. This is particularly welcome when it comes to describing complex systems which are far from being fully understood.

5.2   Heat Conduction

5.2.1   Fourier's Law

The entropy production density due to heat transfer is ([6], Chap. XVI of [7,8]):

$$\sigma = \mathbf{q} \cdot \nabla \left(\frac{1}{T} \right)$$

where $\mathbf{q}$ is defined as a vector that satisfies the following relationship:

$$\nabla \cdot \mathbf{q} = -\frac{\partial}{\partial t} \left(\rho \frac{|\mathbf{v}|^2}{2} + \rho u + \frac{1}{2\mu_0} |\mathbf{B}|^2 + \frac{1}{2}\varepsilon_0 |\mathbf{E}|^2 \right)$$
$$- \nabla \cdot \left[\rho \mathbf{v} \left(\frac{|\mathbf{v}|^2}{2} + h \right) - \mathbf{v} \cdot \sigma' + \frac{\mathbf{E} \wedge \mathbf{B}}{\mu_0} \right]$$

provided that the net electric charge vanishes and the gravitational potential does not depends on time (generalization to unsteady gravitational potential is straightforward).

Since $\sigma \propto \nabla \left(\frac{1}{T} \right)$, it is tantalizing to choose $\nabla \left(\frac{1}{T} \right)$ as a thermodynamic force. The corresponding thermodynamic flux is $\mathbf{q}$. Remarkably, $\nabla \cdot \mathbf{q}$ is the time derivative of no physical quantity, unless all terms but ρ and u vanish identically on the R.H.S. of the definition of $\nabla \cdot \mathbf{q}$.[8] In this case, it is customary to introduce a relationship between $\mathbf{q}$ and ∇T rather than $\mathbf{q}$ and $\nabla \frac{1}{T}$:

$$q_i = -\chi_{ij} \frac{\partial T}{\partial x_j} \quad ; \quad i, j = 1, 2, 3$$

---

[7] For example, the least dissipation principle in Sect. 4.3.2 applies to a system with fixed thermodynamic fluxes in the framework of LNET.

[8] Back in the XIX century, Carnot would have suggested that $\mathbf{q}$ is related (through some form of heat balance) to the time derivative of the total mass of Lavoisier's 'caloric' inside a given volume. But Joule's experiments show that no caloric exists, in contrast with Carnot's own ideas.

This relationship defines the (i, j)-th component of a thermal conductivity tensor of rank 3. If χ_{ij} does not depend on T then the underlying physics is linear. Accordingly, $T\sigma = \chi_{ij}\frac{\partial T}{\partial x_i}\frac{\partial T}{\partial x_j}$. Accordingly, the antisymmetric part $\chi_{ij}^a = \frac{1}{2}\left(\chi_{ij} - \chi_{ji}\right)$ of χ_{ij} provides σ with no contribution, and indeed experiments confirm that we may neglect it altogether. In particular, if the system is isotropic then $\chi_{ij} = \chi\delta_{ij}$ and 'Fourier's law' follows:

$$\mathbf{q} = -\chi\nabla T$$

Correspondingly, $T^2\sigma = \chi|\nabla T|^2$. Irreversibility of heat transport means $\sigma > 0$, hence $\chi > 0$. Moreover, if we want LNET to hold then the Onsager phenomenological coefficient L_{11} in the linear relationship $\mathbf{q} = L_{11}\nabla\left(\frac{1}{T}\right)$ between $\mathbf{q}$ and $\nabla\left(\frac{1}{T}\right)$ must be constant. But Fourier's law requires $\mathbf{q} = \chi T^2\nabla\left(\frac{1}{T}\right)$, hence $\chi = \frac{L_{11}}{T^2} = \frac{const.}{T^2}$. No known material satisfies this scaling law for χ. In steady state, indeed, such material would lead to the following equation for $T(\mathbf{x})$ in steady state: $0 = \nabla \cdot \mathbf{q} = \text{const.}\nabla \cdot \left(\frac{\nabla T}{T^2}\right) = \text{const.}\nabla \cdot \nabla\left(\frac{1}{T}\right) = \text{const.}\Delta\left(\frac{1}{T}\right),$[9] hence $\Delta\left(\frac{1}{T}\right) = 0$, to be compared to the familiar law $\Delta T = 0$ valid for steady $T(\mathbf{x})$ in systems with uniform χ. Generally speaking, LNET does not apply to heat transfer. This conclusion does not rule out the validity of LNET symmetry in particular problems where $\mathbf{q}$ is proportional to the time derivative of total energy, as e.g. in some problems with discontinuous systems. The trouble with $\mathbf{q}$ is that even if the conditions for validity if LNET are satisfied, neither $\mathbf{q}$ nor $\nabla \cdot \mathbf{q}$ are proportional to any time derivative, and $\mathbf{q}$ is therefore not eligible as thermodynamic flux, e.g. whenever convective losses or electromagnetic radiation or viscosity are present (see also Sect. 4.2.8). This fact prevents also any different choice of thermodynamic force and flux from supporting the validity of LNET.[10] All the more reason the MinEP of LNET fails to provide us with a correct description of heat transfer in steady shear flow [14].

---

[9] Here $\Delta\phi \equiv \nabla \cdot \nabla\phi$ is the so-called 'Laplacian' of the scalar field ϕ.

[10] Historically, for example, LNET impact on the research in controlled nuclear fusion has been unfavourable: Onsager and Machlup's principle of Sect. 4.1.5 has been postulated when describing heat transport in toroidal plasmas [9], and as a result, overoptimistic estimates of plasma confinement properties followed. A 'plasma' is a medium a) which is made of both positively and negatively charged, unbound particles b) and where the net electric charge is roughly zero. Stars and lightning are familiar examples of plasmas. When moving, charged plasma particles generate electric currents, and any movement of a charged plasma particle affects and is affected by the fields created by all other charges. The impact of the resulting collective behaviour may overcome the impact of collisional interactions, which are usually responsible for the relaxation of LTE. This is, e.g. the case when large external forces act on the plasma. When it comes to plasmas, accordingly, the assumption of LTE everywhere at all times deserves special attention. In the XIX century, moreover, Rayleigh issued an early warning against the utilization of the dissipation function which plays a fundamental role in LNET least dissipation principle when it comes to *systems in which the cause of the dissipation, or of part of it, is the conduction and radiation of heat* [10]. As for magnetically confined plasmas involved in controlled nuclear fusion research, both theory [11,12] and experiments [13] have shown that LNET is usually violated.

5.2.2   Stability Versus Fourier's Law

Suppose LNET holds in a system with fixed boundary conditions, where no convection and no radiation occur and heat transport occurs well inside the system because of conduction only, according to the law ([6], Sect. V.4.a of [15, 16]):

$$\nabla \cdot \mathbf{q} + \rho c_p \frac{\partial T}{\partial t} = 0$$

where c_p specific heat at constant pressure per unit mass and $\mathbf{q}$ vanishes on the boundary of the volume enclosing the system. Then $L_{11} = $ const. and we have[11]:

$$\frac{d}{dt} \int \sigma d\mathbf{x} = \frac{d}{dt} \int \mathbf{q} \cdot \nabla \left(\frac{1}{T}\right) d\mathbf{x} = \frac{d}{dt} \int L_{11} |\nabla \left(\frac{1}{T}\right)|^2 d\mathbf{x} = 2 \int L_{11} \nabla \left(\frac{1}{T}\right) \cdot \frac{\partial}{\partial t} \nabla \left(\frac{1}{T}\right) =$$

$$= 2 \int \mathbf{q} \cdot \frac{\partial}{\partial t} \nabla \left(\frac{1}{T}\right) d\mathbf{x} = 2 \int \mathbf{q} \cdot \nabla \frac{\partial}{\partial t} \left(\frac{1}{T}\right) d\mathbf{x} = 2 \int \nabla \cdot \left[\mathbf{q} \frac{\partial}{\partial t} \left(\frac{1}{T}\right)\right] d\mathbf{x} +$$

$$-2 \int (\nabla \cdot \mathbf{q}) \frac{\partial}{\partial t} \left(\frac{1}{T}\right) d\mathbf{x} = 2 \oint \frac{\partial}{\partial t} \left(\frac{1}{T}\right) \mathbf{q} \cdot d\mathbf{a} - 2 \int (\nabla \cdot \mathbf{q}) \frac{\partial}{\partial t} \left(\frac{1}{T}\right) d\mathbf{x} =$$

$$-2 \int (\nabla \cdot \mathbf{q}) \frac{\partial}{\partial t} \left(\frac{1}{T}\right) d\mathbf{x} = +2 \int \frac{(\nabla \cdot \mathbf{q})}{T^2} \frac{\partial T}{\partial t} d\mathbf{x} = -2 \int \frac{\rho c_p}{T^2} \left(\frac{\partial T}{\partial t}\right)^2 d\mathbf{x} \leq 0$$

i.e., the non-negative amount of entropy $\int \sigma d\mathbf{x} = \int \mathbf{q} \cdot \nabla \left(\frac{1}{T}\right) d\mathbf{x}$ produced per unit time in the bulk of the system by heat conduction is a Lyapunov function, and is, therefore, a minimum in stable steady states. Straightforward algebra shows that the Euler-Lagrange equation of this variational principle (where the Lagrangian and the Lagrangian coordinate are $L_{11} |\nabla \left(\frac{1}{T}\right)|^2$ and $\frac{1}{T}$, respectively) is just $\Delta \left(\frac{1}{T}\right) = 0$. The latter equation describes therefore stable steady states in problems were heat conduction is the only irreversible process, provided that LNET holds. Since it is not confirmed by experiments, we conclude that LNET does not apply to heat conduction.

5.3   Minimum Entropy Production

5.3.1   Joule Heating: Kirchhoff's Principle

The contribution of Joule heating to the internal entropy production density (we refer to Chap. 3 of [6] and to [8, 17, 18]) is:

$$\sigma = \frac{\mathbf{j}_{el} \cdot (\mathbf{E} + \mathbf{v} \wedge \mathbf{B})}{T}$$

---

[11] Here we invoke Gauss' theorem of divergence and perform an integration by parts.

The quantity $\mathbf{E} + \mathbf{v} \wedge \mathbf{B}$ is the electric field in the reference frame which moves at velocity $\mathbf{v}$. In steady state, this field is $= -\nabla \phi_{el}$, ϕ_{el} electrostatic potential. Moreover, let us introduce the electric charge density ρ_{el}; the electric charge balance reads, therefore

$$\nabla \cdot \mathbf{j}_{el} + \frac{\partial \rho_{el}}{\partial t} = 0$$

Finally, the relationship between $\mathbf{j}_{el}$ and $\mathbf{E} + \mathbf{v} \wedge \mathbf{B}$ ('Ohm's law') is often linear:

$$\mathbf{j}_{el} = \sigma_\Omega \left(\mathbf{E} + \mathbf{v} \wedge \mathbf{B} \right)$$

where σ_Ω is the electrical conductivity.[12] It follows that the Joule heating power density is $T\sigma = \frac{|\mathbf{j}_{el}|^2}{\sigma_\Omega}$. Joule heating is an irreversible process, then $\sigma > 0$ and $\sigma_\Omega > 0$.

If the electric charge carriers (usually, electrons) are acted upon by a local electric field $\mathbf{E} + \mathbf{v} \wedge \mathbf{B}$, then in steady state their acceleration is damped by collisions at frequency, say, ν_{coll} . If $\mathbf{v}_{el}$, m_{el} , n_{el} and q_{el} are the velocity, the mass, the particle density and the electric charge of charge carriers, respectively, then the definition of $\mathbf{j}_{el}$ and compensation between the accelerating and the damping force[13] lead to:

$$\mathbf{j}_{el} = n_{el} q_{el} \mathbf{v}_{el} \quad ; \quad q_{el} \left(\mathbf{E} + \mathbf{v} \wedge \mathbf{B} \right) = m_{el} \nu_{coll} \mathbf{v}_{el}$$

Dot multiplication of both sides of the latter equation by $\frac{n_{el} q_{el}}{\nu_{coll}}$ leads to:

$$\sigma_\Omega = \frac{n_{el} q_{el}^2}{m_{el} \nu_{coll}}$$

As for LNET, two cases are possible.

If $\nabla T = 0$, then the balance of electric charge makes the divergence of $\frac{\mathbf{j}_{el}}{T}$ to be proportional to a time derivative (namely, of ρ_{el}).[14] Moreover, the local electric field $\mathbf{E} + \mathbf{v} \wedge \mathbf{B}$ is $= -\nabla \phi_{el}$, i.e. is proportional to a gradient (of the electrostatic potential). Finally, the relationship between them is linear, and the relevant characteristic time is the characteristic time $\frac{1}{\nu_{coll}}$ of collisions. Accordingly, we may select $\frac{\mathbf{j}_{el}}{T}$ and $\mathbf{E} + \mathbf{v} \wedge \mathbf{B}$ as thermodynamic flux and force, respectively, and we expect that a variational principle related to LNET holds in steady state. (Should we rather select $\mathbf{j}_{el}$

---

[12] Here we limit ourselves to the case of scalar electric conductivity for mathematical simplicity. Should we discuss the general case of tensorial electric conductivity, nothing essential would change in the following.

[13] We have tacitly assumed that there is only one species of charge carriers, and that ν_{coll} does not depend on $\mathbf{v}_{el}$. Both assumptions may be dropped at the expense of heavier algebra; nothing essential changes.

[14] Of course, partial and total time derivatives coincide in the frame of reference locally at rest, where $\mathbf{v} = 0$.

and $\frac{\mathbf{E}+\mathbf{v}\wedge\mathbf{B}}{T}$ as thermodynamic flux and force, respectively, nothing would change). Minimisation of entropy production $\int \sigma d\mathbf{x}$ leads to Kirchhoff's principle:

$$\int \frac{|\mathbf{j}_{el}|^2}{\sigma_\Omega} d\mathbf{x} = \min \quad \text{with the constraint} \quad \nabla \cdot \mathbf{j}_{el} = 0$$

where we have dropped the uniform quantity T and the constraint is given by the electric charge balance in steady state. Kirchhoff's principle prescribes constrained minimization of Joule heating power; since the temperature gradient vanishes, this is equivalent to a constrained minimization of the amount of entropy produced per unit time by Joule dissipation: it is an example of a minimum entropy production principle, a MinEP. In contrast with the MinEP of LNET, however, Kirchhoff's principle holds for Joule dissipation and for vanishing temperature gradient only.

In contrast, if $\nabla T \neq 0$ then the divergence of $\frac{\mathbf{j}_{el}}{T}$ is no more proportional to a time derivative. Correspondingly, $\frac{\mathbf{E}+\mathbf{v}\wedge\mathbf{B}}{T}$ is no more proportional to a gradient. Then, no LNET-related variational principle holds anymore. Such a principle rules Joule heating only if $\nabla T = 0$. Kirchhoff minimum power principle is discussed, for example, in Ref. [19]. Kirchhoff derived this result in 1848 for rigidly fixed conductors only (see Ref. [8] and Problem 21.3 in Ref. [17]). Kirchhoff's principle has been used in early studies of gaseous discharges [20,21] without further theoretical justification. For further discussion see Sects. 5.3.10 and 6.2.9.

5.3.2 Electric Arc

In particular, radiative transport may flatten ∇T in the bulk of radiation-cooled, free-flowing electric arcs. By 'free-flowing' we mean that the arc is in contact with no solid walls but the electrodes, so that the exchange of matter between the arc bulk and the external world is reduced. If the external world supplies the arc with a constant electric current i_{arc} and a voltage drop v_{arc} is obtained, then Kirchhoff's principle reduces to minimization of Joule power $i_{arc} \cdot v$ at constant i_{arc}, i.e. to minimization of v_{arc} at constant i_{arc}. Such minimization has been proposed by Steenbeck in order to explain experimental observations [22]. A discussion of the relationship between Steenbeck's result and MinEP is due to Peters [23]. Our discussion shows that v_{arc} is quite insensitive to arc temperature. An independent investigation of SF6 arcs confirms this result—see Fig. 8 and Sect. V of [24]. For further discussion, see Sects. 5.6.5 and 6.2.8.

5.3.3 A Tale of Two Resistors

As an example, let us consider a resistor 1 (with electric resistance R_1 and at constant temperature T_1) where an electric current I_1 flows as a voltage V_1 is applied and a resistor 2 (with electric resistance R_2 and at constant temperature T_2) where an electric current I_2 flows as a voltage V_1 is applied. (Admittedly, this is rather an example of discontinuous system. However, we anticipate here that this fact leaves the conclusions of our discussion unaffected in the following). The amounts of Joule power in 1 and 2 are $\frac{V_1^2}{R_1}$ and $\frac{V_2^2}{R_2}$, respectively. The amounts of entropy produced by Joule dissipation in 1 and 2 are $\frac{V_1^2}{R_1 T_1}$ and $\frac{V_2^2}{R_2 T_2}$, respectively. The total amount of entropy produced inside the system is

$$P = \frac{V_1^2}{R_1 T_1} + \frac{V_2^2}{R_2 T_2}$$

Let us write down P with the help of thermodynamic flows and thermodynamic forces. By definition, I_1 and I_2 are the total time derivatives of the electric charge flowing across 1 and 2, respectively, then we are allowed to select them as thermodynamic flows. Now, Ohm's law applied to 1 and 2 gives $I_1 = \frac{V_1}{R_1}$ and $I_2 = \frac{V_2}{R_2}$. Then, P reduces to:

$$P = X_i Y_i = L_{ij} X_i X_j \;\; ; \;\; i, j = 1, 2 \;\; ; \;\; X_1 = \frac{V_1}{T_1} \;\; ; \;\; X_2 = \frac{V_2}{T_2}$$

$$Y_1 = I_1 \;\; ; \;\; Y_2 = I_2 \;\; ; \;\; L_{11} = \frac{1}{R_1 T_1} \;\; ; \;\; L_{22} = \frac{1}{R_2 T_2} \;\; ; \;\; L_{12} = L_{21} = 0$$

Onsager's symmetry is trivially satisfied. Validity of LNET requires that the L_{ij}'s are constant, then we take both R_1 and R_2 to be constant below. What about variational principles? Clearly, unconstrained minimization of P leads just to the trivial solution $V_1 = V_2 = 0$: no Joule heating, no electric conduction. If the resistors are connected in parallel, then the total electric current $I_{TOT} = I_1 + I_2$ is constant, and this fact puts a constraint on the minimization. Let us look for an extremum of P:

$$0 = dP = d\left(\frac{V_1^2}{R_1 T_1} + \frac{V_2^2}{R_2 T_2}\right) = 2\frac{V_1}{R_1 T_1}dV_1 + 2\frac{V_2}{R_2 T_2}dV_2 = 2\frac{V_1}{T_1}d\left(\frac{V_1}{R_1}\right) + 2\frac{V_2}{T_2}d\left(\frac{V_1}{R_1}\right) =$$

$$= 2\frac{V_1}{T_1}dI_1 + 2\frac{V_2}{T_2}dI_2 = 2\frac{V_1}{T_1}dI_1 + 2\frac{V_2}{T_2}d\left(I_{TOT} - I_1\right) = \left(\frac{V_1}{T_1} - \frac{V_2}{T_2}\right) \cdot 2dI_1$$

where we have taken advantage of the constraint of constant I_{TOT} (i.e. $dI_{TOT} = 0$). Should a steady state of our system of two resistors in parallel correspond to a minimum of the entropy production P inside the system, then $\frac{V_1}{T_1} = \frac{V_2}{T_2}$. For $T_1 \neq T_2$, however, the result is blatantly *wrong*.[15] This is not surprising provided that we describe our system of two resistors as a discontinuous system, as one of the main assumptions underlying the LNET proof of MinEP in such systems is violated:

---

[15] Historically, this is the birthplace of the skepticism concerning the search of a general-purpose variational principle in non-equilibrium thermodynamics [8].

the constraint $I_1 + I_2 = $ constant implies that $X_i = $ constant and J_i vanishes for no value of i. In contrast, if $T_1 = T_2$ the description of a steady state provided by the condition $dP = 0$ reduces to the familiar, correct result $V_1 = V_2$. In this case, however, T is just a multiplicative constant that can be factorized away and the constrained minimization of P reduces to Kirchhoff's principle. We check that the steady state is actually a minimum of the total Joule heating power by showing that its second differential is ≥ 0

$$
d^2 \left(\frac{V_1^2}{R_1} + \frac{V_2^2}{R_2} \right) = d \left[2V_1 d \left(\frac{V_1}{R_1} \right) + 2V_2 d \left(\frac{V_2}{R_2} \right) \right] =
$$

$$
= \frac{2 (dV_1)^2}{R_1} + \frac{2 (dV_2)^2}{R_2} + 2V_1 d^2 \left(\frac{V_1}{R_1} \right) + 2V_2 d^2 \left(\frac{V_2}{R_2} \right) \geq
$$

$$
\geq 2V_1 d^2 \left(\frac{V_1}{R_1} \right) + 2V_2 d^2 \left(\frac{V_2}{R_2} \right) = 2V_1 d^2 I_1 + 2V_2 d^2 I_2 =
$$

$$
= 2V_1 d^2 I_1 + 2V_2 d^2 (I_{TOT} - I_1) = (V_1 - V_2) \cdot 2d^2 I_1 = 0
$$

We conclude that the electric current is always going to flow along the path of lesser resistance. This result is all too familiar to people working in the prevention of short circuits.

A similar discussion holds for two resistors in series, i.e. with constant total voltage $V_{TOT} = V_1 + V_2$. In particular, the L_{ij}'s are the same as the case with two resistors in parallel. Let $T_1 = T_2 = T$. The condition $dP = 0$ leads to

$$
0 = d \left(\frac{V_1^2}{R_1} + \frac{V_2^2}{R_2} \right) = \frac{2V_1 dV_1}{R_1} + \frac{2V_2 dV_2}{R_2} = \frac{2V_1 dV_1}{R_1} + \frac{2V_2 d (V_{TOT} - V_1)}{R_2} = \left(\frac{V_1}{R_1} - \frac{V_2}{R_2} \right) \cdot 2dV_1
$$

hence $\frac{V_1}{R_1} - \frac{V_2}{R_2} = 0$, i.e. $I_1 = I_2$, just as expected. The sign of $d^2 P$ denotes a minimum:

$$
d^2 P = \frac{2}{T} \left(\frac{V_1 d^2 V_1}{R_1} + \frac{V_2 d^2 V_2}{R_2} + \frac{(dV_1)^2}{R_1} + \frac{(dV_2)^2}{R_2} \right) \geq \frac{2}{T} \left(\frac{V_1 d^2 V_1}{R_1} + \frac{V_2 d^2 V_2}{R_2} \right) =
$$

$$
= \frac{2}{T} \left[\frac{V_1 d^2 V_1}{R_1} + \frac{V_2 d^2 (V_{TOT} - V_1)}{R_2} \right] = \frac{2}{T} \left(\frac{V_1}{R_1} - \frac{V_2}{R_2} \right) d^2 V_1 = 0
$$

But if $T_1 \neq T_2$ then the same argument lead to the wrong result $\frac{I_1}{T_1} = \frac{I_1}{T_2}$.

5.3.4   Back to Ohm

We may obtain further information from Kirchhoff's principle. Let $\mathbf{j}_{el}$ and $-2\phi_{el}$ be the Lagrangian coordinate and the Lagrange multiplier, respectively (As for the Lagrange multipliers, see Sect. A.3). After integration by parts, therefore, Kirchhoff's principle and Gauss' theorem of divergence give:

$$0 = \delta \int \left(\frac{|\mathbf{j}_{el}|^2}{\sigma_\Omega} - -2\phi_{el}\nabla \cdot \mathbf{j}_{el} \right) d\mathbf{x} = 2 \int \left(\frac{\mathbf{j}_{el} \cdot \delta\mathbf{j}_{el}}{\sigma_\Omega} - -\phi_{el}\nabla \cdot \delta\mathbf{j}_{el} \right) d\mathbf{x} =$$

$$= 2 \int \left[\frac{\mathbf{j}_{el} \cdot \delta\mathbf{j}_{el}}{\sigma_\Omega} - \nabla \cdot \left(\phi_{el}\delta\mathbf{j}_{el} \right) + \delta\mathbf{j}_{el} \cdot \nabla\phi_{el} \right] d\mathbf{x} = -2 \oint \phi_{el}\delta\mathbf{j}_{el} d\mathbf{a} +$$

$$+2 \int \delta\mathbf{j}_{el} \cdot \left(\frac{\mathbf{j}_{el}}{\sigma_\Omega} + \nabla\phi_{el} \right) = 2 \int \delta\mathbf{j}_{el} \cdot \left(\frac{\mathbf{j}_{el}}{\sigma_\Omega} + \nabla\phi_{el} \right)$$

as the surface integral vanishes because $\delta\mathbf{j}_{el} = 0$ on the boundary. The relationship above holds for arbitrary $\delta\mathbf{j}_{el}$, hence $\frac{\mathbf{j}_{el}}{\sigma_\Omega} = -\nabla\phi_{el} = \mathbf{E} + \mathbf{v} \wedge \mathbf{B}$, i.e. we retrieve Ohm's law. We have tacitly assumed that $\delta\sigma_\Omega = 0$ in the computation above. This assumption can e.g. justified under the assumption of uniform temperature, as $\sigma_\Omega = 0 = \sigma_\Omega = O(T)$ in most cases. For $\phi_{el} \to \phi_{el} + \delta\phi_{el}$ we retrieve the constraint $\nabla \cdot \mathbf{j}_{el} = 0$. MinEP agrees with Kirchhoff's principle for $T_1 = T_2$ only. If $T_1 \neq T_2$, then Kirchhoff may still hold (it holds in each resistor separately as far as Ohm's law holds and σ_Ω is fixed in each resistor), in contrast with LNET. We discuss the stability of the state described by Kirchhoff's principle below.

5.3.5   An Auxiliary Relationship

We may rewrite Kirchhoff's principle in different forms. Let us replace $\mathbf{j}_{el}$ with $\mathbf{B}$ as Lagrangian coordinate. When performing such a change, the constraint too must involve the new Lagrangian coordinate. Maxwell's equations of electromagnetism include $\nabla \cdot \mathbf{B} = 0$ and lead to $\nabla \wedge \mathbf{B} = \mu_0\mathbf{j}_{el}$ in the nonrelativistic limit $c \to \infty$. We define a Lagrange multiplier ξ and write $\int \left[\frac{|\nabla \wedge \mathbf{B}|^2}{\mu_0^2\sigma_\Omega} + \xi\nabla \cdot \mathbf{B} \right] d\mathbf{x} = \min$, i.e.[16]:

$$0 = \delta \int \left[\frac{|\nabla \wedge \mathbf{B}|^2}{\mu_0^2\sigma_\Omega} + \xi\nabla \cdot \mathbf{B} \right] d\mathbf{x} = \int \left[\frac{2(\nabla \wedge \mathbf{B}) \cdot (\nabla \wedge \delta\mathbf{B})}{\mu_0^2\sigma_\Omega} + \xi\nabla \cdot \delta\mathbf{B} \right] d\mathbf{x} =$$

$$= \frac{2}{\mu_0^2\sigma_\Omega} \int \delta\mathbf{B} \cdot (\nabla \wedge \nabla \wedge \mathbf{B} + \nabla\xi') \quad ; \quad \left(\xi' \equiv -\frac{\mu_0^2\sigma_\Omega\xi}{2} \right)$$

---

[16] Again, we invoke Gauss' theorem of divergence, vanishing $\nabla\sigma_\Omega$ everywhere as well as vanishing $\delta\mathbf{B}$ on the boundary. We invoke the identity $\nabla \cdot (\mathbf{a} \wedge \mathbf{b}) = \mathbf{b} \cdot \nabla \wedge \mathbf{a} - \mathbf{a} \cdot \nabla \wedge \mathbf{b}$ for arbitrary vectors $\mathbf{a}$ and $\mathbf{b}$, and take $\mathbf{a} = \nabla \wedge \mathbf{B}$, $\mathbf{b} = \delta\mathbf{B}$.

for arbitrary $\delta \mathbf{B}$, hence $\nabla \wedge \nabla \wedge \mathbf{B} + \nabla \xi' = 0$. We get rid of the unknown quantity ξ' by taking the curl of both sides[17] and obtain a relationship involving $\mathbf{B}$ only:

$$\nabla \wedge \Delta \mathbf{B} = 0$$

The usefulness of this result will be clear below. For the moment, we may say that we have found a condition satisfied by the generic magnetic field which is produced by a distribution of electric currents which satisfies Kirchhoff's principle. Finally, we remark that the same relationship could also be obtained by taking the curl of Ohm's law in the form $\frac{\mathbf{j}_{el}}{\sigma_\Omega} + \nabla \phi_{el} = 0$ and in the nonrelativistic limit as far as $\nabla \sigma_\Omega = 0$.

5.3.6 What if Joule Heating is Negligible?

The Joule power density $T\sigma = \frac{|\mathbf{j}_{el}|^2}{\sigma_\Omega}$ becomes negligible either if $\mathbf{j}_{el} \to 0$ or if $\sigma_\Omega \to \infty$, i.e. whenever Ohm's law $\mathbf{E} + \mathbf{v} \wedge \mathbf{B} = \frac{\mathbf{j}_{el}}{\sigma_\Omega}$ reduces to

$$\mathbf{E} + \mathbf{v} \wedge \mathbf{B} = 0$$

i.e. the electric field vanishes in the local frame of reference moving at speed $\mathbf{v}$. We refer to this relationship as 'Ohm's law for negligible Joule heating' below. Let us take the curl of both sides of this relationship. Maxwell's equation $\nabla \wedge \mathbf{E} = -\frac{\partial \mathbf{B}}{\partial t}$ leads, therefore, to

$$\frac{\partial \mathbf{B}}{\partial t} - \nabla \wedge (\mathbf{v} \wedge \mathbf{B}) = 0$$

which is to be solved together with the usual condition $\nabla \cdot \mathbf{B} = 0$. The latter two relationships are identical to the relationships $\frac{\partial (\nabla \wedge \mathbf{v})}{\partial t} - \nabla \wedge [\mathbf{v} \wedge (\nabla \wedge \mathbf{v})] = 0$ and $\nabla \cdot (\nabla \wedge \mathbf{v}) = 0$ satisfied by the vorticity $\nabla \wedge \mathbf{v}$ in an inviscid, unmagnetized fluid.[18] It is well known that these equations for vorticity lead to well-known Kelvin's theorem of circuitation, i.e. the circuitation $\int_\gamma \mathbf{v} \cdot d\mathbf{l}$ of $\mathbf{v}$ along any closed line γ in the fluid is constant. If γ encloses a closed surface Λ, then Stokes' theorem ensures the circuitation to be equal to the flux of vorticity across Λ: $\int_\gamma \mathbf{v} \cdot d\mathbf{l} = \oint_\Lambda (\nabla \wedge \mathbf{v}) \cdot d\mathbf{a}$.[19] The analogy displayed above between the equations for vorticity and the equations

---

[17] And by invoking both $\nabla \cdot \mathbf{B} = 0$ and the identities $\nabla \wedge \nabla \wedge \mathbf{a} = \nabla (\nabla \cdot \mathbf{a}) - \Delta \mathbf{a}$ and $\nabla \wedge \nabla w = 0$ for arbitrary vector $\mathbf{a}$ and scalar w with $\mathbf{a} = \mathbf{B}$ and $w = \xi'$.

[18] In order to obtain this relationship, take the curl of both sides of Euler's equation, and recall the identities $\nabla \wedge \nabla f = 0$ and $\nabla \left(\frac{\mathbf{a} \cdot \mathbf{a}}{2}\right) = (\mathbf{a} \cdot) \mathbf{a} - \mathbf{a} \wedge \nabla \wedge \mathbf{a}$ for arbitrary scalar quantity f and vector $\mathbf{a}$.

[19] Kelvin's theorem of circuitation follows from both the equation for $\frac{\partial (\nabla \wedge \mathbf{v})}{\partial t}$ and from the identity $\frac{d}{dt} \int_\Lambda \mathbf{w} \cdot d\mathbf{a} = \int_\Lambda \frac{\partial \mathbf{w}}{\partial t} \cdot d\mathbf{a} + \oint_\gamma (\mathbf{w} \wedge \nabla \wedge \mathbf{w}) \cdot d\mathbf{l}$ for arbitrary $\mathbf{w}$ applied to the case $\mathbf{w} = \nabla \wedge \mathbf{v}$.

for the magnetic field imply that the magnetic flux across a closed surface in the fluid is constant: $\oint_\Lambda \mathbf{B} \cdot d\mathbf{a} = \text{const.}$

Moreover, if the fluid is incompressible ($\nabla \cdot \mathbf{v} = 0$) then $\frac{\partial \mathbf{B}}{\partial t} - \nabla \wedge (\mathbf{v} \wedge \mathbf{B}) = 0$ gives[20]:

$$\frac{d\mathbf{B}}{dt} = (\mathbf{B} \cdot \nabla) \mathbf{v}$$

We are going to proof that a solution of this equation satisfies the so-called 'frozen-field-lines' condition (α_B suitably defined scalar quantity with $\nabla \alpha_B = 0$):

$$\mathbf{B} = \alpha_B \cdot \mathbf{v}$$

Physically, the frozen-field-lines condition means that the magnetic flux across any surface perpendicular to a streamline is constant. Together with Ohm's law for negligible Joule heating, the frozen-field-lines condition implies $\mathbf{E} = 0$: the electric field vanishes in the lab frame of reference.[21] Historically, the term 'frozen' comes from the fact that if the magnetic field lines (which $\mathbf{B}$ is tangent to) coincide (say, at time $t = 0$) with the stream lines (which $\mathbf{v}$ is tangent to), then the former will keep on sticking to the latter for $t > 0$.

In order to proof the frozen-field-lines condition, let us write $\mathbf{v} = \frac{d\mathbf{r}}{dt}$ and let us assume that the magnetic field line coincides initially with a stream line (i.e., $\mathbf{B}$ is initially parallel to $\mathbf{v}$). Moreover, let A and B be two nearby points on the same stream line, with positions $\mathbf{r}_A(t)$ and $\mathbf{r}_B(t) = \mathbf{r}_A(t) + \varepsilon(t)$, respectively. After a time dt, A and B have moved to A' and B', respectively (see Fig. 5.1).

The velocity $\mathbf{v}$ in A has changed by an amount $\mathbf{u}$ such that $\mathbf{r}_{A'} = \mathbf{r}_A + \mathbf{u}dt$. The velocity $\mathbf{v}$ in B has changed by an amount $\mathbf{u} + d\mathbf{u}$ such that $\mathbf{r}_{B'} = \mathbf{r}_B + (\mathbf{u} + d\mathbf{u})\,dt$. The vector ε has changed by an amount $d\varepsilon$ such that $\mathbf{r}_{B'} = \mathbf{r}_{A'} + \varepsilon + d\varepsilon$. Together, the latter three relationships lead to $d\varepsilon = d\mathbf{u}dt$. But $d\mathbf{u}$ is just the increment of $\mathbf{v}$ along ε, hence $d\mathbf{u} = (\varepsilon \cdot \nabla)\mathbf{v}$. Together, the latter two relationships give (after division by dt) $\frac{d\varepsilon}{dt} = (\varepsilon \cdot \nabla)\mathbf{v}$. Now, this equation is identical to the equation $\frac{d\mathbf{B}}{dt} = (\mathbf{B} \cdot \nabla)\mathbf{v}$ above. Consequently, if $\mathbf{B} = \alpha_B \cdot \varepsilon$ at $t = 0$ then $\mathbf{B} = \alpha_B \cdot \varepsilon$ is a solution at all times > 0. Since both ε and $\mathbf{v}$ are tangent to the same stream line at all times, they share the same direction at all times; since $\mathbf{v}$ and $\mathbf{B}$ are initially parallel, they share the same direction at $t = 0$; then, ε and $\mathbf{B}$ share the same direction at $t = 0$, i.e. $\mathbf{B} = \text{const.} \cdot \varepsilon$ at $t = 0$. It follows that $\mathbf{B} = \text{const.} \cdot \varepsilon$ is a solution at all times > 0, hence that $\mathbf{B} = \alpha_B \cdot \mathbf{v}$ is a solution at all times > 0.

Remarkably, the assumptions underlying the proof of the frozen field line condition (incompressibility, negligible Joule heating) are compatible with the assumptions underlying Kirchhoff's principle (Ohm's law, flat temperature gradient, steady

---

[20] Here we invoke both the conditions $\nabla \cdot \mathbf{v} = 0$ and $\nabla \cdot \mathbf{B} = 0$, the definition $\frac{d}{dt} = \frac{\partial}{\partial t} + \mathbf{v} \cdot \nabla$ and the identity $\nabla \wedge (\mathbf{a} \wedge \mathbf{b}) = \mathbf{a}\nabla \cdot \mathbf{b} - \mathbf{b}\nabla \cdot \mathbf{a} + (\mathbf{b} \cdot \nabla)\mathbf{a} - (\mathbf{a} \cdot \nabla)\mathbf{b}$ for arbitrary $\mathbf{a}$ and $\mathbf{b}$ with $\mathbf{a} = \mathbf{v}$ and $\mathbf{b} = \mathbf{B}$

[21] Here we invoke the identity $\mathbf{a} \wedge \mathbf{a} = 0$ for arbitrary vector $\mathbf{a}$, with $\mathbf{a} = \mathbf{v}$.

Fig. 5.1 Evolution of a
stream line

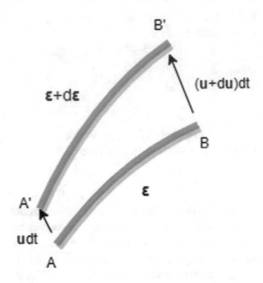

state). In incompressible fluids with uniform temperature and negligible Joule heating
in steady state, therefore, both relationships $\mathbf{B} = \text{const.} \cdot \mathbf{v}$ and $\nabla \wedge \nabla \wedge \mathbf{B} + \nabla \xi' = 0$ hold. After replacing the former in the latter we obtain $\nabla \wedge \nabla \wedge \mathbf{v} + \nabla \lambda = 0$, with
$\lambda = \frac{\xi'}{\alpha_B}$. The same arguments discussed above show that the latter equation is but
the Euler-Lagrange equation of the variational principle

$$\int \left[|\nabla \wedge \mathbf{v}|^2 + \lambda \nabla \cdot \mathbf{v} \right] dx = \min \quad \text{i.e.} \quad \int |\nabla \wedge \mathbf{v}|^2 dx \quad \text{with the constraint} \quad \nabla \cdot \mathbf{v} = 0$$

It is the fact that $|\nabla \wedge \mathbf{v}|^2 \geq 0$ which allows the integral to be a minimum; λ acts as a
Lagrangian multiplier, and incompressibility provides the constraint. Analogously,
if we take the curl of both sides of $\nabla \wedge \nabla \wedge \mathbf{v} + \nabla \lambda = 0$ then the same arguments
invoked above lead to:

$$\nabla \wedge \Delta \mathbf{v} = 0$$

We could as well derive this relationship from the corresponding relationship regard-
ing $\mathbf{B}$ in Sect. 5.3.5 together with $\mathbf{B} = \alpha_B \mathbf{v}$. However, no electromagnetic quantity
appears anymore. Our result may therefore even describe $\mathbf{v}$ in incompressible fluids
with vanishing temperature gradient and no Joule heating even in the limit of vanishing electromagnetic quantities everywhere (i.e., $\alpha_B \to 0$). Alternatively, we may
consider such a limit as a particular case of the solutions satisfying the frozen-field-
line condition.

5.3.7   Viscosity: Korteweg–Helmholtz' Principle

The entropy production density due to viscosity is ($i, k = 1, 2, 3$)

$$\sigma = \frac{\sigma'_{ik}}{T} \frac{\partial v_i}{\partial x_k}$$

(we refer to [6,16,25,26]). Following [16] , here we assume that both η and ζ do not depend on the stresses applied to the fluid. This is the case of 'Newtonian' fluids.[22] It can be shown that $\sigma'_{ik} = \eta \left(\frac{\partial v_i}{\partial x_k} + \frac{\partial v_k}{\partial x_i} - \frac{2}{3}\delta_{ik}\nabla \cdot \mathbf{v} \right) + \zeta\delta_{ik}\nabla \cdot \mathbf{v}$ and that the viscous heating power density is [26]: $T\sigma = \frac{\eta}{2} \left(\frac{\partial v_i}{\partial x_k} + \frac{\partial v_k}{\partial x_i} - \frac{2}{3}\delta_{ik}\nabla \cdot \mathbf{v} \right)^2 + \zeta\left(\nabla \cdot \mathbf{v}\right)^2 = \left(\frac{4}{3}\eta + \zeta\right)\left(\nabla \cdot \mathbf{v}\right)^2 + \eta|\nabla \wedge \mathbf{v}|^2 + 2\nabla \cdot \left[(\mathbf{v} \cdot \nabla)\mathbf{v} - \mathbf{v}(\nabla \cdot \mathbf{v}) \right]$. The coefficients η and ζ are usually referred to as 'dynamic viscosity' and 'bulk viscosity', respectively in the literature.

Since viscous heating is an irreversible process, $\sigma > 0$, hence both $\eta > 0$ and $\zeta > 0$. The viscous term in the equation of motion is $\frac{\partial \sigma'_{ik}}{\partial x_k} = \Delta v_i + \left(\zeta + \frac{\eta}{3}\right) \frac{\partial}{\partial x_i}(\nabla \cdot \mathbf{v}) + O(\nabla\eta, \nabla\zeta)$. If $\nabla\eta = 0$ and $\nabla\zeta = 0$ then Navier-Stokes' equation with no electromagnetic fields reads:

$$\rho\frac{\partial \mathbf{v}}{\partial t} + \rho(\mathbf{v} \cdot \nabla)\mathbf{v} = -\nabla p + \eta\Delta\mathbf{v} + \left(\zeta + \frac{\eta}{3}\right)\nabla(\nabla \cdot \mathbf{v})$$

If the fluid is incompressible ($\nabla \cdot \mathbf{v} = 0$) then Navier-Stokes' equation reduces to:

$$\frac{\partial \mathbf{v}}{\partial t} + (\mathbf{v} \cdot \nabla)\mathbf{v} = -\frac{\nabla p}{\rho} + \nu\Delta\mathbf{v}$$

where the quantity $\nu \equiv \frac{\eta}{\rho}$ is called 'kinematic viscosity'.

Here σ'_{ik} is linear in the spatial derivatives of the components of $\mathbf{v}$, but is equal to the time derivative of no physical quantity (even if $\nabla\eta = 0$ and $\nabla\zeta = 0$); generally speaking, therefore, no LNET-related variational principle is expected to hold.

All the same, a variational principle may still hold for viscous dissipation. In order to discuss this point, we take the curl of both sides of Navier-Stokes for incompressible fluids and obtain the following equation[23] for the vorticity $\nabla \wedge \mathbf{v}$:

$$\frac{\partial}{\partial t}(\nabla \wedge \mathbf{v}) - \nabla \wedge [\mathbf{v} \wedge (\nabla \wedge \mathbf{v})] = \nu\Delta(\nabla \wedge \mathbf{v})$$

---

[22] By definition, a Newtonian fluid is a fluid in which the viscous stresses arising from its flow, at every point, are linearly correlated to the local strain rate, i.e. the rate of change of its deformation over time.

[23] We invoke the identities $\nabla \wedge \nabla \wedge \mathbf{a} = \nabla(\nabla \cdot \mathbf{a}) - \Delta\mathbf{a}$, $(\mathbf{a} \cdot \nabla)\mathbf{a} = \nabla\left(\frac{|\mathbf{a}|^2}{2}\right) - \mathbf{a} \wedge \nabla \wedge \mathbf{a}$, $\nabla \cdot (\nabla \wedge \mathbf{a}) = 0$ and $\nabla \wedge \nabla w = 0$ for arbitrary vector $\mathbf{a}$ and scalar w.

(For vanishing viscosity we retrieve the vorticity balance which leads to Kelvin's circuitation theorem). This balance of vorticity reduces to a form similar to the equation $\frac{\partial \mathbf{B}}{\partial t} - \nabla \wedge [\mathbf{v} \wedge \mathbf{B}] = 0$ for the magnetic field in a fluid with no Joule heating, provided that $\nu\Delta (\nabla \wedge \mathbf{v}) = 0$. In turn, this is only possible if either no viscous dissipation occurs at all (i.e., $\nu \equiv 0$) or $\Delta (\nabla \wedge \mathbf{v}) = 0$. We rewrite the latter relationship as[24]:

$$\nabla \wedge \Delta \mathbf{v} = 0$$

This is one of the equations satisfied by $\mathbf{v}$ in incompressible fluids with negligible Joule heating and $\nabla T = 0$. Correspondingly, we may repeat the arguments discussed above step by step and show that $\mathbf{v}$ satisfies the variational principle

$$\int |\nabla \wedge \mathbf{v}|^2 d\mathbf{x} = \min \quad \text{with the constraint} \quad \nabla \cdot \mathbf{v} = 0$$

If $\nabla\eta = 0$, then the expression displayed above for the viscous heating power density makes this variational principle to reduce to the variational principle[25]

$$\int \sigma'_{ik} \frac{\partial v_i}{\partial x_k} d\mathbf{x} = \min \quad \text{with the constraint} \quad \nabla \cdot \mathbf{v} = 0$$

According to Ref. [25] this variational principle is due to both Korteweg and Helmholtz, and will therefore be referred to as 'Korteweg–Helmholtz' principle' in the following.

Let us check that Korteweg–Helmholtz' principle agrees with the equation of motion. We start from $\delta \int [\eta|\nabla \wedge \mathbf{v}|^2 + \lambda\nabla \cdot \mathbf{v}] = 0$ (λ Lagrange multiplier). For $\mathbf{v} \to \mathbf{v} + \delta\mathbf{v}$ we write[26]: $0 = \delta \int [\eta|\nabla \wedge \mathbf{v}|^2 + \lambda\nabla \cdot \mathbf{v}] = \int [2\eta (\nabla \wedge \mathbf{v}) \cdot \nabla \wedge \delta\mathbf{v} + \lambda\nabla \cdot \delta\mathbf{v}] d\mathbf{x} = \int \delta\mathbf{v} \cdot (2\eta\nabla \wedge \nabla \wedge \mathbf{v} - \nabla\lambda) d\mathbf{x}$ for arbitrary $\delta\mathbf{v}$, hence $2\eta\nabla \wedge \nabla \wedge \mathbf{v} - \nabla\lambda = 0$; we take the curl of both sides and retrieve $\nabla \wedge \Delta \mathbf{v} = 0$ just as expected, provided that $\nabla\eta = 0$ (i.e. $\nabla\nu = 0$ as $d\rho = 0$ for incompressibility). If $\lambda \to \lambda + \delta\lambda$ then we retrieve $\nabla \cdot \mathbf{v} = 0$.

It is worthwhile to stress a point that will surface again and again in the following. Per se, a condition of extremization does not necessarily provide complete information on the relaxed state. It is rather just one of its features, which is useful for further investigation. Here, in Helmholtz' words [27] *for extremely small velocities and constant flows, the currents in a viscous fluid are so distributed, that the loss of kinetic energy due to friction becomes a minimum, provided that the velocities at*

---

[24] Here we invoke the identity $\Delta (\nabla \wedge \mathbf{a}) = \nabla \wedge \Delta\mathbf{a}$ for arbitrary vector $\mathbf{a}$.

[25] Here we apply Gauss' theorem of divergence. Moreover, in the variation, we neglect surface integrals on the boundary, as boundary conditions are fixed.

[26] Gauss' theorem of divergence; $\delta\mathbf{v} = 0$ at the boundary; $\nabla \cdot \mathbf{v} = 0$; moreover, $\nabla \wedge \nabla w = 0$, $\nabla \cdot (w\mathbf{a}) = \mathbf{a} \cdot \nabla w + w (\nabla \cdot \mathbf{a})$; $\nabla \wedge \nabla \wedge \mathbf{a} = \nabla (\nabla \cdot \mathbf{a}) - \Delta\mathbf{a}$ and $\nabla \cdot (\mathbf{a} \wedge \mathbf{b}) = \mathbf{b} \cdot \nabla \wedge \mathbf{a} - \mathbf{a} \cdot \nabla \wedge \mathbf{b}$ for arbitrary vectors $\mathbf{a}$, $\mathbf{b}$ and arbitrary scalar w.

the boundary of the fluid are considered as fixed. In his own words [28], Korteweg's contribution is that *in any simply connected region, when the velocities along the boundary are given, there exists, as far as the squares and products of the velocities may be neglected, only one solution of the equations for the steady motion of an incompressible viscous fluid, and that this solution is always stable.* We stress the following points:

- *the velocities along the boundaries [...] are given,* i.e. boundary conditions are always required, as it may be expected;
- as for these boundary conditions, Helmholtz claims (with no proof) that *as for a body floating in such fluid, is is also possible to extend this theorem as follows: a floating body is in equilibrium in a slow, steady current of a flowing fluid when the friction in the steady current is a minimum even if the value of the velocity of the fluid along the surface of the floating body changes in the same way it would change should any of the possible motions of the body actually occur* [27]. In other words, the minimum condition is to be satisfied not only with respect to changes in the flow around the body, but also to the position of the body itself. Minimization affects, therefore, the geometry of the problem.
- the result holds *as the squares and products of the velocities may be neglected.* (Helmholtz speaks of *extremely small velocities*). This corresponds to low values of *Re*. Korteweg himself vividly explains the consequences of this condition: *When, on the contrary, the squares and higher powers of the velocities are taken into account, I have my reasons for supposing that—even in the case of a sphere moving with uniform velocity—no state of steady motion can be reached, and the motion must finally become unstable.* In other words, if *Re* increases then turbulence becomes possible, and this extremum condition no longer applies.

Korteweg–Helmholtz' principle for viscous, incompressible Newtonian fluids with $\nabla \nu = 0$ and $\nabla T = 0$ is the counterpart of Kirchhoff's principle for resistive, incompressible fluids with $\nabla \sigma_\Omega = 0$ and $\nabla T = 0$. Since $\nabla T = 0$, minimization of viscous power is equivalent to minimization of the amount of entropy produced per unit time through viscosity, i.e. Korteweg–Helmholtz' principle too is a MinEP. In contrast with the MinEP of LNET, Korteweg–Helmholtz' principle holds for viscous dissipation and $\nabla T = 0$ only. Admittedly, both Refs. [27,28] refer to temperature nowhere. However, the viscosity coefficient is taken as constant and uniform. If it depends on temperature, this is possible whenever $\nabla T = 0$. Being the viscous contribution to the total heating power density P_h proportional to the squares of the components of the velocity gradients, which are neglected here, the low-*Re* assumption corresponds to negligible viscous power density everywhere across the fluid, which in turn is compatible with vanishing temperature gradient as no other heating mechanism is considered. In the general case $\nabla T \neq 0$, the work of [29] casts doubt on the validity of the MinEP of LNET in fluid dynamics. The connection between Korteweg–Helmholtz' and Kirchhoff's principles on one side and the condition $\nabla T = 0$ on the other side is not trivial and will be discussed in the following. For further discussion, see Sects. 5.3.10 and 6.2.9.

5.3.8   Maximum Economy: Yardangs, Rivers and the Human Blood

We have discussed minimization of viscous heating power for Newtonian fluids. But even if human blood is non-Newtonian, when it comes to the circulation of the blood *the general principle [...] is that the form and the arrangement of the blood vessels are such that the circulation proceeds with a minimum of effort, and with a minimum of wall-surface, the latter condition leading to a minimum of friction and being therefore included in the first* and, more generally *we assume that the cost of operating a physiological system is a minimum, what we mean by the cost being measurable in calories and ergs, units whose dimensions are equivalent to those of work*—see Chap. XV of [30].

This minimization, which is based on Murray's tenet of 'maximum economy' [31],[27] is rigorously proven nowhere, but *all the experience and the very instinct of the physiologist tells him it is true; he comes to use it as a postulate [...] and it does not lead him astray* [30] and leads to correct predictions concerning the branching and the size of arteries. Since the temperature of the human body is constant and uniform, minimization of the amount of power dissipated through viscosity is equivalent to minimization of the amount of entropy produced per unit time by viscous dissipation.[28]

Even if $Re \gg 1$, the idea that minimum friction rules the development of structures in a system with negligible ∇T is not limited to physiology. In hydrology, for example, MinEP has been postulated in Ref. [32] in order to describe the fractal pattern of river basins. In desertic areas, it has been empirically observed that yardangs (desertic landforms sculpted by wind erosion) evolve spontaneously into streamlined minimum-drag forms. When viewed from above, yardangs resemble the hull of a boat—a benchmark of reduced fluid dynamic resistance. Softer rocks are eroded and removed by the wind, and the harder material remains. The resulting pattern is the outcome of both geology (the original distribution of rocks) and fluid mechanics (the motion of the wind) [33].[29]

5.3.9   Porous Media

Problems involving flows across porous media are of paramount practical relevance (think e.g. of oil wells, aquifer pollution, etc.), involve both viscosity and capillarity and have also been described with the help of a postulated MinEP-like principle [34].

---

[27] In this paper we find the thought-provoking words *physiology is a problem in maxima and minima*.

[28] Of course, the actual relaxed flow in a non-Newtonian fluid may differ from the corresponding flow in the Newtonian case, as the viscous power depends differently on velocity.

[29] For further discussion see Sect. 6.2.9.

A porous medium is defined as a solid permeated by an interconnected network of pores filled with a fluid (water, here and below). Darcy's law states that

$$\mathbf{q}_w = -K_w \nabla H_w \quad ; \quad \mathbf{q}_w \equiv \mathbf{v}\varphi_p \quad ; \quad H_w \equiv z + \frac{p}{\rho g}$$

in saturated, isotropic porous media. Here $\mathbf{v}$, g, p, z, ρ, $\varphi_p \equiv \frac{V_p}{V_{tot}}$, V_p and V_{tot} are the water velocity $\mathbf{v}$, the absolute value of the acceleration of gravity, the pressure, the vertical coordinate, the water mass density, the porosity of the medium (which is supposed uniform below), the volume of voids and the total volume, respectively; $\mathbf{q}_w$, H_w, $K_w > 0$ and $\frac{1}{K_w}$ are dubbed 'volumetric water flux', 'hydraulic head', 'hydraulic conductivity' and 'hydraulic resistivity', respectively. (If the medium is anisotropic K is replaced by a tensor; nothing essential changes in the following). Here 'saturated' means that all the pores are filled and conducting, so that hydraulic conductivity is maximal. Two remarks are of interest: firstly, $\rho g H_w$ is a contribution to the enthalpy h per unit mass in the energy balance of Sect. 4.2.7 in a uniform, constant, vertical gravitational field—the obvious example being Earth's field. Secondly, Darcy's law is formally analogous to Ohm's law in a rigid conductor at rest (Sect. 5.3.1): $\mathbf{q}_w$, $\frac{1}{K_w}$ and H_w play the role of electric current density, electric resistivity and electrostatic potential, respectively. This analogy suggests that some variational principle is lurking.

In Darcy's original formulation $\frac{dK_w}{dt} = 0$, $\frac{dK_w}{d|\mathbf{q}_w|} = 0$ and Darcy's law is linear. Just like its electrical counterpart, however, K_w depends on a quantity ('permeability') which depends on the porous medium.[30] If the permeability is too low, then Darcy's law is no more linear, and K_w is not negligible only if $|\mathbf{q}_w|$ exceeds a threshold. Another example of nonlinearity occurs at large $|\mathbf{q}_w|$: too much water poured into the system, indeed, tends to destroy the porous structure of the solid and to form microfractures; the latter may coalesce ('fingering') into larger fractures which act as preferential channels for the water flow across the system. In this case, $\frac{dK_w}{d|\mathbf{q}_w|} > 0$.

Self-consistent treatment of large water flow across a porous medium requires that we drop the assumption of the saturated medium. In an 'unsaturated' medium, not all the pores are filled with water and conducting. In unsaturated soil, for example, some pores become air-filled: unsaturated soil is a water-air two-phase flow system. Larger pores in soil are filled with air; smaller pores hold water. By describing a pore that allows water to flow as a capillary tube, it is possible to show that a capillary tube with a radius equal to or less than a critical radius is filled with the wetting fluid

---

[30] Formally, the permeability is the product of the dynamic viscosity of the fluid, the thickness of the porous medium and the average flow velocity calculated as if the fluid was the only phase present in the porous medium, divided by the applied pressure difference. Permeability is a specific property characteristic of the solid skeleton and the microstructure of the porous medium itself, independently of the nature and properties of the fluid flowing through the pores of the medium. This allows to take into account the effect of temperature on the viscosity of the fluid flowing through the porous medium and to address other fluids than pure water, e.g., concentrated brines, petroleum, or organic solvents. Permeability has the dimension of an area.

(water). Otherwise, it is filled with non-wetting fluid (air). Similar results hold for other couples of fluids (e.g. water and oil).

Formally, Darcy's law holds in unsaturated media too, but K is lower than in the saturated case (as the number of pores available for water transport is lower) and depends on saturation, which is basically the ratio between the volume of water-filled pores and the volume of all pores. In the definition of H_w, moreover, p is replaced by the 'capillary pressure' p_c. Capillary pressure is the pressure between two immiscible fluids in a thin tube resulting from the interactions of forces between the fluids and solid walls of the tube. In most cases, it depends also on saturation.

Viscosity, capillarity and gravity act together. On microscopic scales, viscosity rules viscous heating. On a scale not larger than a critical radius of a pore, capillarity allows wetting and may even bring water upwards, depending on the gradient of p_c (i.e., of saturation). On a much larger scale, gravity rules, drags water backwards and allows fingering.

Let our porous medium achieve a relaxed state under suitable constant boundary conditions (e.g. a constant amount of water supplied from above by rain). Liu (see Sect. 2.4 of Ref. [34]) has postulated that if the contribution of kinetic energy to the energy balance of water is negligible (as is usually the case) then the relaxed state satisfies the variational principle

$$\int \Delta E_c d\mathbf{x} = \min \quad ; \quad \text{with the constraint of fixed} \int |\nabla H_w|^2 d\mathbf{x}$$

where Liu defines the 'energy expenditure rate per unit volume' $\Delta E_c \equiv \nabla \cdot (\mathbf{q}_w H_w)$. In Liu's own words, *the energy expenditure rate [...] is equal to the energy carried by water flowing into the volume minus the energy carried by water flowing out of the volume. Obviously, a smaller energy expenditure rate corresponds to a smaller water flow resistance.*

As a result, the variational principle provides a dependence of K_w on $|\mathbf{q}_w|$ in agreement with observations, provided that the impact of capillarity on K_w is negligible, i.e. whenever the system is ruled by gravity. (The constraint plays a crucial role; the unconstrained version of the variational principle does not work—see the discussion of Eq. 1 in Ref. [35]). It turns out that *for a gravity-dominated unsaturated flow process, relative permeability is not only a function of [...] saturation, but also a power function of water flux.* In other words, K increases with increasing $|\mathbf{q}_w|$ as a power law, in agreement with physical intuition about fingering: the larger the water flow, the stronger the tendency to form larger and larger fractures, the easier the motion of water in the resulting channels backwards. In the words of Liu: *a nonlinear natural system that is not isolated and involves positive feedback tends to minimize its resistance to the flow process that is imposed by its environment* [35]. Inclusion of the further constraint of constant energy (see Sect. 2.7 of Ref. [34]) leaves this result basically unaffected.

No justification (but comparison *a posteriori* with observations) is provided for the variational principle. Remarkably, indeed, this variational principle provides no prediction in agreement with observations if capillarity rules the system. In Ref. [35], this failure is due to the fact that the only system which is subject to optimization process *are not isolated and involve positive feedbacks* (like fingering). Liu's discussion deserves particular attention because it is one of the very few examples where the far from equilibrium relaxed states which are *not* correctly described by the postulated variational principle are explicitly and clearly identified, thus facilitating a critical examination.[31]

5.3.10 Stability Versus Kirchhoff's and Korteweg–Helmholtz' Principles

The analogy between Kirchhoff's and Korteweg–Helmholtz case outlined in Sects. 5.3.1 and 5.3.7, respectively, allows us to limit ourselves to Kirchhoff's case. If $\nabla T = 0$ then $T = T_0$ and we may take $\mathbf{X} = \mathbf{E} + \mathbf{v} \wedge \mathbf{B}$ and $\mathbf{Y} = \frac{\mathbf{j}_{el}}{T_0}$ as the vector of thermodynamic forces and thermodynamic flows, respectively. Moreover, LNET holds for Joule heating, hence $0 \geq \frac{d_X P}{dt} = \frac{d_Y P}{dt} = \frac{1}{2}\frac{dP}{dt}$ and a stable steady state enjoys Lyapunov stability with P as Lyapunov function. In the following, we are going to check under which condition the inequality is satisfied, i.e. under which condition the existence of a stable steady state agrees with $\nabla T = 0$. To this purpose, we limit ourselves to the nonrelativistic limit where $\nabla \wedge \mathbf{B} = \mu_0 \mathbf{j}_{el}$, take the curl of both sides of this equation, invoke Ohm's law, take into account that $\nabla \wedge \mathbf{E} = -\frac{\partial \mathbf{B}}{\partial t}$ and obtain the so-called 'magnetic diffusion equation':

$$\frac{\partial \mathbf{B}}{\partial t} - \nabla \wedge (\mathbf{v} \wedge \mathbf{B}) = \frac{1}{\mu_0 \sigma_\Omega} \Delta \mathbf{B}$$

If $\sigma_\Omega \to \infty$ then the magnetic diffusion equation reduces to the condition for the conservation of magnetic flux $\frac{\partial \mathbf{B}}{\partial t} - \nabla \wedge (\mathbf{v} \wedge \mathbf{B}) = 0$, just like the vorticity balance $\frac{\partial}{\partial t}(\nabla \wedge \mathbf{v}) - \nabla \wedge [\mathbf{v} \wedge (\nabla \wedge \mathbf{v})] = \nu \Delta (\nabla \wedge \mathbf{v})$ reduces in the $\nu \to 0$ limit to the condition for the conservation of the flux of vorticity (the circuitation) $\frac{\partial}{\partial t}(\nabla \wedge \mathbf{v}) - \nabla \wedge [\mathbf{v} \wedge (\nabla \wedge \mathbf{v})] = 0$. Both ν and $\frac{1}{\mu_0 \sigma_\Omega}$ have the dimensions of a diffusion coefficient, i.e. $(\text{length})^2 \cdot (\text{time})^{-1}$. Now, let l be a typical length of the system. We define the dimensionless 'Reynolds' number' $Re \equiv \frac{|\mathbf{v}|l}{\nu}$ and the dimensionless 'magnetic Reynolds' number' $Re_M \equiv |\mathbf{v}|l\mu_0\sigma_\Omega$. If $Re \gg 1 \, (\ll 1)$ then the typical time scale $\frac{l^2}{\nu}$ of viscous dissipation is $\gg (\ll)$ the inertial time scale $\frac{l}{|\mathbf{v}|}$. If $Re_M \gg 1 \, (\ll 1)$ then the typical time scale $l^2 \mu_0 \sigma_\Omega$ of Joule dissipation is $\gg (\ll)$ the

---

[31] See Sect. 6.2.10 for further discussion.

inertial time scale $\frac{l}{|\mathbf{v}|}$. Let us check whether the sign of $\frac{d_Y P}{dt}$ is actually non-positive, as predicted by LNET. We obtain[32] :

$$
\begin{aligned}
\frac{d_Y P}{dt} &= \frac{1}{T_0} \int (\mathbf{E} + \mathbf{v} \wedge \mathbf{B}) \cdot \frac{d\mathbf{j}_{el}}{dt} dx \\
&= \frac{1}{T_0} \int (\mathbf{E} + \mathbf{v} \wedge \mathbf{B}) \cdot \frac{\partial \mathbf{j}_{el}}{\partial t} dx + O\,(Re_M) \\
&= \frac{1}{\mu_0 T_0} \int (\mathbf{E} + \mathbf{v} \wedge \mathbf{B}) \cdot \nabla \wedge \frac{\partial \mathbf{B}}{\partial t} dx + O\,(Re_M) \\
&= \frac{1}{\mu_0 T_0} \int \frac{\partial \mathbf{B}}{\partial t} \cdot \nabla \wedge (\mathbf{E} + \mathbf{v} \wedge \mathbf{B})\, dx + O\,(Re_M) \\
&= \frac{1}{\mu_0 T_0 \sigma_\Omega} \int \frac{\partial \mathbf{B}}{\partial t} \cdot \nabla \wedge \mathbf{j}_{el} dx + O\,(Re_M) \\
&= \frac{1}{\mu_0^2 T_0 \sigma_\Omega} \int \frac{\partial \mathbf{B}}{\partial t} \cdot \nabla \wedge \nabla \wedge \mathbf{B} dx + O\,(Re_M) \\
&= -\frac{1}{\mu_0^2 T_0 \sigma_\Omega} \int \frac{\partial \mathbf{B}}{\partial t} \cdot \Delta \mathbf{B} dx + O\,(Re_M) \\
&= -\frac{1}{\mu_0 T_0} \int |\frac{\partial \mathbf{B}}{\partial t}|^2 dx + O\,(Re_M)
\end{aligned}
$$

Clearly, $\frac{d_Y P}{dt} \leq 0$ for arbitrary perturbation only for $Re_M \to 0$. The configuration satisfying Kirchhoff's principle enjoys Lyapunov stability provided that $Re_M \to 0$ and that Joule heating only occurs. Physically, the motion of the fluid is so slow that incompressibility is satisfied and all inhomogeneities in $\mathbf{B}$ smear out throughout the system, so that the Joule heating power density $\propto |\mathbf{j}_{el}|^2 \propto |\nabla \wedge \mathbf{B}|^2$ is small and the system attains a state with $\nabla T = 0$. The latter allows Kirchhoff's principle to describe the relaxed state.

We may appreciate the generality of Kirchhoff's principle when considering the experimental results reported in Ref. [36]. A prototypical example of dissipative structure (Sect. 5.8.9), Belousov–Zhabotinsky reaction is a spontaneously oscillating reaction of inorganic chemistry. Inspired by the excess potential approach of Sect. 6.1.3, the authors performed accurate measurements of the oscillating electrostatic potential in the cell where this chemical reaction occurs[33] shows that the Joule power is a slowly but monotonically decreasing function of time, in agreement with the idea that the Joule power acts as a Ljapunov function of the system regardless of the complexity of its oscillations.

---

[32] We take into account both Ohm's law and the following relationships: $\mathbf{v} \propto O\,(Re_M)$, $\frac{d}{dt} \equiv \frac{\partial}{\partial t} + \mathbf{v} \cdot \nabla$, $\nabla \wedge \frac{\partial}{\partial t} = \frac{\partial}{\partial t} \nabla \wedge$, $\nabla \wedge \mathbf{B} = \mu_0 \mathbf{j}_{el}$, $\nabla \cdot \mathbf{B} = 0$, $\nabla \wedge \nabla \wedge \mathbf{a} = \nabla (\nabla \cdot \mathbf{a}) - \Delta \mathbf{a}$ and $\nabla \cdot (\mathbf{a} \wedge \mathbf{b}) = \mathbf{b} \cdot \nabla \wedge \mathbf{a} - \mathbf{a} \cdot \nabla \wedge \mathbf{b}$ for arbitrary vectors $\mathbf{a}$, $\mathbf{b}$.

[33] And where the mean macroscopic velocity of the fluid mixture vanishes, so that Re_M is negligible

The analogy with Korteweg–Helmholtz is straightforward, provided that we replace Re_M with Re. In Korteweg's own words [28] *when in a given region occupied by viscous incompressible fluid there exists at a certain moment a mode of motion which does not satisfy the equations of motion, then—the velocities along the boundary being maintained constant the change which must occur in the mode of motion will be such (neglecting squares and products of velocities) that the dissipation of energy by external friction is a constraint, decreasing till it reaches the value corresponding to the steady solution of the equation of motion and the mode of motion becomes identical with such solution.* As a consequence, *the mode of motion represented* by the solution of the equations of motion *is always stable as far as squares and products of velocities may be neglected.* To show this point, *let the mode of motion be disturbed by any cause whatever. Then the dissipation of energy by internal friction is necessarily increased. But as soon as the cause of disturbance ceases it must decrease again till it reaches the value* corresponding to the steady solution of the equation of motion, and then the latter *mode of motion is restored.* Korteweg–Helmholtz' principle of Sect. 5.3.7 describes the relaxed state provided that $Re \ll 1$ and that viscous heating only occurs. We conclude that incompressible fluids are always going to flow across the path of lesser resistance.[34]

Finally, there is still a subtle connection with thermodynamics hidden in Helmholtz' work. Korteweg and Helmholtz obtain the minimum condition in different ways under the same $Re \ll 1$ assumption. Korteweg [28] shows that all incompressible flows which satisfy the boundary conditions but do not satisfy Navier-Stokes' equation correspond to an amount $\int \sigma'_{ik} \frac{\partial v_i}{\partial x_k} d\mathbf{x}$ of dissipated viscous power larger than the same amount in the flow which satisfies Navier-Stokes equation.[35] Helmholtz [27] writes down a balance of mechanical energy $\frac{dK}{dt} = P_{ext} - \int \sigma'_{ik} \frac{\partial v_i}{\partial x_k} d\mathbf{x}$, where K and P_{ext} are the kinetic energy and the mechanical work done per unit time by the external world on the fluid; then, he shows that in steady state the quantity $W' \equiv P_{ext} - \frac{1}{2} \int \sigma'_{ik} \frac{\partial v_i}{\partial x_k} d\mathbf{x}$ is a minimum. The balance of mechanical energy ensures that $W' = \frac{dK}{dt} + \frac{1}{2} \int \sigma'_{ik} \frac{\partial v_i}{\partial x_k} d\mathbf{x}$. In steady state $W' = \frac{1}{2} \int \sigma'_{ik} \frac{\partial v_i}{\partial x_k} d\mathbf{x}$

---

[34] The fact that $Re \ll 1$ is essential (think e.g. of honey flowing across small pipes with inner walls of different roughness). If $Re \gg 1$ stable flows with complex distributions of temperature may even prefer the path of greatest resistance. For example, tornado winds may (unfortunately) be stronger near the ground, where friction is larger. Reference [37] is a fascinating review of problems in fluid dynamics where Korteweg–Helmholtz' principle works—and of other problems where it fails.

[35] Reference [38] discusses a generalization of this result, as its equation (54) postulates minimization of the volume integral of T times a linear combination of $\sigma'_{ik} \frac{\partial v_i}{\partial x_k} d\mathbf{x}$ and $\frac{\chi}{2} |\nabla T|^2$ for an incompressible fluid with uniform dynamic viscosity and thermal conductivity. The Lagrangian density (Sect. A.2) is T^2 times the linear combination of the amounts of entropy produced per unit time and volume by viscous dissipation and heat transport. The Lagrangian is related to the time derivative ('entransy dissipation') of the product of internal energy and temperature ('entransy'). The Lagrangian coordinates are T and the components of $\mathbf{v}$. The Euler-Lagrange equations are the energy balance of the fluid (with heat conduction and viscous heating as the only energy transport process and energy source, respectively) and the Navier–Stokes' equation in the $Re \to 0$ limit. Basically, it is a combination of Korteweg–Helmholtz' principle and Fourier's law.

and minimization of $\int \sigma'_{ik} \frac{\partial v_i}{\partial x_k} d\mathbf{x}$ follows. But then Korteweg–Helmholtz' principle takes on another meaning. The definition of W' gives: $P_{ext} = W' + \frac{1}{2} \int \sigma'_{ik} \frac{\partial v_i}{\partial x_k} d\mathbf{x}$ where both quantities tend to a minimum in steady state, hence the work done per unit time by the external world on the fluid is a minimum. This is in agreement with Le Châtelier's principle: in a stable state, the flow must be in the direction of the applied force, accommodating itself to it rather than opposing it. Taking into account that the work done by the fluid onto the external world has the opposite sign, we could rephrase our conclusion by saying that the work done per unit time by the fluid on the external world is a maximum is steady state.[36]

5.3.11 Convection at Moderate Ra

Here we take advantage of the results of Sect. 5.3.1 and 5.3.7 and provide a qualitative discussion of a variational principle for convection cells, provided by Chandrasekhar in his rigorous, seminal discussion of the heat transport processes occurring in stars [39].

Convection occurs in a gravitating fluid under the effect of a temperature gradient: if the dimensionless Rayleigh number Ra (an increasing function of the absolute value of $|\nabla T|$) exceeds a threshold value Ra_{thr} then buoyancy drives the spontaneous formation of convective motions, counteracted by dissipative effects like viscous heating and (in electrically conducting, magnetized fluids) Joule heating. A small mass element of a fluid in a constant gravitational field and with a nonuniform temperature gradient may undergo a rotational motion in convective cells because of buoyancy. The notion of a steady state is therefore meaningful in an averaged sense only, where the averaged is performed on both space and time. In all cases, in a steady state, the averaged amount of energy ε_ν transformed into heat per unit volume and time by viscous heating and Joule heating is equal to the averaged amount of energy ε_g supplied per unit volume and time by the buoyancy to the fluid

$$\varepsilon_\nu = \varepsilon_g$$

Unless $|\nabla T|$ is very large, $\frac{T}{|\nabla T|} \gg$ the linear size of a small mass element. Consequently, the temperature jump inside a small mass element is negligible, at least if $|\nabla T|$ is not too large across the system, or, in other words, if $Ra - Ra_{thr}$ is assumed to be not too large. Moreover, even if convection may locally affect the fluid density, we neglect the impact of this perturbation on the gravitational field, which may therefore be assumed to be constant at all times; it follows that—as shown in Sect. 4.2.7—gravity leaves entropy unaffected. Together, these assumptions allow us to apply both Korteweg–Helmholtz' principle and Kirchhoff's principle locally, hence: $\varepsilon_\nu = \min$. Together with the relationship above, this leads to:

$$\varepsilon_\nu = \min \quad \text{with the constraint} \quad \varepsilon_\nu = \varepsilon_g$$

---

[36] See note in Sect. 6.1.11.

Now, convection is ruled by the interplay of gravity and ∇T. Accordingly, if convection cells are present then ε_g is an increasing function of $|\nabla T|$, i.e.: $\frac{d\varepsilon_g}{|\nabla T|} > 0$. Since $Ra \propto |\nabla T|$, the variational principle above reduces therefore to a variational principle which we refer to as 'Chandrasekhar's principle' in the following:

$$Ra = \min \quad \text{with the constraint} \quad \varepsilon_\nu = \varepsilon_g$$

Quoting Chandrasekhar [39] *instability occurs at the minimum temperature gradient at which a balance can be maintained between the kinetic energy dissipated by viscosity and the internal energy released by the buoyancy force.* Remarkably, we killed more birds with one stone, as Chandrasekhar's principle applies equally well to fluids with viscous heating only, Joule heating only or with a mixture of both. This fact makes it uniquely useful when it comes to convection problems in geophysics and stellar physics. In all cases, it allows the selection of solutions of the equations of motion for the convective cells which are stable against perturbations and are therefore likely to be observed in Nature.

For $Ra - Ra_{thr} > 0$ small but finite, Busse [40] analyzes the equations of motion of an unmagnetized fluid with the help of the expansion in powers of the amplitude ϵ of convective velocity and shows (under the reasonable assumption that ϵ is an increasing function of Ra) that a stable, steady solution of the convection problem corresponds to a minimum of $\int_0^{\epsilon_0} Ra(\epsilon) \, d\epsilon$ where ϵ_0 is an increasing function of the convective heat flux. The fact that ϵ_0 is a given quantity means that the convective heat flux is given. Being $Ra - Ra_{thr}$ an increasing function of ϵ (which vanishes for vanishing ϵ), Busse's result generalizes Chandrasekhar's one, but with the constraint of given heat flow. We shall discuss the thermodynamic foundation of Chandrasekhar's and Busse's results in Sect. 6.2.2.

5.3.12 Turbulent Flow Between Fixed Parallel Surfaces

Basically, Chandrasekhar in Sect. 5.3.11 requires minimization of $|\nabla T|$ ($\propto Ra$) at a given amount of dissipated power. We are going to show that—formally at least—this requirement agrees with Malkus' analysis [41] of the steady-state turbulent shear flow in a viscous, incompressible fluid moving with mean velocity $U = U(z)$ in the $\mathbf{x}$ direction between fixed parallel surfaces of infinite extent $z = +z_0$ and $z = -z_0$ in the Cartesian frame of coordinates $\{x, y, z\}$ and with $U(+z_0) = U(-z_0) = 0$. Here 'mean' is defined as an average in the $\mathbf{y}$ direction. Equations (1.12) and (1.14) of Ref. [41] maximize the total power dissipated by viscosity per unit mass with fixed $U_m \equiv \frac{1}{2z_0} \int_{-z_0}^{+z_0} U \, dz$. In order to show the relationship between Chandrasekhar's and Malkus' statements, it is enough to show that $|\nabla T|$ increases with increasing U_m, so that we are allowed to replace the constraint in Malkus' analysis with the constraint of fixed temperature gradient. A lemma of variational calculus, the reciprocity principle for isoperimetric problems (see e.g. Sect. IX.3 of Ref. [42]), ensures that the solution of Malkus' variational problem is also the solution of the 'reciprocal' variational problem of a minimum temperature gradient with given viscous power, i.e. Chandrasekhar's problem.

Physically, of course, there are many fundamental differences between Chandrasekhar's and Malkus' problems. Firstly, Chandrasekhar's argument is a rigorous proof in a system where both Joule and viscous dissipation may occur, while Malkus' one is a suggestion in a system where no Joule dissipation occurs. Secondly, $|\nabla T|$ is given by the different temperatures of the top and the bottom plate in convection, while in Malkus' problem the temperature of the plates at $z = +z_0$ and $z = -z_0$ may be taken as equal (say, to $T_{boundary}$).

According to Ref. [43], however, Chandrasekhar's theorem and Malkus' suggestion are but two faces of the same problem from the point of view of thermodynamics; a thermodynamic proof of the former (to be found in Sect. 6.2.2) may therefore justify also the latter. The line of reasoning is as follows.

Viscous dissipation produces some heating in the fluid and the local value of $|\nabla T|$ vanishes exactly nowhere at all times because of the competing effect of viscous heating and of heat transport due to the turbulent motions; speaking of $|\nabla T|$ is still therefore meaningful, provided that we refer to some average on turbulent time- and spatial scale at least. We may e.g. invoke the same averaging procedure invoked in Ref. [41]. Admittedly, the actual values of both viscous power and $|\nabla T|$ are usually neglected in most practical applications. However, what is crucial here is their mutual dependence.

If we suppose, with no loss of generality, that $U \geq 0$ everywhere, then $U_m \geq 0$. In particular, $U_m = 0$ if and only if $U = 0$ everywhere, which in turn is equivalent to vanishing viscous heating everywhere, which is equivalent to $|\nabla T| = 0$ everywhere because $T(+z_0) = T(-z_0) = T_{boundary}$ at all times. Correspondingly, $U_m > 0$ if and only if a suitably averaged $|\nabla T|$ is > 0, i.e. the latter quantity is an increasing function of U_m as required, in a neighbourhood of $U_m = 0$ at least. Broadly speaking, the larger U_m the larger $|\frac{\partial U}{\partial z}|$ (because $U(+z_0) = U(-z_0) = 0$), the larger the viscous heating power, the larger the averaged $|\nabla T|$ (because $T(+z_0) = T(-z_0) = T_{boundary}$). A similar behaviour is expected in the more interesting case when the Reynolds number $Re \equiv \frac{z_0 U_m}{\nu}$ exceeds the threshold value Re_{thr} corresponding to the onset of turbulence, as this onset corresponds to a sharp increase in viscous dissipation: the larger $Re - Re_{thr}$, the larger $U_m - \frac{\nu Re_{thr}}{z_0}$, the larger the dissipated power, the larger the averaged $|\nabla T|$, so that once again the latter quantity is an increasing function of U_m in a right neighbourhood of Re_{thr} at least.

Finally, we stress the point that care must be taken with the selection of the relevant constraints and boundary conditions. For example, Malkus added some additional constraints on both the mean and the fluctuating velocity field, which have been heavily criticized by Reynolds and Tiederman [44] on the basis of comparison with experiments.

5.4   Bejan's 'Constructal Law'

More recently, Bejan [45] looked for a solution of the problem of locating and
connecting the roads in a region Ω in order to minimize the time Δt required by one
person to reach a given, common destination M in Ω (and, once arrived there, to leave
Ω), given the number $\dot{N}$ of persons distributed uniformly throughout Ω which reaches
M per unit time. Simple, straight roads connecting M with the starting points of each
traveller do not solve the problem, as people reach M starting from any point in Ω
and the travellers would literally walk one upon the other. The nature of the problem
remains basically unchanged if we allow different points in Ω to be a different height.
Thus, Ω can be either a flat area, a mountainous region or a skyscraper. According
to Bejan, the solution is a tree-like, approximately self-similar, fractal-like network
of roads similar to the existing ones in the real world.

A similar treatment is relevant to the problem [46] of locating and connecting
the heat conductors in a region Ω filled with electronic circuits. The aim of these
heat conductors is to minimize the maximum value of temperature achieved inside Ω
because of the Joule heating associated with the operation of the electronic circuits.
Since the temperature of the external world is difficult to control, the best strategy
is to minimize the absolute value of the temperature gradient across Ω. The heat
conductors (basically, wires of metal) do the job through the conduction of heat
towards a heat sink located at a given point M. The amount of heat produced per unit
time by Joule heating during the operation of the electronic circuits is also given and
is supposed to be uniformly distributed uniformly throughout Ω. The solution is a
tree-like, approximately self-similar, fractal-like network, similar to the network of
roads above, to the ever-bifurcating tree of arteries in the human lung and to other
fractal-like structures in Nature [47], e.g. the meanders in the mouth of a river, the
formation of crack patterns in solids subjected to volumetric cooling by convection
[48] and the branching tips of lightning.

Speaking of electricity, a similar result is retrieved in [49], where the tree-like
shape of a dielectric breakdown (with a typical 'Lichtenberg's figure') is recon-
structed through an iterative solution of Laplace equation for the electrostatic potential (i.e. of the variational problem of Kirchhoff) near the tip of each branch of the
discharge.

In all cases, solutions are obtained after many steps of a recursive algorithm,
each step achieving an ever better approximation to the required minimization. In
every problem addressed by Bejan and coworkers, the variational description of the
final state, to which Bejan's algorithms converge intuitively, corresponds to the best
removal (allowed by the constraints) of obstacles for the transport of something
(people, heat, electric charge...) across the system. There is an analogy with an
electrical circuit described by Ohm's law, where Kirchhoff's principle prescribes
that the electric current flows along the path of lesser resistance in the relaxed state.

The fact that many details of various fractal-like structures are successfully
described with the help of iterative algorithms converging to configurations that
satisfy some constrained minimization both in engineering design and in natural
sciences led Bejan to put forward his 'constructal law' [47]: *for a finite-size flow*

system to persist in time (to survive) its configuration must evolve in such a way that it provides an easier access to the currents that flow through it. As pointed out in Refs. [34,35], fingering (Sect. 5.3.9) is a vivid example of physical process where spontaneous evolution occurs towards stronger and stronger reduction of obstacles to the flow of water across a porous medium. Since obstacles often act with the help of some kind of friction, it makes sense to classify the constructal law as a MinEP for the purposes of our discussion. Admittedly, however, this point is debatable, as we shall see below.

The analogies among different phenomena show that the constructal law applies to different dissipative processes: viscous dissipation, Joule heating, heat conduction, etc. The constructal law acts as a necessary condition for the relaxed state. It is also a prescription for the evolution both of engineers' design activity (which achieves an ever better optimization after many steps of an iterative procedure) and of complex natural systems. As such, the constructal law seems a true, universal, general-purpose, fundamental principle of systems far from thermodynamic equilibrium.

Far from gaining universal consensus, the constructal law has been heavily criticized. Critics highlight its limits of applicability [50], its vagueness [51] and its alleged universality [52]. For example, according to [51] the constructal law speaks of 'flow', but it is never explained how to decide which flow should be used to model a given 'flow system'. As a result, a major issue that is not addressed with the constructal law is how to apply it objectively. Correspondingly, the ambiguities in the utilization of the word 'flow' seem to imply that the meaning of the word 'access' (ubiquitous in Bejan's papers) is also ambiguous. In our formalism, we may ask how to write a Lyapunov function for every problem allegedly addressed by the constructal law.

Even authors who invoke the constructal law when dealing with vascular structures in biology [53] realize that the *major criticism of the constructal theory is that it has not been derived based on First Principles [...] Hence, the existence of global design of macroscopic (finite size) systems that govern evolution is not without controversy. The complication lies in the fact that mechanical forces that determine structural changes act locally rather than globally.* For further discussion, see Sect. 6.2.4.

5.5 Zipf's Principle of Least Effort

5.5.1 Of Words and Bells

Even if usually postulated in a more or less qualitative way, the idea that a system eventually evolves towards a configuration that minimizes the dissipation inside the system is widely popular, well beyond the domain of physics and engineering. Admittedly, the present Section is concerned with problems outside the domain of physics and engineering we wander through in the rest of this book, and is therefore a digression the reader may as well overlook. However, the wide popularity of the results discussed here makes them worth discussing.

In steady states where the amount of energy lost per unit time through dissipation is compensated by the work done per unit time by human beings, for example, the so-called *principle of least effort* ('PLE') has been proposed (long before Bejan's 'constructal law' of Sect. 5.4) as a general-purpose criterion driving human behaviour in many fields, from linguistics to urban planning.

After a first, preliminary formulation by Ferrero (*men try to make the minimum mental effort* [54]), PLE and its consequences have been systematically, even if qualitatively, investigated by Zipf: *any person* [...] *will strive to solve his problems in such a way as to minimize the total work that he must expend in solving both his immediate problems and his probable future problems. That in turn means that the person will strive to minimize the probable average rate of his work-expenditure over time. And in so doing he will be minimizing his effort.* [...] *Least effort, therefore, is a variant of least work* [55].[37]

If e.g. a word is used more frequently, then the associated effort can be reduced by shortening the phrase which contains such a word. One can observe it easily in greetings. However, thereby merely the speaker's effort gets reduced but the process must not result in a total disappearance of the phrase. This is, of course, the concern of the hearer who gets decoding problems because of reduced redundancy. Thus the hearer forces the speaker to a compromise. The requirement of least effort of all participants in the process is a source of self-regulation: as a result, the lexicon of a language is made of many words which are rarely used, as well as of a small bunch of few words which are utilized very frequently [57]. Even if nobody has designed it on purpose from the beginning, the resulting configuration (the coexistence of a few commonly used words with many, rarely used words) of the system (the lexicon) is highly ordered (we shall come back to this point below), is the unintended consequence of distinct, competing attempts to minimize some effort (minimization of speakers' and hearer's effort shrink and enlarge the lexicon, respectively), and is stable.

To understand this stability, we refer to a thought experiment described in Ref. [55] outside the domain of linguistics. Let us take one blackboard and many, $n \gg 1$ bells, the i-th bell being at a distance $d_i > 0$ from the blackboard with $d_i < d_j$ for $i < j$ and $i, j = 1, \ldots n$. The blackboard is ruled with n columns, the i-th column corresponding to the i-th bell (we suppose the blackboard to be wide enough to contain all n columns). At the initial time, we station a lazy demon at the blackboard. A spell makes the demon act as a bell ringer; in particular, the demon must ring one bell each second of time, and after he has finished ringing the bell once he must return to the blackboard to record that fact in the bell's column. Thus in order to ring one bell ten times, or ten bells once each, he will in either case make ten trips back and forth in the space of ten seconds, and will have ten marks therefore on the blackboard. We give the demon the order to ring all bells at least once and to stop after he has

---

[37] According to these words, the validity of PLE is not likely to be somehow weakened if the number of persons actually interacting with our effort-minimizing person is low. Together, indeed, both PLE and the principle of indifference of Sect. 6.1.9 are invoked [56] when it comes to describing the behaviour of small social groups—an example of the small systems of Sect. 1.

rung the n-th and farthest bell once. Once the constraints above are satisfied, the demon is free to choose which bell to ring and when. In each one of his one-second trips, of course, the lazy demon will run to each bell over the shortest distance, in order to minimize the effort. When starting a new trip, moreover, we may suppose he always rings the easiest remaining bell first, while postponing as long as possible the more distant and hence more difficult bells. When doing so, however, the drawback is that the demon will be forced to run faster and faster, and therefore to work at an ever-increasing rate, as he proceeds farther and farther from the blackboard, if he is to complete each round trip within the prescribed second. And in so doing he will be unevenly distributing his work over time with the risk of collapsing before he gets the n-th bell rung. This uneven distribution of expended work over time is in contrast with the idea of a steady state; it is also forbidden by PLE, which prescribes minimization of the *probable average rate of his work-expenditure over time* where the word *probable* here is a shortcut for the words *which is considered to be likely to occur from the demon's point of view*. After a while, our clever demon will realize what the optimal strategy is like: from second to second he will try to counterbalance the cumulative amount of expended work with the frequency of his visits to the farthest bells; in other words, every time the demon rings a distant bell ($i \approx n$) he will have to ring a succession, or cluster, of nearer bells ($i \ll n$). Correspondingly, the distribution of marks on the blackboard evolves towards a configuration where most marks are concentrated in a few low-i columns and a minority is distributed in all other columns. It is the never-ending demon's attempt to minimize his own effort that stabilizes the distribution. This distribution is the outcome of two competing, effort-minimizing tendencies, acting on a short term and a long term, respectively. The former leads the demon to prefer the low-i bells, in order to minimize the effort at each trip separately; the latter prevents sudden, future and undesired growth of effort by judiciously distributing the tasks of ringing different bells over time. Stabilization is never to be taken for granted once for all: it is rather the provisional outcome of an ever-going process of continuous adjustment driven by the clash between competing processes of effort minimization; the time series of the number of marks on the i-th column on the blackboard at given $i < n$ tends to be made of intermittent fluctuations near a constant value which depends on i; the collection of these constant values form the 'final' or 'relaxed' configuration of the system.

The final configuration of the demon's marks on the blackboard is analogous to the distribution of the words in the lexicon, and for the same reason, i.e. the coexistence of competing, effort-minimizing tendencies. Again, these tendencies usually act on different time scales; this is particularly clear in the case of written communication. The shorter the time the sender takes to write a letter, the shorter the letter, the more irksome its reading on the addressee's side. The more irksome this reading, the longer the time it takes to understand the letter's content (possibly after looking up words in the dictionary, etc.). Next time, the addressee is going to ask the sender for a less cryptic letter; after many letters, the willingness of both to minimize each one's effort makes the shared lexicon include a few words of frequent use and many, much less utilized words.

Our discussion suggests that the same PLE describes relaxation to similar configurations in widely different systems, provided that it applies to many processes acting on different time scales in the same system. As for our bells, a common feature of these configurations is the occurrence of clusters of bells with $i \ll n$. Now, let us raise the number n of bells while the demon keeps on going back and forth; we may ask what happens to the distribution of marks on the blackboard. Generally speaking, the larger n, the longer the time it takes to accomplish the demon's mission. If we keep on raising n so that relaxation occurs well before the demon has finished his work, then the stability discussed above implies that the overall shape of the relaxed distribution changes no more with increasing n. As n gets larger and larger, we reach the point where the cluster includes as many marks as the whole initial distribution. Being equal in both the total number of marks and the shape, the cluster is indistinguishable from the initial distribution. Consequently, the cluster contains a subcluster, just like the original distribution contained a cluster, and with the same structure. Of course, we may repeat the argument again and again. This shows that the relaxed configuration is *scale-invariant*, i.e. its structure looks the same no matter at which scale we look at it. Scale invariance implies that there is no privileged spatial scale in the dynamics underlying relaxation; since the propagation speed of information is finite in all systems, it follows that there is also no privileged time scale. This conclusion fits nicely the fact that distinct processes occurring on different time scales rule relaxation on an equal footing.

Let us denote with w_i the positive-definite amount of work expended by the demon for each trip back and forth between the blackboard and the i-th bell. Generally speaking, w_i is an increasing function of d_i. In the following we assume $w_i = \alpha \cdot d_i$, $\alpha = $ const. for simplicity; we shall discuss this assumption below. Let us denote with N_i the total number of back and forth trips done by the demon between the blackboard and the i-th bell—or, equivalently, the total number of marks on the i-th column on the blackboard—when the demon has finished its work (i.e., when N_n becomes $= 1$). Then, the total amount of work spent by the demon is $R \equiv N_i w_i = \alpha N_i d_i$ and the sum is extended to all values of i. The demon tries to minimize R as much as possible, but of course, the actual value of R depends on the detailed history of the demon's motion and we do not expect the demon to attain an absolute minimum first time. Now, demons are notoriously sensitive, and the failure to do the job with the minimum possible effort cuts our demon to the quick. Now, he volunteers to repeat the whole job again and again, in order to lower R. At his request, we allow him to start by trying $k = 1, \dots K$ times. Let us define the value $R_k = (N_i d_i)_k$ of R at the k-th demon's attempt, and $M \equiv \inf_{k=1,\dots K} R_k$.[38] Now, the definition of R helps us to write down the demon's roadmap towards minimization of R: he should try to ring each i-th bell with the same absolute minimum expended amount of work $\frac{M}{n}$ regardless of i; if all amounts of work reach this common absolute minimum, indeed, R too gets minimized. Having this goal in mind, we allow the demon to try K more times, obtain a new value of M, and so on. We conclude that the relaxed configuration satisfies (approximately at least) the condition of equal amount of work

---

[38] Note that $M > 0$ as all w_i's and N_i's are > 0.

expended for each bell, i.e. $N_i \propto \frac{1}{d_i}$. The farther the bell from the blackboard, the smaller the number of the demon's visits.

The fact that $N_i \propto \frac{1}{d_i}$ and not, say, $\propto \frac{1}{d_i^2}$ follows from our assumption $w_i = \alpha \cdot d_i$ above, i.e. is a consequence of the dynamics. This is an example of the fact that PLE—just like other variational principles discussed in this book—cannot provide a full description of the relaxed state without further information. Generally speaking, it is often possible to retrieve a known behaviour of the relaxed state of a system just by minimizing some suitably defined functional, provided that the minimization is subject to suitably chosen constraints; many examples are listed e.g. in Ref. [58]. Meaningful predictions are therefore possible only if some working hypothesis justifies our choice of both the variational principle and its constraints, so that unsuccessful comparison of the resulting predictions with observations may possibly falsify our initial hypothesis.

In particular, let us assume that $d_i = \beta \cdot i$, $\beta = $const. This is just another particular assumption concerning the structure of our system of bells, and cannot be obtained from PLE. In this case the condition $N_i \propto \frac{1}{d_i}$ leads to *Zipf's law*:

$$N_i \propto \frac{1}{i}$$

i.e. the number N_i of demon's visits to the i-th bell is exactly i times smaller than the number of visits to the most visited bell with $i = 1$. Here i is just an integer ('rank'): the larger i, the less frequent the ringing of the bell. Zipf's law describes with surprising accuracy the actual distribution of words in the lexicon of many languages and in long literary texts like Homer's *Iliad* and Joyce's *Ulysses* [55]: if the words are ordered from the most frequent to the less utilized and are given a rank i in such a way that the larger the rank, the less utilized the word, then the number N_i of occurrences of the word with rank i is i times smaller than the number N_1 of occurrences of the $i = 1$ word. The larger i, the smaller N_i, the larger the relative weight of the unavoidable fluctuations; we expect therefore that Zipf's law breaks down at very large i.[39] It is easy to see that Zipf's law is scale-invariant. If $n \gg 1$, indeed, we make a small error if we consider i as a continuous variable. The scaling transformation $i \rightarrow \kappa \cdot i$ with constant $\kappa > 0$ leaves therefore Zipf's law unaffected, as the scaling factor κ is absorbed in the proportionality constant.

5.5.2 City Air Makes You Free

The literature on Zipf's law is enormous, and a dedicated discussion goes beyond the scope of this book.[40] Its undisputed success is due to the fact that it applies

---

[39] For example, the most common word ($i = 1$) in English, which is *the*, occurs $\approx$ one-tenth of the time in a typical text; the next most common word ($i = 2$), which is *of*, occurs $\approx$ one-twentieth of the time; and so forth. The law breaks down for $i > 1000$.

[40] For an introduction, see Ref. [59].

to problems in many different fields. For example, in many countries (from India to the USA) the population of the largest city is about twice the population of the next largest, three times the population of the third largest, and so forth [60]. There are 2 competing processes. Firstly, the attempt to concentrate people where mass production of consumer goods occur, namely the big cities, in order to minimize the effort of transporting commuters between their home and their workplace. Secondly, the attempt to minimize the effort of obtaining the raw materials (food, minerals) required by mass production from the environment makes people live in small communities immediately near the sources of these raw materials (e.g. at the farm or at the mine pit). Following Ref. [61], we rephrase the PLE of Sect. 5.5.1 for a system of many cities.

To start with, we describe a generic city A as a system that provides the population with an amount W_I of resources per unit time while receiving (delivering) an amount Q_I (Q_{II}) of resources per unit time from (to) either other cities or the environment. In the simplest case, there is only 1 city in the world; then, Q_I and Q_{II} are the amount of resources taken from the environment and the amount of waste into the environment per unit time, respectively.[41] For simplicity, we write $W_I = k_I(Q_I - Q_{II})$ with k_I constant, and define the 'efficiency' $\eta_A \equiv \frac{W_I}{Q_I}$. It follows immediately that $\eta_A = k_I(1 - \frac{Q_I}{Q_{II}})$. The actual value of η_A stands for the processes (economical, political...) occurring inside A. The analogy with the textbook description of thermodynamic cycles is purely formal: even if both Q_I, Q_{II} and W_I are assumed to be positive-definite (so that η_A too is > 0), these quantities are not directly related to energy and there is no energy conservation to be invoked; both $Q_I - Q_{II}$ and k_I may e.g. be < 0.

Now, let us consider a world made of 2 cities, say A and B. As for A, nothing changes except that Q_{II} is not wasted but is delivered to B: it represents the net amount of resources exchanged per unit time between A and B (through commerce, war etc.). As for B, it receives Q_{II} from A, provides the population with an amount W_{II} of resources per unit time and releases an amount Q_{III} of resources per unit time to the environment. Again, we write $W_{II} = k_{II}(Q_{II} - Q_{III})$ and $\eta_B \equiv \frac{W_{II}}{Q_{II}}$. Indeed, we may consider the system of the 2 cities as one unity, which takes a net amount of resources Q_I from the environment per unit time and releases a net amount of waste Q_{III} to the environment, while providing the population with an amount $W_I + W_{II}$ of resources per unit time. For mathematical simplicity, we assume $k_I = k_{II} = k$; this is equivalent to assume that the production processes in the two cities are basically the same. Let us introduce the overall efficiency $\eta_{A,B} \equiv \frac{W_I + W_{II}}{Q_I}$, which keeps into account the interaction between A and B; the larger $\eta_{A,B}$, the lower the amount of resources required by the system made of A and B for the production of a given amount of resources for the population. Straightforward algebra shows that:

---

[41] Here by 'resources' we mean food, money, services... anything of economic value. We postulate that there is a common measurement unit that allows a quantitative estimate of both Q_I, Q_{II} and W_I.

$$1 - \frac{1}{k}\eta_{A,B} = \left[1 - \frac{1}{k}\eta_A\right]\left[1 - \frac{1}{k}\eta_B\right]$$

Generalization to the realistic cases of $N_{TOT} \gg 1$ cities with a total population $X_{TOT} \gg 1$ is straightforward. Let us denote with N_l the number of cities with population X such that $x_l \equiv l \cdot \Delta \leq X < x_l + \Delta$, where $l = 1, \ldots N_{int}$, $\Delta \equiv \frac{X_{TOT}}{N_{int}}$ and $N_{int} \gg 1$ number of intervals. Of course, $\Sigma_l N_l = N_{TOT}$ and $N_l x_l = X_{TOT}$, where the sum is extended to all values of l. The probability $p_l \equiv \frac{N_l}{N_{TOT}}$ that a city belongs to this class satisfies the conditions $p_l > 0$ and $\Sigma_l p_l = 1$. Furthermore, let us denote with Q_l and W_l the amount of resources received and delivered to the population, respectively, by these n_l cities per unit time. Now, the actual value of the corresponding efficiency $\eta_l \equiv \frac{W_l}{Q_l}$ stands for the processes occurring inside the system of our N_l cities, including the exchanges among them, the production process etc. Generally speaking, therefore, η_l will depend on N_l, or, equivalently, on p_l; we may therefore write $\eta_l = \eta(p_l)$. What about the interaction of our N_l cities with the N_m ($m \neq i$) cities with population $x_m \leq X < x_m + \Delta$ which deliver an amount W_m to the population? Formally, we may still define an overall efficiency $\eta_{l,m} \equiv \frac{W_l + W_m}{Q_l}$, which stands for the interactions between the two groups of cities. Admittedly, this time the cities of the two groups may also interact with all other cities outside these groups. All the same, however, we may say that $\eta_{l,m}$ depends on the number of the cities of the first group which actually interacts with cities of the second group; this number being proportional to $N_l N_m \propto p_l p_m$ we may write $\eta_{l,m} = \eta(p_l p_m)$. We follow the same line of reasoning outlined above and write:

$$1 - \frac{1}{k}\eta(p_l p_m) = \left[1 - \frac{1}{k}\eta(p_l)\right]\left[1 - \frac{1}{k}\eta(p_m)\right]$$

Direct substitution shows that a solution is $1 - \frac{1}{k}\eta(p_l) = p_l^b$, i.e.:

$$\eta(p_l) = k(1 - p_l^b)$$

where b is some real number. It is possible to proof that this solution is unique.[42]

---

[42] To this purpose, let us start from the fact that $f(y) \equiv 1 - \frac{1}{k}\eta(y)$ is a continuous, positive-definite function of a positive-definite variable $y \equiv p_l$ such that $f(xy) = f(x)f(y)$. (Here $f(y) > 0$ because all Q_l's are > 0). Indeed, let us define $g(y) = \ln f(e^y)$. Then $g(x + y) = \ln f(e^{(x+y)}) = \ln f(e^x e^y) = \ln\left[f(e^x)f(e^y)\right] = \ln f(e^x) + \ln f(e^y) = g(x) + g(y)$. This result, together with the trivial identity $1 = \Sigma_k \frac{1}{n}$, $k = 1, \ldots n$ for arbitrary integer n, leads to $g(1) = g(\Sigma_k \frac{1}{n}) = \Sigma_k g(\frac{1}{n}) = ng(\frac{1}{n})$, hence $g(\frac{1}{n}) = \frac{1}{n}b$ where we have defined $b \equiv g(1)$. For arbitrary integer p, therefore, it follows that $g(\frac{p}{n}) = \frac{p}{n}b$. Generally speaking, for any arbitrary positive real number y two successions p_m and n_m ($m = 1, 2 \ldots$) of integers exist such that $y = \lim_{m \to \infty} \frac{p_m}{q_m}$. Furthermore, continuity of $f(y)$ implies continuity of $g(y)$. It follows that $g(y) = g(\lim_{m \to \infty} \frac{p_m}{q_m}) = \lim_{m \to \infty} g(\frac{p_m}{q_m}) = \lim_{m \to \infty} \frac{p_m}{q_m}b = yb$. Let us define $z \equiv \ln y$. We have $f(y) = f(e^z) = e^{g(z)} = e^{bz} = e^{b \ln y} = y^b$. Substitution of the definitions of $f(y)$ and y gives precisely $1 - \frac{1}{k}\eta(p_l) = p_l^b$.

PLE requires that the effort in obtaining a given amount of resources of the population is minimized—or, in other words, that the amount of resources required in order to maintain the system is a minimum. Alternatively, we may say that people are going to organize themselves in order to maximize the amount of resources available to the population for a given amount of effort, i.e. the efficiency. Since everyone shares this same desire and since the efficiency may be different in different cities, we require maximization ('MaxEff') of the averaged efficiency $< \eta >\equiv \Sigma_l p_l \eta(p_l)$. Since $\Sigma_l p_l = 1$, the above formula for $\eta(p_l)$ gives $< \eta >= k\Sigma_l(1 - p_l^{b+1})$. Constraints are: a) the normalization condition $\Sigma_l p_l = 1$; b) a given total population $X_{TOT} = \Sigma_l N_l x_l$, so that $\Sigma_l p_l x_l = \frac{X_{TOT}}{N_{TOT}}$ and $\Sigma_l p_l \cdot l = \frac{X_{TOT}}{N_{TOT} \cdot \Delta}$. Accordingly, we write:

$$\frac{d}{dp_q}\left[k\Sigma_l(1 - p_l^{b+1}) + \xi(\Sigma_l p_l - 1) + \mu(\Sigma_l p_l \cdot l - \frac{X_{TOT}}{N_{TOT} \cdot \Delta})\right] = 0$$

where ξ and μ are Lagrange multipliers, $q = 1, \ldots N_{int}$ and $\frac{dp_l}{dp_q} = 1$ if $i = q$, $= 0$ otherwise. If $b \neq 1$ we can write the solution as follows:

$$p_l = (C_1 + C_2 l)^{\frac{1}{b}}$$

where the values of the 2 quantities $C_1 \equiv \frac{\xi}{k(1+b)}$ and $C_2 \equiv \frac{\mu}{k(1+b)}$ are determined by the 2 constraints $\Sigma_l p_l = 1$ and $\Sigma_l p_l x_l = \frac{X_{TOT}}{N_{TOT}}$. The fact that $p_q > 0$ and the requirement that p_q is well-defined for arbitrary values of b and X_{TOT} (hence of N_{int}) imply $C_1 > 0$ and $C_2 > 0$, respectively. The fact that $\Sigma_l p_l = 1$ for arbitrarily large N_{int} implies $b < 0$, as the sum diverges otherwise.

The facts that $b < 0$, $C_1 > 0$ and $C_2 > 0$ imply $\frac{dp_l}{dl} < 0$, i.e. there are more small cities than large cities (we have tacitly considered l as a continuous variable here). The facts that $\eta(p_l) = k(1 - p_l^b)$, $\eta(p_l) > 0$ and $p_l \leq 1$ imply that k and b have the same sign. Since $b < 0$ implies $k < 0$, $\frac{d\eta(p_l)}{dp_l} < 0$ so that $\frac{d\eta}{dl} = \frac{d\eta(p_l)}{dp_l}\frac{dp_l}{dl} > 0$, i.e. the larger the city, the larger its efficiency.

Let us rewrite our results in terms of rank $i = 1, \ldots N_{TOT}$, which replaces l in the role of the new independent variable. Formally at least, we may repeat our arguments above step by step and obtain again $< \eta >= k'\Sigma_i(1 - p_i^{b'+1})$; generally speaking,[43] $k' \neq k$ and $b' \neq b$. The rank i increases monotonically from $i = 1$ for the largest, $l = N_{int}$ city to $i = N_{TOT}$ for the smallest, $l = 1$ city. Then η, which is an increasing function of l, is a decreasing function of i. Accordingly, p_i is an increasing function of i as there are more small cities than large cities. Apart from that, when it comes to play the role of independent variable i is as allowable as l. Then, MaxEff maximization still reads ($r = 1, \ldots N_{TOT}$):

---

[43] As for the exact meaning of the notation p_i, see below.

$$\frac{d}{dp_r}\left[k'\Sigma_i(1 - p_i^{b'+1}) + \zeta(\Sigma_i p_i - 1) + \lambda(\Sigma_i i \cdot p_i - Z)\right] = 0$$

where λ and ζ play the role of Lagrange multipliers and Z is a constant quantity, whose actual value is not relevant in the following. The solution is:

$$p_i = (D_1 + D_2 i)^{\frac{1}{b'}}$$

where D_1 and D_2 are suitable constant quantities and $b' > 0$ as p_i increases with increasing i.

Now, we make a small error when considering both l and i as continuous variables whenever both N_{int} and N_{TOT} are so large that we may assume $N_{int} \to \infty$ and $N_{TOT} \to \infty$. The fact that p_i and p_l refer to the same quantity (the distribution of cities[44]) links the rank i of a city to its population X_i, which in turn corresponds approximately[45] to $l = \frac{X_i}{\Delta}$, so that $C_1 + C_2 l = C_1 + \frac{C_2}{\Delta} \cdot X_i$. Then, we require that $(C_1 + \frac{C_2}{\Delta} \cdot X_i)^{\frac{1}{b}} = (D_1 + D_2 i)^{\frac{1}{b'}}$. As for the largest cities (C_1 negligible for large l) this relationship reduces to:

$$X_i \propto \frac{1}{(i_0 + i)^\gamma}$$

for the population X_i of the city with rank i; here $i_0 \equiv \frac{D_1}{D_2}$ and $\gamma \equiv -\frac{b}{b'} > 0$ for $b < 0$ and $b' > 0$. This relationship, often dubbed as *Zipf-Mandelbrot law*, has been proposed by Mandelbrot [62] in the framework of information theory for the number N_i of a word with rank i in a lexicon (see Sect. 5.5.1). Remarkably, Zipf-Mandelbrot law has been originally obtained as a result of information theory and is concerned with language, not cities. This result definitely strengthens Zipf's original idea that a common variational principle underlies the self-organization of widely different systems. Scale invariance is retrieved for negligible i_0. If both $i_0 \to 0$ and $\gamma \to 1$, then Zipf's law (Sect. 5.5.1) is obtained and the original result of Zipf (the population of the largest city in a country is about i times the population of the city of rank i) is retrieved.[46] Even if less frequent, largest cities are more efficient than smaller cities, i.e. provide people with more resources for a given effort (including financial investment, environmental impact, etc.): as the old saying goes, *city air makes you free*.

---

[44] Since i and l are continuous variables, $p_i di$ and $p_l dl$ stand for the probability of finding a city in the interval with lower bound i, upper bound $i + di$ and with lower bound l, upper bound $l + dl$, respectively.

[45] Rigorously speaking, $l \cdot \Delta \le X_i < \Delta$; but we take here a negligible $\Delta \propto \frac{1}{N_{int}}$.

[46] Again, the larger i, the smaller X_i, the larger the relative weight of fluctuations, the more likely the deviation from Zipf. Indeed, the smaller X_i, the smaller l, the flatter p_l. Power laws hold for large values of the independent variable only.

Coexistence of effort-minimization processes on many time scales, scale invariance (at least for large values of the independent variable) and occurrence of intermittent fluctuations even after the distribution of the components of the system has
settled down seem to be common features of relaxed states described by PLE. But
the values of constant quantities like i_0 and γ involved in the scaling laws depend on
the detailed dynamics and cannot be obtained by PLE only.

5.5.3 Pareto

The freedom when choosing the independent variable hints at a fact we have not yet
put in evidence: our computation of p_l and the underlying MaxEff principle are not
limited to cities, and the variable x_l is not necessarily bound to stand for population
[61]. Should we e.g. divide the total population X_{TOT} in N_{int} classes where N_l is
the number of persons with income X such that $x_l \equiv l \cdot \Delta \leq X < x_l + \Delta$, where
$l = 1, \ldots N_{int}$ and $\Delta \equiv \frac{X_{TOT}}{N_{int}}$, then $p_l \equiv \frac{N_l}{N_{TOT}}$ would again satisfy a relationship
like $p_l = (C_1'' + C_2'' l)^{\frac{1}{b''}}$ with constant quantities $C_1'' > 0, C_2'' > 0$ and $b'' < 0$. Let us
introduce the number $N_>(Y) = N_{TOT} \int_Y^{+\infty} p_l dl$ of persons whose income is larger
than a given value Y (as usual by now, we consider l as a continuous variable). For
large enough income (i.e. for $C_1'' \ll C_2'' l$) we obtain *Pareto's distribution*:

$$N_>(Y) \propto \frac{1}{Y^{-\frac{1}{b''}-1}}$$

This relationship rules the distribution of wealth in many societies, according to
Pareto [63]. We cannot compute the actual value of b'' with the help of MaxEff
only. All the same, PLE ensures stability; in other words, Pareto's distribution of
wealth—which describes a society with many poor people and a few rich people, as
p_l is a monotonically decreasing function of l—is expected to resurface again and
again—in agreement with Pareto's original intuition.

5.5.4 A Tale of Two Cities

The PLE of Sect. 5.5.1 provides us with useful information even when it comes to
describing the interaction between two cities, say A and B, in more detail [60]. Let
us assume that a dweller of A sends some goods to a dweller of B for a whole year.
(Correspondingly, the inhabitant of B may ship goods to the correspondent in A, the
situation being perfectly symmetrical; we focus on shipping from A to B here with
no loss of generality). To this purpose, she makes use of trucks; let us denote with
d_{AB} the distance to travel between A and B. At the end of the year, the amount of
money for the fuel is proportional both to d_{AB} and to the total number of trips in the
year, i.e. to the total amount of shipped goods. According to PLE (and to common
sense), the sender will try to minimize the cost (e.g. through suitable time-scheduling
of the shipping, careful choice of the most reliable trucks and the cheapest diesel

oil, etc. Minimization of d_{AB} among all available routes connecting A and B is also an obvious choice). After optimization has been successfully achieved, the total cost per year is still proportional both to d_{AB} (which is supposed to be minimized, here and in the following) and to the quantity of goods shipped from A to B in one year, but its value has been fixed once for all at its minimum possible value by the minimization process. In other words, the total amount of shipped goods is inversely proportional to d_{AB}, the constant of proportionality being fixed. Being the problem perfectly symmetrical with respect to the permutation $A \leftrightarrow B$, the same result applies also to the amount of goods shipped in one year from B to A. We may say that the amount of goods exchanged between one inhabitant in A and one inhabitant in B in one year is still inversely proportional to d_{AB}. Now, our choice of a time interval of one year is purely matter of convenience: we may as well speak of shipping in one month, or a quarter of the year. Moreover, means of transportation may include trains, ships, etc. Generally speaking, therefore, we conclude that the amount of goods exchanged between one inhabitant in A and one inhabitant in B in one year is inversely proportional to d_{AB}, the constant of proportionality being fixed.

In the more realistic case where A and B have $N_A > 1$ and $N_B > 1$ inhabitants, respectively, each inhabitant of A is equally likely to send goods to inhabitants of B, and vice-versa. The total amount of goods exchanged per unit time between the two cities is the sum of the amounts of goods exchanged between 1 inhabitant of A and 1 inhabitant of B, and the sum is to be performed on all possible couples made of 1 inhabitant of A and 1 inhabitant of B each. The number of these couples is equal to $N_A N_B$.[47] Now, the relationship of inverse proportionality discussed above holds for each one of these amounts separately. Crucially, moreover, the constant of proportionality that appears in such relationship is the same for all couples, because it is equal to the minimum found through PLE for one couple and this minimum is the same for all couples (same oil price, same means of transportation, etc.).

According to PLE, therefore, we conclude that the total amount of goods exchanged per unit time between two cities A and B is directly proportional to the product $N_A N_B$ of their populations and inversely proportional to the length d_{AB} of the cheapest transportation distance d_{AB} between them. Similar arguments apply if we replace the total amount of goods exchanged per unit time with the total number of persons who move[48] between A and B. This result is usually dubbed 'gravity model', due to its formal similarity with the Newtonian formula for the gravitational potential between two masses. A better fit of the gravity model with observations in urban planning is found in the literature:

$$T_{AB} = k \frac{N_A N_B}{d_{AB}^\beta}$$

---

[47] Even if only a fraction aN_A of inhabitants of A ($a < 1$) is actually involved in trading with B, and a fraction bN_B of inhabitants of B ($b < 1$) do business with A, nothing changes in our discussion provided that we replace $N_A N_B$ with $abN_A N_B$.

[48] Either for business, or tourism, or honeymoon...

Here T_{AB} the total number of persons who travel between cities A and B per unit time, k is a constant (usually set $= 1$ through suitable definition of units, a choice we adopt below) and β is a dimensionless constant, usually in the range between $\beta = 1$ [60] and $\beta = 2$ [64]; its value is found empirically.

5.5.5   Travels with Entropy

Generalization of the two-cities gravity model of Sect. 5.5.4 to the problem of T persons who travel between $i = 1, \ldots n$ 'origin' cities (each with O_i departures) and $j = 1, \ldots m$ 'destination' cities (each with D_j arrivals) per unit time is straightforward (the problem where the same cities are both 'origin' and 'destination' is a particular case with $n = m$). If T_{ij} is the number of travels from i to j per unit time and d_{ij} is the distance between the i-th origin city and the j-th destination city, then the gravity model for each couple (i, j) gives just:

$$T_{ij} = \frac{N_i N_j}{d_{ij}^{\beta}}$$

with the proviso that the following self-consistency relationships hold:

$$\Sigma_{ij} T_{ij} = T$$
$$\Sigma_j T_{ij} = O_i$$
$$\Sigma_i T_{ij} = D_j$$

The question is supposedly settled. However, a problem arises. Let us double the distance between two given cities, say $O_{i=1}$ and $D_{j=1}$. If only these two cities exist, then the two-city gravity model unambiguously predicts that $T = T_{11}$ gets multiplied by $2^{-\beta}$. But in case of more cities, people may just change destination in agreement with PLE, as there is plenty of available, alternative choices. How do people redistribute? The lack of unambiguous answer reflects the fact that many possible configurations are possible, i.e., the number of ways of distributing the travelling people over the O_i's and the D_j's is usually $\gg 1$, all of them satisfying the self-consistency relationships above. In other words, even if our description of this transport network relies on the knowledge of the travels between just one origin and one destination, this knowledge is not enough for us to grasp the behaviour of the network.

All the same, PLE is still helpful. To start with, let us compute the number $W = W(T_{11}, T_{12}, \ldots T_{nm})$ of possible travellers' combinations when the number of travels per unit time from the i-th origin city to the j-th destination city is T_{ij}. According to Ref. [65], we write

$$W = \frac{T!}{\Pi_{ij} T_{ij}!}$$

where $n! \equiv 1 \cdot 2 \cdot 3 \cdot \ldots n$ (with $0! = 1$ by definition) and Π_{ij} is the product over all i and j.[49] If a measure c_{ij} of the effort[50] a travel from the i-th origin city to the j-th destination city calls from a traveller is available, then the total effort C for all travels per unit time is

$$C = \Sigma_{ij} T_{ij} c_{ij}$$

Here we postulate no link between c_{ij} and the distance d_{ij} between the i-th origin city to the j-th destination city. We shall discuss this point in more detail below. In the spirit of PLE, we look for the configuration which minimizes the total effort C for given W, or, equivalently, which maximizes W for given C. Looking for maxima of W is perfectly equivalent to the much simpler problem of looking for maxima of $\ln W = \ln T! - \Sigma_{ij} \ln T_{ij}!$, as the logarithm is a monotonically increasing function of its argument. Formally, $\ln W$ resembles Boltzmann's definition of entropy in statistical mechanics[51]; historically, this is the reason why the present discussion is usually referred to as *entropy model* in urban planning research. Looking for maxima of the logarithm is much simpler, because we may invoke *Stirling's approximation* $\ln n! \approx n \ln n - n$. so that $\ln W \approx T \ln T - T - \Sigma_{ij} T_{ij} \ln T_{ij} + \Sigma_{ij} T_{ij}$. The self-consistency relationships above concerning T, the O_i's and the D_j's provide us with further constraints. In conclusion, we look for extrema of the quantity [67]:

---

[49] In order to understand this formula, imagine that four persons, Alice, Bob, Charlie and Daisy spend the holidays in Italy ($T = 4$). Alice and Bob live in New York ($i = 1$), Charlie and Daisy live in Miami ($i = 2$). For budgetary reasons, each of them can choose only one low-cost flight to an Italian destination, either Rome ($j = 1$) or Venice ($j = 2$). Let all of them book their seats in advance. To make an order, the clerk at the travel agency drops all air tickets in boxes, labelled 'New York–Rome', 'Miami–Venice' and the like. As a result, the older the booking, the deeper the ticket in the box. One possible set of holidays is: Alice in Rome and Bob, Charlie and Daisy in Venice. In this case, the boxes 'New York–Rome', 'New York–Venice', 'Miami–Rome' and 'Miami–Venice' contain $T_{11} = 1$, $T_{21} = 0$, $T_{12} = 1$, $T_{22} = 2$ tickets, respectively. Another possible set of holidays is: Bob and Daisy in Rome and Alice and Charlie in Venice. Correspondingly, $T_{11} = T_{21} = T_{12} = T_{22} = 1$ ... and so on. (Note that in all cases $T_{11} + T_{21} + T_{12} + T_{22} = T$). The choice of both the destination and of the booking date of each of our tourists does not depend on the choice of others. Then, after listing all possibilities in order to compute the total number of possible sets of holydays— homework for the reader—it turns out that this number is equal to $T! = 1 \cdot 2 \cdot 3 \cdot 4 = 24$. In other words, and not surprisingly, the total number of possible holidays is just the total number of possible combinations of our four tourists. Now, the number W of the possible distributions of seats on the flights is always $< T!$, as the order of the booking is not relevant. When looking at the case of T_{22} seats on the flight Miami-Venice, for example, it turns out that the case with Charlie booking before Daisy (i.e., with Charlie's ticket on the top) and Daisy booking before Charlie (i.e., Daisy' ticket on the top) are counted separately when computing $T!$, but of course, both of them correspond to the same T_{22}. In order to prevent unduly multiple counting of the tickets in the Miami-Venice box when computing W, therefore, we have to divide $T!$ by the number $T_{22}!$ of the possible combination of the T_{22} tickets in this box. After repeating the same for all boxes, we obtain the expression for W.

[50] Including, e.g. the price of the ticket, the possible jet lag, the trouble in packing, etc.

[51] This is one of the examples hinted at in Ref. [66] quoted in Sect. 1.

$$L \equiv T \ln T - T - \Sigma_{ij} T_{ij} \ln T_{ij} + \Sigma_{ij} T_{ij} +$$

$$+\Sigma_i \lambda_i (O_i - \Sigma_j T_{ij}) + \Sigma_j \gamma_j (D_j - \Sigma_i T_{ij}) + \chi (C - \Sigma_{ij} T_{ij} c_{ij})$$

where both the λ_i's, the γ_j's and χ are Lagrange multipliers. The Euler-Lagrange equation with respect to the Lagrangian coordinate T_{pq} ($p = 1, \ldots n, q = 1, \ldots m$) is just $\frac{\partial L}{\partial T_{pq}} = 0$, i.e.[52]: $\ln T_{pq} = -\lambda_p - \gamma_q - \chi c_{pq}$, or, equivalently:

$$T_{pq} = \exp(-\lambda_p) \exp(-\gamma_q) \exp(-\chi c_{pq})$$

Remarkably, this result coincides *exactly* with the result $T_{pq} = \frac{N_p N_q}{d_{pq}^\beta}$—which follows directly from the two-cities theory of Sect. 5.5.4 and is therefore also a consequence of PLE—provided that $\lambda_p = -\ln N_p, \gamma_q = -\ln N_q, \chi = \beta$ (these relationships provide us with the values of the Lagrange multipliers) and

$$c_{pq} = \ln d_{pq}$$

This means that self-consistence of PLE in urban planning—i.e., an agreement between the gravity model and the entropy model—unambiguously defines the correct measure of the effort required by travel: the logarithm of the distance.

This result raises some interesting issues. Firstly, we may ask if the entropy model holds also for problems outside the domain of urban planning. The answer is affirmative, as an entropy model exists even for the distribution of the words in a lexicon and is allegedly linked to Zipf's law [68]. Secondly, we recall that PLE is concerned with minimization of the effort as it is perceived by the agent who has to support it (remember the demon of Sect. 5.5.1). The fact that the measure of effort to be minimized scales logarithmically with the effort's physical cause (here, the distance between two cities) agrees nicely with *Weber–Fechner's law* of physiology, which states that the relationship between stimulus and perception is logarithmic. Finally, we may as well act in reverse: in other words, we may just start from $c_{pq} = \ln d_{pq}$, look for a maximum of the number W of configurations (in agreement e.g. with Bejan's approach of Sect. 5.4) and retrieve the gravity model, thus providing us with a confirmation of PLE.

5.6   Maximum Entropy Production

5.6.1   Muffled Intuitions

In discontinuous systems, LNET implies that the total amount of entropy produced per unit time by all irreversible processes in the system is twice the Rayleigh's dissipation function of Sect. 4.1.3: $\frac{dS}{dt} = 2f$. Moreover, if the thermodynamic forces

---

[52] Here we take into account that the T_{ij}'s are independent from each other.

are kept fixed then the least dissipation principle holds: $\frac{dS}{dt} - f = \max$. These two relationships imply that a stable steady state corresponds to a constrained maximum of Rayleigh's dissipation function, the constraint being given by fixed thermodynamic forces. Depending on the problem, LNET leads also to a maximum entropy production principle ('MEPP'), not just to a MinEP. Many researchers have been vigorously working towards the establishment of MEPP beyond the domain of LNET. In continuous systems, LNET deals with the entropy production P due to some (not all!) irreversible processes occurring within the bulk of the system. Even if LNET does not apply, there are variational principles which minimize the entropy production due to selected irreversible processes in the bulk for particular problems, like Kirchhoff's principle of Sect. 5.3.1, Korteweg–Helmholtz' principle of Sect. 5.3.7 and Chandrasekhar's principle of Sect. 5.3.11.

Remarkably, researchers in many different fields independently posit that MEPP describes relaxed states of many systems far from thermodynamic equilibrium and beyond the domain of LNET. In a system where a MEPP holds, a relaxed state far from thermodynamic equilibrium corresponds to a maximum of the amount of entropy produced per unit time. Remarkably, most examples of MEPP available in the literature usually involve irreversible phenomena occurring on the boundary of the system [43,69]. In contrast, MinEP in LNET is usually related to dissipative phenomena occurring in the bulk of the system. Bejan himself claims [47] that the constructal law is in full agreement with Malkus' findings discussed in Sect. 5.3.12 [41] when it comes to turbulence in fluids. Words in pure constructal jargon like *Nature takes the easiest and most accessible paths and, hence, processes are accomplished very quickly in a minimum time* are found in the work of a staunch supporter of MEPP [70]. According to Reis [71], both MEPP and MinEP are particular corollaries of Bejan's constructal law of Sect. 5.4. According to Liu [35], if Bejan is right then relaxation minimizes all obstacles on the path of the amount of heat Q flowing per unit time across a physical system entering from one hotter boundary at temperature T_h and coming out through the opposite cooler boundary at temperature $T_l < T_h$, and maximizes therefore Q; if T_l and T_h are fixed, then the amount $Q(\frac{1}{T_l} - \frac{1}{T_h})$ of entropy of the Universe produced per unit time is maximized. All the same, no generally accepted proof of MEPP is yet available[53] ; see e.g. [73].

Unfortunately, physical intuition may lead to contrasting conclusions. On one side, we know that fluids spontaneously follow the path of lesser resistance, and the same holds for the electric current (a well-known example is a short-circuit). This familiar behaviour suggests that dissipation (hence entropy production) tends to *minimization*. On the other side, we know that a physical system spontaneously evolves towards a state of maximum entropy, and it is only natural to imagine that it is eager to do so as quickly as possible, thus accelerating the production of entropy as

---

[53] Even if no less than 11 different definitions of MEPP are taken into account [72].

Fig. 5.2 S_{TOT} *versus* t

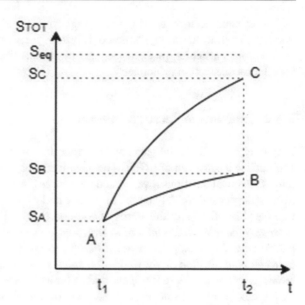

much as possible: accordingly, the natural course of things should *maximize* entropy production.[54]

In order to clarify this point, let us plot entropy versus time. Let S_A and S_{eq} be the values of the total entropy S_{TOT} at $t = t_A$ and at thermodynamic equilibrium, respectively—see Fig. 5.2. According to the Second Principle of thermodynamics, entropy is ever-increasing in the march towards equilibrium, and we expect $S(t_1 > t_A) > S_A$. At $t = t_1$, we ask ourselves if it is more likely to find the system at B or C. Statistical mechanics provides us with the answer: the probability of observing the system at B is $\propto \exp\left(\frac{S_B - S_{eq}}{k_B}\right)$, and the probability of observing the system at C is $\propto \exp\left(\frac{S_C - S_{eq}}{k_B}\right)$ (with the same proportionality constant). Accordingly, if $S_{eq} > S_C > S_B$ then C is more probable than B. Consequently, the evolution of the system from $t = t_A$ to $t = t_1$ is more likely to follow a path from A to C than from A to B. It follows that the increment of S_{TOT} from $t = t_A$ to $t = t_1$ is more likely to take the value $S_C - S_A$ than the value $S_B - S_A$. Thus, the value of $\frac{dS_{TOT}}{dt}$ is more likely to be $\approx \frac{S_C - S_A}{t_1 - t_A}$ than $\approx \frac{S_B - S_A}{t_1 - t_A}$, so that in steady state $\frac{dS_{TOT}}{dt} \approx$ max.

This argument surfaces again and again in the literature, more or less implicitly and in various forms, over the decades.[55] The trouble is that it is unacceptably qualitative. The maximum of $\frac{dS_{TOT}}{dt}$, indeed, is defined just in some statistical sense,

---

[54] In the words of Ref. [74], the most important conclusion of MEPP is that *there is life on Earth, or the biosphere as a whole, because ordered living structures help dissipate the energy from the sun on our planet more quickly as heat than just absorption of light by rocks and water.*

[55] Reference [75] provides us with an example. After a sophisticated proof of maximization of entropy in a fluid described by Navier–Stokes equation, the logical step from maximization of entropy to maximization of entropy production is apparently explained just by the words *Maxi-*

as the repeated occurrence of the wordings 'more likely' in the lines above makes it clear. Far from thermodynamic equilibrium, the word 'statistical' has a precise meaning for particular, problem-dependent models like e.g. in the case of LNET-based least dissipation principle.[56]

5.6.2 Maximum Versus Minimum

Moreover, it remains to be seen how to reconcile a MEPP with the MinEP both inside and outside the domain of LNET. The question remains unsettled to date. In the system of chemical reactions investigated by Endres [74] the choice between MEPP and MinEP depends ultimately on the boundary and the initial conditions of the simulation (see Sect. 6.1.9). In the words of Kleidon et al. [76] *While MinEP applies near thermodynamic equilibrium and to transient effects, MEPP is the selection principle that decides among the possible steady states—hence, there is no contradiction between minimization or maximization since we deal with optimization under different conditions.*[57] Here it is clear that as far as we are concerned with a MEPP we may speak of 'rule of selection' rather than of a true variational principle.[58] Given the fact that the problems in atmospheric physics discussed in Ref. [76] obviously deal with oscillating states (think e.g. of the circadian cycle), Kleidon et al.'s words suggest that extremization criteria in non-equilibrium thermodynamics may, e.g. be useful when dealing with oscillating states.[59]

As for fluids and plasmas, a physical—even if qualitative—argument supporting the point of view of Ref. [76] is provided by Kawazura et al. in Ref. [78]: diffusion—a process of diminishing inhomogeneities—works stronger on smaller scales, and the measure of dissipation is unbounded for small-scale fluctuations. Moreover, both viscous dissipation and Joule dissipation act on small scales, and tend to destroy the macroscopic order of the fluid they act upon (we have e.g. seen that magnetic field lines are frozen, i.e. are transported without being destroyed, by the motion of a magnetized fluid provided that $Re_M \gg 1$). Eventually, heating (i.e., conversion of ordered flow to random motion) is related to small-scale processes. The latter may, therefore, drive relaxation towards a state described by some MinEP, with some dissipation-related Lyapunov function ruling the relaxation. On a quite different time scale, however, large-scale processes like, e.g. convection may change the pattern of $\mathbf{v}$, $\mathbf{j}_{el}$ and $\mathbf{B}$ across the system, and therefore the actual value of entropy

---

mization for each value of t maximizes, in turn, the entropy production rate. Generally speaking, however, maximization of a quantity a does not necessarily imply maximization of $\frac{da}{dt}$.

[56] Remarkably, in his first papers on least dissipation Onsager wrote precisely of 'most probable paths'.

[57] Independently, Struchtrup et al. [77] support Kleidon et al.'s point of view [76] by postulating that the boundary conditions in a problem of one-dimensional heat transfer in a gas are chosen in such a way to minimize the maximum over all positions x's of the entropy production density $\sigma(x)$.

[58] This conclusion in agreement with the discussion of Sect. 5.1.

[59] See Sect. 5.8.1 for further discussion.

production rate at all times. The relaxed state (if any exists) will be determined by the macroscopic boundary conditions imposed by the external world which keep the system far from thermodynamic equilibrium and act on macroscopic spatial scales and correspondingly long time scales. In this conceptual framework, MEPP can act as a selection rule: among many possible final relaxed states, the system as a whole will evolve towards the state with the largest entropy production. MEPP and MinEP coexist because they act on different time scales.[60]

Basically, the discussion above relies on the fact that the evolution of a system as a whole towards a relaxed state described by MEPP is but a succession of small steps where each small component of the system evolves in a very short time scale towards a local, quasi-steady state described by MinEP. In other words, it requires that the evolution of the system on the large scale is much slower than the evolution of each subsystem on a small scale. Unfortunately, this is the weak point of the argument. In fact, there is plenty of examples where large-scale dynamics is much *faster* than the predicted time scale of small-scale dynamics. When magnetic confinement is lost for any reason in plasmas for controlled nuclear research, for example [79], the plasma undergoes a catastrophic disruption in a few milliseconds, which is a much shorter time than the typical diffusion time of heat in the plasma (up to 1 s).[61] One could argue that such disruption is but a gross instability, while the relaxed states allegedly described by MEPP are stable by definition; but then the relevance itself of MEPP is at stake, as it may hold only provided there is no disruption and fails, therefore, to provide us precisely with information about disruption. Lack of a priori justification of MEPP in stable states is particularly painful here, as it implies a lack of much desired information about the loss of stability.[62]

A different approach to the open issue of coexistence of MinEP and MEPP is discussed by Glimm et al. [75]. (The authors consistently refer to MinEP and MEPP and to 'Prigogine's' and 'Ziegler's principle', having in mind the results of Sects. 4.3.10 and 6.1.10, respectively). Following Glimm et al.'s line of reasoning, let us *consider the irreversible experiment of dropping a stone from the tower of Pisa. The entropy, which must increase due to the irreversibility of the experiment, is not found in the stone. Considered in isolation, the stone is an open system and its entropy increase is minimized (zero) according to Prigogine's principle. The entropy does increase within the air disturbed by the falling stone. This disturbance accounts for the aerodynamic (turbulent) drag on the falling stone and, if included in the model, the system has a maximum rate of increase, according to Ziegler's principle. Therefore, we see the Prigogine–Ziegler distinction not as a controversy regarding laws of physics but as alternate modelling strategies in the construction of a physical model. [...] For this reason, the two opposite principles coexist.*[63] From the point of view

---

[60] In more abstract terms, a similar point of view is discussed also in Ref. [69]. See also Endres' results [74] in the note of Sect. 6.1.9.

[61] Disruptions are a major obstacle in the search of controlled nuclear fusion.

[62] In contrast, the thermodynamic relationship $C_v < 0$ ensures that self-gravitating bodies with no interaction but gravity are intrinsically unstable (Sect. 3.4.1).

[63] A similar point of view is put forward also in Ref. [80].

of applications to fluid dynamics, the following words are crystal-clear: *In the case of observational or experimental turbulence, the dominant dissipation and entropy production occur in boundary layers. Rather than modelling boundary layer entropy production explicitly, carefully controlled measurements of turbulence are generally located in regions of space far from such boundaries. This convention explains the fact that turbulence measurements generally support Ziegler's principle and not Prigogine's principle. Thus, the main result of the programme proposed here, in support of Ziegler's principle, is that classical fluid turbulence is a closed system.* The last words are somehow surprising at their face value. However, their meaning is made clear by a discussion of the coexistence of different phenomena satisfying different extremum conditions in a steady relaxed state in Ref. [76]. In the convective motion of a turbulent fluid between parallel walls at different temperatures, for example, viscous dissipation is strong near the walls, while turbulent energy flows almost with no viscous dissipation in the fluid bulk, far from the walls. Entropy is produced both in the boundary layer (because of viscosity) and in the bulk (because of irreversible heat transport). If the control volume excludes (includes) the boundary layers, then the ruling contribution to the entropy balance changes, and either MEPP or MinEP may hold. Different principles hold in different regions.[64]

5.6.3 A Thought Experiment

A physically intriguing approach to the issue of maximization of entropy production is due to Sawada [81]. Here we discuss in detail Sawada's thought experiment. Even if tantalizing, the argument is far from rigorous, for the reasons explained below. All the same, its simplicity and generality deserve careful attention.

There are three bodies, (1), (2) and (0)—see Fig. 5.3. Bodies (1), (2) and (0) have linear size L_1 , L_2 and L_0 , thermal conductivity χ_1, χ_2 and χ_0, and heat capacity c_1, c_2 and c_0 respectively. Moreover, bodies (1), (2) and (0) are in thermal contact with a heat reservoir at temperature T_1 , with a heat reservoir at temperature T_2 and with a heat reservoir at temperature T_0, respectively. Heat reservoirs are thermally insulated from each other (Fig. 5.3), no thermal contact across thickened boundaries is ever allowed. Finally, body (1) may be in thermal contact with the body (0). Body (0) may be in thermal contact with the body (2). Body (1) and (2) are never in thermal contact with each other. Heat conduction only is considered. This implies that:

- The relaxation time for temperature across the whole body (1) is $\tau_1 \approx \frac{c_1 L_1^2}{\chi_1}$
- The relaxation time for temperature across the whole body (2) is $\tau_2 \approx \frac{c_2 L_2^2}{\chi_2}$
- The relaxation time for temperature across the whole body (0) is $\tau_0 \approx \frac{c_0 L_0^2}{\chi_0}$

and that for a perturbation of temperature with wavelength λ

---

[64] Further examples are referred to in Sects. 6.2.18, 6.2.19 and 6.2.21.

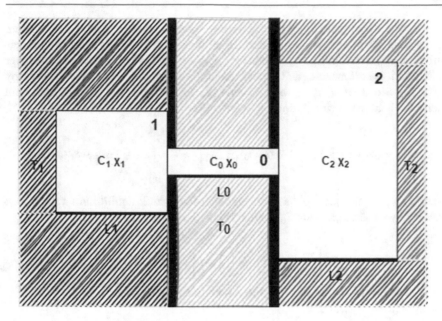

Fig. 5.3 Sawada's thought experiment (see text)

- The relaxation time in the body (1) is $\tau_{\lambda 1} \approx \frac{c_1 \lambda^2}{\chi_1}$
- The relaxation time in the body (2) is $\tau_{\lambda 2} \approx \frac{c_2 \lambda^2}{\chi_2}$
- The relaxation time in the body (0) is $\tau_{\lambda 0} \approx \frac{c_0 \lambda^2}{\chi_0}$

We assume $\chi_1, \chi_2 \gg \chi_0$ and $L_1, L_2 \gg \lambda \approx L_0$. It follows that $\tau_{\lambda 1}, \tau_{\lambda 2} \ll \tau_{\lambda 0} \approx \tau_0 \ll \tau_1, \tau_2$. Moreover, let the three bodies (1), (2) and (0) be thermally decoupled from each other at $t < t_0$. Thus, (1) is in thermal equilibrium with its own heat reservoir, and the same holds for (2) and (0). Entropies of reservoirs are constant. The entropy of the Universe at $t = t_0$ is:

$$S_{TOT}(t = t_0) = S_1(t = t_0) + S_2(t = t_0) + S_0(t = t_0)$$

Now, let us put in thermal contact body (0) with body (1) and body (0) with body (2) in the time interval $t_0 < t < t_0 + \Delta t$, where we choose Δt such that:

$$\tau_{\lambda 1}, \tau_{\lambda 2} \ll \tau_{\lambda 0} \approx \tau_0 \ll \Delta t \ll \tau_1, \tau_2$$

The inequalities $\tau_{\lambda 1}, \tau_{\lambda 2} \ll \tau_{\lambda 0} \approx \tau_0 \ll \tau_1, \tau_2$ discussed above ensure that such Δt exists. Then, at $t = t_0 + \Delta t$ let us decouple again the bodies for a further interval Δt, i.e. for t such that $t_0 + \Delta t < t < t_0 + 2\Delta t$; and let us put the bodies on thermal contact once more for t such that $t_0 + 2\Delta t < t < t_0 + 3\Delta t$, and so on, again and again...

We want to compute $S_{TOT}\left[t = t_0 + \left(n + \frac{1}{2}\right)\Delta t\right]$, where n is a positive integer. Every time we put the bodies in thermal contact we perturb (1) and (2) with perturbation of wavelength $\lambda \approx L_0$. Since $\tau_{\lambda 1}, \tau_{\lambda 2} \ll \Delta t$, both (1) and (2) will recover LTE, and it is still meaningful to speak about the entropy S_1 of (1) and the entropy S_2 of (2). Since $\Delta t \ll \tau_1, \tau_2$, both (1) and (2) are not in thermal equilibrium with their own heat reservoirs. Accordingly, we write:

$$S_1\left[t = t_0 + \left(n + \frac{1}{2}\right)\Delta t\right] \neq S_1\,(t = t_0) \quad ; \quad S_2\left[t = t_0 + \left(n + \frac{1}{2}\right)\Delta t\right] \neq S_2\,(t = t_0)$$

Since $\tau_0 \ll \Delta t$, the body (0) has recovered thermodynamic equilibrium with its own heat reservoir at temperature T_0, which is just the initial value. Then:

$$S_0\left[t = t_0 + \left(n + \frac{1}{2}\right)\Delta t\right] = S_0\,(t = t_0)$$

Since n is a positive integer, the three bodies are decoupled from each other at $t = t_0 + \left(n + \frac{1}{2}\right)\Delta t$, then:

$$S_{TOT}\left[t = t_0 + \left(n + \frac{1}{2}\right)\Delta t\right] = S_1\left[t = t_0 + \left(n + \frac{1}{2}\right)\Delta t\right] +$$
$$+ S_2\left[t = t_0 + \left(n + \frac{1}{2}\right)\Delta t\right] + S_0\left[t = t_0 + \left(n + \frac{1}{2}\right)\Delta t\right]$$

Together, the relationships above lead to the looked-for result for $S_{TOT}\left[t = t_0 + \left(n + \frac{1}{2}\right)\Delta t\right]$:

$$S_{TOT}\left[t = t_0 + \left(n + \frac{1}{2}\right)\Delta t\right] - S_{TOT}\,(t = t_0) =$$
$$= S_2\left[t = t_0 + \left(n + \frac{1}{2}\right)\Delta t\right] - S_2\,(t = t_0) +$$
$$+ S_1\left[t = t_0 + \left(n + \frac{1}{2}\right)\Delta t\right] - S_1\,(t = t_0)$$

Body (0) has disappeared. Now, the value of n is arbitrary. In the limit of large n, the Second Principle of thermodynamics requires that $\lim_{n \to +\infty} S_{TOT}\left[t = t_0 + \left(n + \frac{1}{2}\right)\Delta t\right] = S_{eq} = \max.$ for arbitrary initial conditions, i.e. for arbitrary $S_{TOT}\,(t = t_0)$, and the L.H.S. goes to a maximum as $n \to +\infty$. In other words, since S_{TOT} evolves towards a maximum and we may choose t_0 at will, the increment of S_{TOT}, namely the L.H.S., is bound to evolve towards a maximum. This is basically the argument referred to in Fig. 5.2. Moreover, since $\tau_0 \ll \tau_1, \tau_2$ we may take $\Delta t \to 0$, provided that the body (0) is small enough. Division of both sides of the relationship above by Δt gives just:

$$\text{max} = \frac{dS_{TOT}}{dt} = \frac{dS_1}{dt} + \frac{dS_2}{dt}$$

at the time $t = t_0$. (We will drop the dependence on t_0 below with no ambiguity).

Since $\tau_{\lambda 1} \ll \tau_1$, the body (1) interacts much more quickly with the body (0) than with the heat reservoir at temperature T_1. Let us denote with $Q_{1 \to 0}$ the amount of heat going from the body (1) to body (0) per unit time. Then

$$T_1 dS_1 = -Q_{1 \to 0} dt$$

Analogously

$$T_2 dS_2 = +Q_{0 \to 2} dt$$

Finally, the inequalities $L_1, L_2 \gg L_0$ allow us to neglect the impact of energy exchange between body (0) and its own heat reservoir on the global energy balance. The latter reduces, therefore, to

$$Q_{1 \to 0} = Q_{0 \to 2} \equiv Q$$

It follows that:

$$\text{max} = \frac{dS_{TOT}}{dt} = Q \left(\frac{1}{T_2} - \frac{1}{T_1} \right)$$

Since the total entropy increases as heat conduction is an irreversible process, the value of the maximum on the L.H.S. is positive. Accordingly, $Q > 0$ corresponds to $T_1 > T_2$. This is to be expected, as Q is just the amount of heat transmitted by (1) to (2) per unit time. If (1) receives heat from the external world and (2) gives back the same amount of heat towards the external world after having received it from (1), then

$$\frac{Q}{T_2} = \oint_{\text{boundary of 2}} \frac{\mathbf{q}}{T} \cdot d\mathbf{a} \quad ; \quad \frac{Q}{T_1} = -\oint_{\text{boundary of 1}} \frac{\mathbf{q}}{T} \cdot d\mathbf{a}$$

Body (1) has received an amount of heat Q across its own boundary from the external world per unit time, hence $\mathbf{q} \cdot d\mathbf{a} < 0$ for (1). In contrast, body (2) gives it back across its own boundary to the external world, hence $\mathbf{q} \cdot d\mathbf{a} > 0$ for (2). The final result is:

$$\text{max} = \frac{dS_{TOT}}{dt} = Q \left(\frac{1}{T_2} - \frac{1}{T_1} \right) = \oint_{\text{boundary of 1+2}} \frac{\mathbf{q}}{T} \cdot d\mathbf{a}$$

The generality of the result follows from three facts. Firstly, we may identify the domain of the surface integral on the R.H.S. with the boundary of the whole system with negligible error, as $L_1, L_2 \gg L_0$; body (0) disappears altogether from the result, and plays therefore just the role of an auxiliary intermediate support of Sawada's line of reasoning. Secondly, the discussion relies on no detailed information concerning the physical nature of the bodies involved and on the physical mechanism of transport embedded in **q**. Finally, our result describes a configuration that is both steady ($\frac{dS_{TOT}}{dt}$ achieves a maximum), is far from thermodynamic equilibrium (transmission of heat, an irreversible phenomenon, occurs and is the only source of change in both S_1 and S_2) and is stable (as a consequence of the Second Principle of thermodynamics). According to Ref. [81], a steady, stable, far from equilibrium state of a system where only heat transport occurs corresponds to a maximum of the entropy produced by the irreversible transport of heat across the system's boundary towards the external world. This result seems the generalization of the result on shock waves of Ref. [82], which is concerned with convection only—see Sect. 5.6.9 below.

Unfortunately, Sawada's argument is not a rigorous one, for two reasons. The first reason is rather formal. When it comes to compute the time derivative of S_{TOT} we have to take into account the fact that the derivative on time is by definition a limit for $\Delta t \to 0$. Unfortunately, we cannot rigorously achieve this limit, as both $\tau_{\lambda 1}$ and $\tau_{\lambda 1}$ provide us with a positive lower bound on Δt. The second reason is more fundamental. The Second Principle of thermodynamics ensures relaxation to a state of maximum S_{TOT}, but provides no information about the relaxation time. In Sawada's argument, however, $\frac{dS_{TOT}}{dt} \to$ max precisely when (and because) $S_{TOT} \to$ max, i.e. relaxation of the system to a stable state occurs precisely when the system relaxes to thermodynamic equilibrium. Consequently, there is no stable steady state far from thermodynamic equilibrium, where $S_{TOT} = $ const., $T_1 = T_2$ and there is no net flux of heat across the boundary.

Should we reject Sawada's result altogether? Not entirely. Generally speaking, a physical system is made of many different parts. In order to write down S_{TOT} explicitly for the system as a whole, we have to assume that each part is at LTE. It follows that whenever the entropy produced per unit time in the bulks of the various parts vanishes, then $\frac{dS_{TOT}}{dt}$ depends only on the physics at the interfaces among different parts. Sawada's arguments suggests that if maximization of $\frac{dS_{TOT}}{dt}$ in stable, steady state is proven somehow, then such proof relies on some constraint of fixed entropy production in the bulk. Moreover, Sawada's point of view agrees with our analysis of the copper wire: if any maximization holds, it is concerned with the boundaries of the system, and the interfaces with the external world. This fits nicely with the results below concerning turbulent flow between fixed parallel surfaces at large Re, convection at large Ra, shock waves at $Ma \gg 1$ (Sect. 5.6.9) and crystal growth in Sects. 5.3.12 and 5.6.9, where the contribution of the bulk to the production of entropy is negligible in all cases.

5.6.4 Again, the Copper Wire

To start with, let us see what our copper wire tells us about the properties of MEPP, if any MEPP is valid. The entropy balance of a copper wire in steady state with no electric current (i.e., $\sigma = 0$) leads to the relationships:

$$\frac{dS_{wire}}{dt} = 0 \quad ; \quad \frac{dS_{TOT}}{dt} = \left(\frac{dS}{dt}\right)_{\text{produced in the wire}}$$

$$\frac{dS_{wire}}{dt} = \left(\frac{dS}{dt}\right)_{\text{produced in the wire}} + \left(\frac{dS}{dt}\right)_{\text{coming into the wire from the external world}}$$

In our discussion of the global form of the entropy balance in a fluid (Sect. 4.2.9), we have seen that the fluid is like a copper wire. Then, we write:

$$\left(\frac{dS}{dt}\right)_{\text{coming into the wire from the external world}} = -\oint_{boundary} \mathbf{j}_s \cdot d\mathbf{a}$$

where $\mathbf{j}_s = \frac{\mathbf{q}}{T} - \frac{\mu_k^0 \mathbf{j}_k}{T}$ and the R.H.S. is $\frac{d_e S}{dt}$, i.e. the amount of entropy coming into the fluid per unit time across the fluid boundary. It follows that:

$$\frac{dS_{TOT}}{dt} = \oint_{boundary} \mathbf{j}_s \cdot d\mathbf{a} \quad ; \quad \text{in steady state}$$

Accordingly, a steady state corresponds to a maximum of $\frac{dS_{TOT}}{dt}$ if and only if it corresponds to a maximum of $\oint_{boundary} \mathbf{j}_s \cdot d\mathbf{a}$. We stress the point that we have provided no proof of the fact that a stable steady state satisfies MEPP. Rather, the copper wire shows that if MEPP holds, then it is related to the boundary terms in the entropy balance.

So far, we have implicitly assumed that the copper wire is at rest. Generalization to the case of a moving wire is straightforward, provided that we require that the entropy per unit volume of wire does not depend on time, i.e. that $\frac{\partial(\rho s)}{\partial t} = 0$. In this case, the entropy balance in local form (Sect. 4.2.8) implies that nothing changes in our discussion provided that we replace $\mathbf{j}_s$ with $\mathbf{j}_s' \equiv \mathbf{j}_s + \rho s \mathbf{v}$.[65] This way, convection enters the matter just like conduction and radiation. In all cases, σ remains zero, hence $\frac{d_i S}{dt}$ vanishes. As a rule of thumb, MEPP seems to work fine when $\frac{d_i S}{dt}$ is negligible. In this case, what is commonly meant by the wordings 'maximization of entropy production' actually refers to 'maximization of the boundary contribution to the entropy production'.

5.6.5   Two Remarkable Exceptions

Christen [83] claims that Steenbeck's principle of Sect. 5.3.2 follows from MEPP with the constraint of given applied voltage, and Zupanovic et al. [84] claim that Kirchhoff's laws ruling a net of electric resistors at the same temperature follow

---

[65] It makes sense to apply a fluid-dynamical formalism to a wire if we think of a floating wire in a channel of streaming water.

from MEPP provided that the input power is given. In both cases the (Joule) heating mechanism acts inside the system, not at the boundaries, hence $\frac{d_i S}{dt} \neq 0$ in contrast with Sect. 5.6.4. We deal with Christen's and Zupanovic et al.'s results in Sects. 6.2.8 and 6.2.14, respectively.

5.6.6 Heat Conduction in Gases

Holyst et al. [85] investigate heat conduction in a a perfect gas with uniform density n_0 at LTE everywhere at all times in slab geometry. The gas is bounded by two parallel plates located at $x = 0$ and $x = d$ and with the same, constant temperature $T_{boundary}$ and is heated by radiofrequency with a constant amount P_{TOT} of heat per unit time. This amount is an input of the problem.

If $P_{TOT} = 0$, then the fluid is at thermodynamic equilibrium with the walls, the temperature of the fluid is uniform and is equal to $T_{boundary}$ everywhere. In this case, we denote the amount of internal energy of the fluid with U_0. As the heating is switched on, the fluid relaxes to some final state after a while. The constant value > 0 of P_{TOT} keeps the steady, relaxed far from thermodynamic equilibrium. In this relaxed state the temperature inside the fluid is larger than $T_{boundary}$ (generally speaking, it depends on the position), and the internal energy U of the fluid is larger than U_0, say by an amount $\Delta U \equiv U - U_0 > 0$. Conservation of energy in steady state requires that an amount Q_{loss} of heat flows from the fluid towards the plates per unit time, and that $Q_{loss} = P_{TOT}$. It is supposed that conduction is the only physical process that is responsible for the transport of heat from the fluid towards the walls. If there is gravity, this is equivalent to assume that $|\nabla T|$ is nowhere too large, so that $Ra < Ra_{thr}$ everywhere.

Different scenarios, with different distributions of additional heating in the fluid between the plates, have been investigated both analytically and numerically. Results show that the relaxed state corresponds to a constrained minimum of $\frac{\Delta U}{Q_{loss}}$ with the constraint of given ΔU. We show in Sect. 6.2.16 that this result agrees with the maximization of $\int \mathbf{q} \cdot \nabla \left(\frac{1}{T}\right) d\mathbf{x}$, rather than with its minimization required by LNET.[66]

5.6.7 Convection at Large Ra

According to Ozawa et al. [86], the maximization of $\oint_{boundary} \frac{\mathbf{q}}{T} \cdot d\mathbf{a}$ applies reasonably well to the viscous ($P_h = \sigma'_{ik} \frac{\partial v_i}{\partial x_k}$), turbulent, time-averaged relaxed fluids

---

[66] For a generalization of the results of Ref. [85], see Sect. 5.8.6.

between parallel walls at different temperatures with Bénard cells[67] in the limit of large Rayleigh number, i.e. $Ra \rightarrow \infty$. In this limit both viscosity and the viscous heating power density becomes negligible in the fluid bulk at least. Time-averaging is performed over many periods of rotation of the convective cells.[68]

A seemingly different extremum condition has been required by Malkus and Varonis [87] in incompressible fluids, i.e. maximization of $\beta \{v_z T\}$ where $\beta \equiv -\frac{\partial \{T\}}{\partial z}$ ($\beta > 0$ if the walls are parallel and horizontal and the bottom wall is hotter), $\{a\}$ is the average over the horizontal plane of the generic quantity a and z is the spatial coordinate in the direction linking the two walls. For further discussion, see Sect. 6.2.19.

5.6.8 The H-Mode

Basically, the same approach is followed in [88] when it comes to describing the 'H-mode'[69] spontaneous onset of a large gradient of temperature[70] near the edge of tokamak[71] plasmas when the flow of heat from the plasma bulk towards the walls of the toroidal chamber which contains the plasma exceeds a threshold value. Given the heat flow, the authors postulate maximization of temperature inhomogeneity, which is proportional to the entropy production.

From the point of view of thermodynamics, we recall that the overall amount of heat produced by both nuclear fusion, Joule heating and auxiliary systems is negligible near the plasma edge; moreover, the amount of power locally produced by viscous heating can be small when compared to the amount of heat coming from the plasma bulk and flowing per unit time across the plasma edge region towards the external world.

In steady state, the entropy is constant, and the entropy balance reads therefore $0 = \frac{dS}{dt} = \frac{d_i S}{dt} + \frac{d_e S}{dt} = \int \mathbf{q} \cdot \nabla \left(\frac{1}{T}\right) d\mathbf{x} - \int \frac{\mathbf{q}}{T} \cdot d\mathbf{a}$.[72] Maximization of the first

---

[67] Rayleigh-Bénard convection is a type of natural convection, occurring in a planar horizontal layer of fluid heated from below, in which the fluid develops a regular pattern of convection cells known as 'Bénard cells'.

[68] For further discussion, see Sect. 6.2.13.

[69] A shortcut for 'high confinement mode'

[70] Usually dubbed as 'internal transport barrier'.

[71] A tokamak is a device in the research on controlled nuclear fusion [79] where magnetic fields confine a hot plasma inside a chamber. In steady state, the heating power released into the plasma, e.g. by Joule heating, auxiliary heating systems or (hopefully) nuclear fusion is equal to the power lost by the plasma towards the walls of the chamber. Ideally, the plasma is colder (hotter) near the (far from the) walls. Usually, the temperature follows a quasi-parabolic profile. When the heating power (hence, the energy flux outwards) exceeds a threshold, then the temperature profile displays a typical pudding-like shape, i.e. a strong gradient near the wall and a flat profile near the centre. Correspondingly, the energy content of the plasma increases dramatically, thus facilitating the occurrence of nuclear fusion reactions. To date (2022), the H-mode seems therefore to be a breakthrough on the roadmap towards successful utilization of fusion energy.

[72] We have invoked Gauss' theorem of divergence.

term (Sect. 6.2.16) corresponds therefore to maximization of the second term (Sect. 6.2.13). Being the plasma energy transport essentially ruled by turbulent convection, the analysis of Ref. [88] is somehow an extension of the work of Ref. [86] to plasma physics.

The nature of the extremum condition depends on the constraint. In the model of H-mode described in Ref. [78], the system selects the state where $\int \mathbf{q} \cdot \nabla \left(\frac{1}{T}\right) d\mathbf{x}$ is maximum when the heat flux $\int \mathbf{q} \cdot d\mathbf{a}$ is fixed, but is minimum whenever the fixed quantity is the temperature of the plasma bulk. The latter condition is equivalent to the condition of a fixed gradient of temperature, because the temperature of the wall, the lowest temperature in the system, is constant. The former and the latter case correspond to maximum temperature gradient at a given heat flux and minimum heat flux at a given temperature gradient. Such duality is in agreement with a theorem discussed in Sect. 6.2.7. For further discussion, see Sect. 6.2.13.

5.6.9 Shock Waves

A shock wave is a type of propagating disturbance that moves faster than the local speed of sound in the medium. Like an ordinary wave, a shock wave carries energy; however, it is characterized by an abrupt, nearly discontinuous change in pressure, temperature and density of the medium. At a given time, a thin region (the wavefront) separates the region where the shock wave has already passed ('region behind the wave') from the region where the shock wave has still to pass ('region in front of the wave'). It is customary to define the jump $[a] \equiv a_{behind} - a_{front}$ of the generic physical quantity a as the difference between the value a_{behind} of a (in the region) behind the wave and the value a_{front} of a in (the region in) front of the wave. When a shock wave passes through matter, entropy increases, i.e. $[s] > 0$. Generally speaking, the faster the shock wave, the larger $|\mathbf{v}|$, the nearer the value of $\mathbf{j}'_s$ to $\rho s \mathbf{v}$, the larger the value of the Mach number Ma (the dimensionless ratio between the propagation speed of the shock wave and the speed of sound), the thinner the wavefront, the smaller the volume V_S occupied by a wavefront of given area A_S, the smaller $\int_{V_S} \sigma d\mathbf{x} \propto O(V_S)$, the more negligible the contribution of σ to the entropy balance. For $Ma \gg 1$, we neglect this contribution altogether, and apply the discussion of Sect. 5.6.4 and write $\frac{dS_{TOT}}{dt} = \oint \mathbf{j}'_s \cdot d\mathbf{a} \approx A_S \left[j'_{sn}\right]$[73]: $\frac{dS_{TOT}}{dt}$ gets maximized if and only if the jump $[\rho s v_n]$ of entropy flux across the wavefront gets maximized. In both

---

[73] The surface integral is the sum of the contribution of the front side and the back side of the shock wave. With respect to the wave front, the fluid moves at velocity $\mathbf{v}$; the latter points towards (away from) the shock wave on the front (back) side, hence the contributions of the front side and of the back side to the entropy balance are < 0 and > 0, respectively; moreover, we write $b_n \equiv \mathbf{b} \cdot \mathbf{n}$ with $\mathbf{b}$ generic vector and $\mathbf{n}$ unit vector perpendicular to the shock wave pointing outwards (i.e. forwards on the front side and backwards on the back side).

classical and relativistic fluid mechanics, indeed, maximization of this jump has been rigorously proven by Rebhan [82] in the $Ma \to \infty$ limit starting from the full set of balance equations for mass, momentum and energy.[74]

5.6.10 Dunes

It is worthwhile to rewrite the condition of maximum $\left[j'_{sn} \right]$ in the particular case of no heat flow and only one chemical species (so that $\left[T^{-1} \cdot \mu_k^0 \cdot j_{kn} \right] = \left[T^{-1} \cdot q_n \right] = 0$) as $\left[\rho s v'_n \right] = \max$. In contrast with shock waves, if there is a discontinuity moving so slowly across the fluid that s remains basically unaffected (i.e. $[s] \approx 0$) then this maximization reduces to $\left[\rho v'_n \right] = \max$., i.e. $\left[(\rho v_n)_{behind} + (\rho v_n)_{front} \right] - (\rho v_n)_{front} = \max$. In the frame of reference where our discontinuity is at rest, this means the square-bracketed quantity is a maximum whenever the mass flow $(\rho v_n)_{front}$ imping-ing on the discontinuity is constant. In the physics of subaqueous dunes, the latter quantity is known as 'gross bedford-normal transport'. Its constrained maximization has been observed experimentally; in other words, given the impinging flow, it has been observed that the dunes spontaneously align themselves in order to maximize the gross bedford-normal transport. Just like in Korteweg–Helmholtz' principle of Sect. 5.3.7, here the geometry adapts itself; but the extremum condition, like the val-ues of Re relevant to the problem are quite different. Moreover, if the impinging flow oscillates periodically with a period much shorter than the typical bedford reconsti-tution time, then the time-averaged gross bedford-normal transport is maximized [89]. Extension to desertic dunes is discussed in Ref. [33].[75]

5.6.11 Detonation Versus Shock Waves

A comparison of the shock waves of Sect. 5.6.9 with the deceitfully similar case of detonation waves is useful when it comes to highlighting the subtleties of the search for a stability criterion. In its simplest ('Chapman–Jouguet') description, a detona-tion wave is propagating shock wave accompanied by exothermic heat release (due e.g. to combustion). This description confines the chemistry and diffusive transport processes to an infinitesimally thin zone. For a plane wave, the balances of mass, energy and momentum provide a single ('Hugoniot') relationship which links the a_{behind}'s with the a_{front}'s.[76] Generally speaking, for a given set of a_{front}'s Hugo-niot's relationship provide an infinite collection of solutions, i.e. of sets of a_{behind}'s, each set corresponding to a different value of the propagation velocity—see Sect. 3.c and note 37 of Ref. [90]. We may ask if thermodynamics provides us with a stability

---

[74] Rebhan has shown that maximization of the jump of entropy flux may even *replace* one conser-vation equation see Eq. (17) of Ref. [82]. For further discussion, see Sect. 6.2.17.

[75] See Sect. 6.2.17 for further discussion.

[76] We adopt the same formalism of Sect. 5.6.9.

criterion that allows us both to select one solution and to predict one value of the propagation velocity, to be compared with experiments. In spite of the geometrical similarity,[77] the thermodynamics of a detonation wave differs from the thermodynamics of a shock wave as the propagation of the former relies on an exothermic heat release. Correspondingly, the amount of entropy produced per unit time by a unit area of detonation wave and released in steady-state propagation to the external world is still equal to $[\rho s v_n]$. Now, however, this entropy production is due to irreversible processes related to exothermic heat release, rather than to the exchange of energy and matter with the external world as in the shock wave of Scct. 5.6.9. In other words, $\frac{d_i S}{dt}$ is far from negligible: it rules.[78] Our discussion in Sects. 5.6.3 and 5.6.4 suggests that no MEPP holds, in contrast with shock waves. Is stability described by minimization of $[\rho s v_n]$? The answer is affirmative. We recall that Hugoniot's relationship provides many possible solutions, and that we expect thermodynamics to select the stable one. It turns out that the only experimentally observed velocity of spontaneous propagation of detonation corresponds to a minimum of the detonation velocity, which is also a minimum of s_{behind}; for a description of the underlying stabilizing mechanism, see Ref. [91]. In other words, given the a_{front}'s the observed stable configuration of a steadily propagating detonation wave corresponds to a minimum of $[\rho s v_n]$, which is an increasing function of both velocity and s_{behind}. Detonation waves seem therefore to provide us with an example of MinEP, rather than MEPP. As stressed in Ref. [90], no LNET and no Onsager symmetry is ever invoked. Here, indeed, entropy is mainly produced inside the system, rather than by transport processes across the boundaries of the system.

5.6.12 Solids

When dealing with continuous systems, we have focussed our attention on fluids so far. However, our arguments are not limited to fluids, as far as the assumption of LTE everywhere at all times holds. For example, when it comes to crystal growth both σ and $\mathbf{q}$ are usually neglected. This fact, together with the relationships $h_k = \mu_k^0 + T s_k$, $\mathbf{j}_s = \frac{\mathbf{q}}{T} - \frac{\mu_k^0 \mathbf{j}_k}{T}$ and $\frac{dS_{TOT}}{dt} = \oint_{boundary} \mathbf{j}_s \cdot d\mathbf{a}$ in steady state, leads to $\frac{dS_{TOT}}{dt} = -\oint_{boundary} \frac{\mu_k^0 \mathbf{j}_k}{T} \cdot d\mathbf{a} = \oint_{boundary} s_k \mathbf{j}_k \cdot d\mathbf{a} - \oint_{boundary} \frac{h_k \mathbf{j}_k}{T} \cdot d\mathbf{a}$. Accordingly, $\frac{dS_{TOT}}{dt}$ is an increasing function of the net amount $\oint_{boundary} s_k \mathbf{j}_k \cdot d\mathbf{a}$ of entropy exchanged per unit time across the boundary because of diffusion. Maximization of the latter amount has been postulated by many authors in different ways when describing crystal growth—see Ref. [92] for a comprehensive bibliography. (The term $\propto h_k$ may, e.g. be of interest in case of exothermal reactions).

---

[77] Both detonation and shock waves are described as very thin, surface-like region of separation between a 'front' region and a 'behind' region

[78] *There is no need to take into account a change in entropy through exchanges of either heat or matter with the environment* [90].

Fig. 5.4 Paltridge's model of
Earth's atmosphere

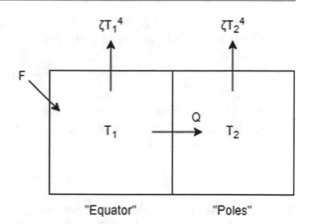

Examples of MEPP concerning the entropy exchange across the boundary of the system may also be found in metallurgy [93] and when it comes to processes of crystallization [94] and solidification [95]. The growth of a crystal is obviously affected by what happens at its boundaries. Correspondingly, the selection criterion discussed in [93] is related to the *conjecture that each lamella must grow in a direction that is perpendicular to the solidification front*. Analogously, the results of [94] are related to the growth velocity via the *rate of entropy production per unit area*, which is a surface quantity. In [92], a crystal grows and changes its surface so as to select a flow that maximizes the entropy production, $\frac{d_e S}{dt}$ being the only contribution to the entropy production which depends on both a flow and the surface of the system. As for solidification, the relevant quantity is the *entropy generation term which depends on the solid–liquid region* [95], i.e. on an interfacial boundary. For further discussion, see Sect. 6.2.20.

5.6.13 Earth's Oceans and Atmosphere

Paltridge's simplified model of the system made of Earth's ocean and atmosphere divides this system into two regions [96], dubbed 'the equator' and 'the poles'. The temperatures of the equator and of the poles are T_1 and T_2 with $\Delta T \equiv T_1 - T_2 > 0$. The equator supplies the poles with an amount Q of energy per unit time[79] with the help of winds, sea currents, etc. The equator and the poles radiate towards the outer space a an amount of power $\propto \zeta T_1^4$ and $\propto \zeta T_2^4$ respectively, where ζ is a proportionality constant $\propto$ Stefan-Boltzmann's constant. Finally, the equator is supposed to receive the whole amount F of power reaching the Earth from the Sun, for simplicity—see Fig. 5.4.

---

[79] Time-averaged over seasonal variations.

We denote with E_1 and E_2 the internal energies of the equator and of the poles, respectively. The balances of these energies read

$$\frac{dE_1}{dt} = F - \zeta T_1^4 - Q \quad ; \quad \frac{dE_2}{dt} = -\zeta T_2^4 + Q$$

Note that $\frac{d_i S}{dt} = 0$. We take $\zeta = 1$, $F = 1$ with no loss of generality,[80] and focus our attention on steady states ($\frac{d}{dt} = 0$) where the balances above reduce to $0 = 1 - T_1^4 - Q$ and to $0 - -T_2^4 + Q$, i.c. to

$$1 - T_1^4 - T_2^4 = 0 \quad ; \quad \Delta T = (1 - Q)^{\frac{1}{4}} - Q^{\frac{1}{4}}$$

This is a system of two equations in two unknown quantities, once Q is given. Thus, Q defines unambiguously and completely the system in a steady state. In particular, if $Q = 0$, i.e. if the poles are completely decoupled from the equator, then $\Delta T = 1$; if $Q = 1$ then $\Delta T = 0$, i.e. the equator and the poles are at the same temperature. Remarkably, the net amount of entropy $Q \left(\frac{1}{T_2} - \frac{1}{T_1} \right)$ produced per unit time by the transport of energy between the equator and the poles is also a function of Q in steady state. In particular, it is a differentiable, non-negative function of Q for $0 \leq Q \leq 1$ and vanishes for both $Q = 0$ and $Q = 1$. Consequently, a maximum of $Q \left(\frac{1}{T_2} - \frac{1}{T_1} \right)$ exists for some value of Q in the range $0 \leq Q \leq 1$. According to Paltridge's argument, such maximum describes a stable steady[81] state. The argument runs as follows.

The continuous line in Fig. 5.5 represents steady states. Let the system be at O1 at the beginning. Let a fluctuation bring the system away from the steady state. Local fluctuations leave ΔT unaffected. Accordingly, they shift the system vertically, either into A1 or into B1. For the moment, let us assume that $|\Delta T|$ is small near both O1, A1 and B1; we shall drop this assumption below. Then, the law which rules the evolution of the system is approximately linear, i.e. $Q \propto \Delta T$, with $Q (\Delta T = 0) = 0$ (represented by the dotted line crossing the origin). Accordingly, starting from B1 the system relaxes to the steady state C1, while starting from A1 the system relaxes to the steady state D1. The middle point O2 between C1 and D1 on the line of steady states is on the right of the initial state O1. On average, after many fluctuations, the system moves towards the central part of the plot, where $Q \left(\frac{1}{T_2} - \frac{1}{T_1} \right)$ is maximum.

What if we start from an initial state on the left side of the plot, near $\Delta T = 1$? Let O3 be the initial state, A3 and B3 are the perturbed unsteady states. Now $|\Delta T|$ is large. We expect therefore that the evolution of the system is ruled by a nonlinear law (represented by the dotted curves linking the origin, A3 and B3). Accordingly, starting from B3, the system relaxes to the steady state C3, while starting from A3, the system relaxes to the steady state D3. The larger $|\Delta T|$, the stronger the nonlinearity,

---

[80] Accordingly, we write all values of energy and temperature in suitable dimensionless units.
[81] Time-averaged over seasonal variations.

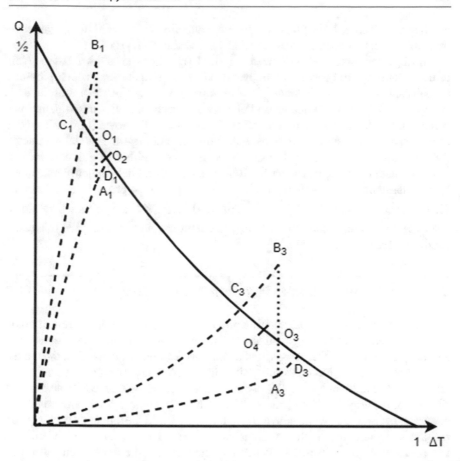

Fig. 5.5 *Q versus* ΔT *in Paltridge's model of Earth's atmosphere—see text*

the more vertical the dotted curve in D3, then in C3. It follows that the middle point
O4 between C3 and D3 on the line of steady states is on the left of the initial state
O1. On average, after many fluctuations, the system moves the central part of the
plot, where $Q \left(\frac{1}{T_2} - \frac{1}{T_1} \right)$ is maximum.

Paltridge's findings are partially confirmed by Ozawa et al.'s [97], who apply
the maximization of $\oint \frac{q}{T} \cdot d\mathbf{a}$ to the general circulation of Earth's global fluid (the
atmosphere and the ocean). In particular, **q** includes no large-scale convection—in
the authors' words it *does not in principle include the advective heat flux* but includes
heat conduction ruled by small-scale turbulence as well as radiation. In particular, the
maximization of $\oint \frac{q}{T} \cdot d\mathbf{a}$ describes well the overall outcome which results from the
radiation power balances of many different regions of the global fluid. Independent
results reported in [98], indeed, suggest that Paltridge's approach works fine as far
as horizontal heat fluxes only are considered, and becomes unrealistic when vertical,
convection-related heat fluxes are taken into account. Similar results in geophysics

are also described in Refs. [76,99], where the attention is focussed on the radiative transport of heat. A more refined model is discussed in Ref. [100].

The arguments above are criticized in Ref. [101], where it is shown that they fail to take properly into account the impact of viscous dissipation,[82] which provides a contribution to the production $\frac{d_i S}{dt}$ of entropy in the bulk of the atmosphere.[83] Nicolis' work [103] casts doubts on Paltridge's conclusions after investigating the behaviour of entropy production in the evolution of a well-known (Lorenz') model of atmosphere: it is shown that—even after time-averaging—the model's attractor (which the system eventually relaxes to) corresponds neither to a maximum nor to a minimum of entropy production. Remarkably, this model explicitly takes into account the contribution of thermal conduction to the entropy production in the bulk. Finally, maximization of $Q\left(\frac{1}{T_2} - \frac{1}{T_1}\right)$ fails to describe relaxed states where either viscosity and thermal conductivity are present in the bulk or complex boundary conditions are assumed [104].[84]

5.7   Lotka and Odum's Maximum Power Principle

Just like the minimum dissipation principles discussed in Sect. 5.5, maximization of dissipation has been appealing well beyond the domains of physics and engineering. As such, it deserves further discussion; again, this Section is a digression the reader may overlook. The results concerning cities in Sect. 5.5.2 rely on Zipf's principle of least effort and hold for $k = \frac{W}{Q_I - Q_{II}} < 0$. In this case, the city A receives an amount $Q_I > 0$ of resources per unit time and feeds the population with an amount W of resources per unit time while delivering to the outer world the amount $Q_{II} > Q_I$ of resources per unit time. Such 'resources' are not bound to be energy-related quantities (i.e., quantities measured in Watt): they may as well be immaterial resources like money, social services (measured, e.g. through their monetary value), etc.

---

[82] And of Coriolis forces which are counteracted by viscous forces.

[83] In Ref. [102], Shutts puts in evidence a conundrum hidden in Paltridge's results: in spite of their *surprising* agreement with observations, they are *invariant to the Earth's rotation rate [...] which is particularly disconcerting in view of the importance attached to these quantities in numerical simulations of climate*. Accordingly, Shutts puts forward a different maximization criterion, namely maximization of a linear combination of the time derivatives of mechanical energy and enstrophy; such quantity differs from the familiar entropy of LTE as it involves no temperature. Usually, enstrophy (i.e. the volume integral of the sum of all squared absolute values of all partial derivatives of all components of $\mathbf{v}$) is a so-called 'rugged invariant', i.e. the typical time scale of enstrophy decay due to viscosity is much longer than the corresponding time scale for energy (see also the note in Sect. 6.1.4). Moreover, if, after time-averaging on suitable time scales at least, the total mechanical energy is constant, then the maximization of the time derivative of mechanical energy is only possible if the rate of dissipation of such energy due to viscosity is also a maximum (the corresponding amount of heat produced by viscosity being carried away through some energy transport process). Thus, for a $Re \gg 1$ fluid like the atmosphere Shutts' result seems to be somehow related to Malkus' one [41] (Sect. 5.3.12).

[84] See Sect. 6.2.13 for further discussion.

We may rightfully ask ourselves what happens when the quantities Q_I, Q_{II} and W are precisely amounts of power, measured in Watts. Conservation of energy implies $W = Q_I - Q_{II}$, i.e. $k = +1$, and the results of Sect. 5.5.2 hold no more.

Even so, the main lesson of Sect. 5.5—namely, a single variational principle provides us with information concerning the behaviour of a network of many similar, interrelated agents—is still likely to hold. Biology and ecology, for example, offer plenty of problems where a large number of interlinked entities (cells, living species...) display collective behaviour. This behaviour is both remarkably resilient against perturbations and definitely far from thermodynamic equilibrium. In the words of Lotka *For the battle array of organic evolution is presented to our view as an assembly of armies of energy transformers-accumulators (plants), and engines (animals); armies composed of multitudes of similar units, the individual organisms. The similarity of the units invites statistical treatment, the development of a statistical mechanics of which the units shall be, not simple material particles in the ordinary reversible collision of the type familiar in the kinetic theory, collisions in which action and reaction were equal; the units in the new statistical mechanics will be energy transformers subject to irreversible collisions* [105].

Let us focus our attention on just one living being, which corresponds to the city of Sect. 5.5.2. According to DeLong, *organisms maximize fitness by maximizing power. With greater power, there is a greater opportunity to allocate energy to reproduction and survival, and therefore, an organism that captures and utilizes more energy than another organism in a population will have a fitness advantage* [106]. The ability to maximize the amount W of energy our living being is able to utilize per unit time is useful to its metabolism, then it is selected by evolution. According to Lotka, *if sources are presented, capable of supplying available energy in excess of that actually being tapped by the entire system of living organisms, then an opportunity is furnished for suitably constituted organisms to enlarge the total energy flux through the system. [...] natural selection will so operate as to increase [...] the rate of circulation of matter through the system, and to increase the total energy flux through the system, so long as there is presented an unutilized residue of matter and available energy [...] natural selection tends to make the energy flux through the system a maximum, so far as compatible with the constraints to which the system is subject [...] the term energy flux is here used to denote the available energy absorbed by and dissipated within the system per unit of time* [107]. Of course, in steady state the larger this *energy flux* the larger the amount of heat flowing across the boundaries of the system towards the external world: in the words of Ref. [108] *energy dissipated in the organization also has an optimum for maximum power where the power output is a dissipative process.*[85] In biology, Lotka's *energy flux* includes the metabolic rate (i.e., the amount of energy per unit of time that a living being needs to keep its body functioning at rest) which is, therefore, to be maximized. The idea that natural selection prefers the living beings which maximize W was initially strengthened by the experimental results reported in Ref. [109]. DeLong's experiments [106] provide us with further

---

[85] See also Sect. 6.1.12.

confirmation: as two species of microorganisms are put in competition in a culture, after a while two outcomes are observed: (a) either the species survives which has the highest metabolic rate (in DeLong's words *a winning species uses more power at its steady-state density than the losing species uses at its steady-state density*) (b) or a stable community where the species coexist is obtained, which has higher metabolic rate than either species could have alone (*if coexistence occurs, it will be characterized by community-level power that exceeds the highest-power competitors when grown in isolation*).[86]

Odum developed Lotka's idea (referred to as 'maximum power principle' ('MPP') or 'maximum rate of useful energy transformation') in a systematic way, and discussed in depth its applications to ecology—for a review of Odum's vastly influential work, see Ref. [111]. In particular, Odum stressed the point that MPP is *not* equivalent to the maximization of efficiency $\eta = \frac{W_l}{Q_l}$ which is rather an outcome of the PLE of Sect. 5.5. Should a living being maximize η, indeed, it would be quite similar to an ideal Carnot cycle. In this case, however, transformations should be infinitely slow: under these conditions, the power output would clearly vanish as it would take an infinite time to a finite amount of work.[87] In the opposite limit of very fast transformations, on the other hand, unavoidable irreversibility spoils η and reduces W. It follows that MPP requires that η achieves an optimal value η_{opt} which is less than the ideal Carnot efficiency η_C. Remarkably, the same is true in heat engines working between a maximum temperature T_{max} and a minimum temperature $T_{min} < T_{max}$. For such engines, it is well known that $\eta_C = 1 - \frac{T_{min}}{T_{max}}$. According to Ref. [112], in the same engines $\eta_{opt} = 1 - \sqrt{\frac{T_{min}}{T_{max}}}$. In order to apply the same analysis to biological systems, it is reasonable to write $T_{max} \approx T_{min}$, or, more precisely, $T_{min} = T_{max}(1 - \varepsilon)$ with $0 < \varepsilon \ll 1$. Substitution of this *ansatz* in the expressions for η_C and η_{opt} leads to[88] $\lim_{\varepsilon \to 0} \eta_{opt} = \frac{\eta_C}{2}$. Indeed, this relationship is confirmed by Odum in many cases. In order to grasp the ecological meaning of this result, Odum postulated the analogy of ecological system with electrical networks [113]. According to this analogy, a battery (or more explicitly the solar electricity generator Odum sometimes used) pushes electrons around a copper circuit in almost exactly the same way that the sun pushes energy (or reduced carbon) around the invisible circuits of an ecosystem. Odum actually built electric networks of resistors, capacitors, etc. which simulated the behaviour of particular ecological systems, and postulated the validity of a generalization of Ohm's law in ecology. With no need of postulating small differences in temperatures, Ohm's law in an electric circuit made of a voltage source (e.g., a battery) with voltage V and inner resistance R_S connected in series with a 'load' resistance R_L leads to the so-called 'maximum transfer power

---

[86] DeLong's experiments are performed under controlled conditions, including light, etc. Things may change if the outcome of the competition includes the depletion of the environmental resources available to the competing species. In this case, a modified version of the principle, namely the 'optimal power principle', has been put forward—see Ref. [110] for a discussion.

[87] The tortoise won the hare; but the fairy tale says nothing of what occurred when the hawk came.

[88] Here we take into account that $\sqrt{1+x} \approx 1 + \frac{x}{2} + \ldots$ for $0 < x \ll 1$.

theorem'. This theorem states that the efficiency η (defined as the ratio of power P_L dissipated by the load to the power developed by the source) achieves one half of its maximum allowable value[89] whenever $P_L = \text{max}$. What the ecological equivalent of current and voltage exactly are is a debated topic.[90]

5.8 Oscillating Relaxed States

5.8.1 Rules of Selection

Even beyond the domain of applicability of LNET, we have shown that in some cases at least it seems possible to describe relaxed states with the help of the constrained minimization or maximization of some thermodynamic quantity. So far, we have described relaxed states which are steady states. However, there are some problems where the system relaxes to a final, far from equilibrium state which is not a steady state, but an oscillating state, i.e. a configuration where all quantities keep on oscillating at some fixed frequency. Now, when deciding whether the final outcome of the relaxation process is a steady state or an oscillating one, the system seems still to follow a criterion involving thermodynamic quantities. Thus, Tykodi postulated MinEP in oscillating states [114].

Generally speaking, the selection of the relevant time scale is crucial. In geophysics, for example, when applied to the seasonal cycle the validity of MEPP is restricted to energy transport due to turbulent convection processes which are fast enough to allow relaxation on time scale $\ll$ the time scale of interest [115]. More precisely, the selected relaxed state still corresponds to a constrained minimum (or maximum) of some thermodynamic quantity, provided that the value of the latter is averaged in time over many oscillations. The minimization (or maximization) acts as a 'selection rule' between a stable steady state and a stable, oscillating state.

We denote with $a_0(\mathbf{x})$, $a_1(\mathbf{x}, t)$ and $\overline{a}(\mathbf{x}, t) = \lim_{t_0 \to \infty} t_0^{-1} \int_t^{t+t_0} a(\mathbf{x}, t') \, dt'$ the steady-state value of the generic quantity a, a fluctuation of a near $a_0(\mathbf{x})$ (occurring on small time scales and short spatial scales) and the time average of a, respectively, here and in the following. Throughout our discussion, we assume that the system evolves towards a relaxed configuration where $\overline{a}$ depends on time no more. In order to assure that the definition of $\overline{a}$ remains meaningful during the relaxation, we must take care that the auxiliary time t_0 in the definition of $\overline{a}$ is much longer than the typical time scale of small-scale fluctuations but remains much shorter than the characteristic time scale of relaxation. In particular, $\overline{a} = a_0(\mathbf{x})$ and $\overline{a_1}(\mathbf{x}, t) = 0$: steady-state values are time averages which smear out fluctuations. Finally, if $|a| < \infty$ at all times then the definition of $\overline{a}$ implies $\overline{\frac{\partial a}{\partial t}} = 0$.

---

[89] On one side, $\eta = \frac{R_L}{R_L + R_S}$ increases monotonically from its minimum value 0 towards its maximum value 1 as $\frac{R_S}{R_L}$ increases from 0 towards $+\infty$. On the other side, $P_L = R_L I^2$ where $I = \frac{V}{R_S + R_l}$ is the electric current. Straightforward algebra shows that $P_L = \text{max}$ for $R_L = R_S$. In this case $\eta = \frac{1}{2}$.

[90] Accordingly, the connection between MPP and MEPP remains rather obscure [72].

Fig. 5.6 Lay-out of the experiment described in Ref. [116]

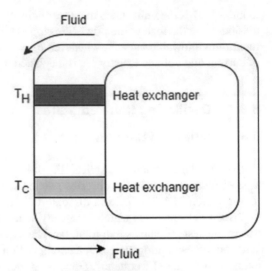

Fig. 5.7 Heat flows in the experiment described in Ref. [116]

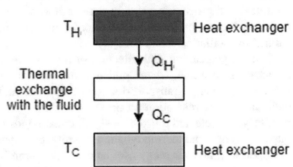

5.8.2 Biwa et al.'s Experiment

Here we describe (Fig. 5.6) the experiment of Biwa et al.'s [116]. Two heat exchangers at temperatures T_H and $T_C < T_H$ exchange an amount Q_H and Q_C of heat per unit time, respectively, with a fluid in a closed circuit (see the flows of heat in Fig. 5.7). Heat exchange between the fluid and the heat exchangers occurs through the walls of the latter; moreover, $\frac{d_i S}{dt} = 0$ in both heat exchangers. The balance of entropy of the system made by the two heat exchangers reads:

$$\Delta S = \frac{Q_C}{T_C} - \frac{Q_H}{T_H} = Q_H \left(\frac{1}{T_C} - \frac{1}{T_H} \right) - \frac{\Delta I}{T_C}$$

where $\Delta I \equiv Q_H - Q_C$ and ΔS is the net amount of entropy that comes out of the fluid into the heat exchangers per unit time, i.e. (-1) times the net amount of entropy $\oint \frac{\mathbf{q}}{T} \cdot d\mathbf{a}$ which comes out of the heat exchangers per unit time (where the domain of integration is the union of the boundary surfaces of the two heat exchangers)

$$\Delta S = - \oint \frac{\mathbf{q}}{T} \cdot d\mathbf{a}$$

The quantities ΔI, T_H and T_C are measured. The control parameter is Q_H. Given these quantities, the above relationships provide us with the value of ΔS.

Experiments show that

1. a threshold Q_{thr} exists, such that:

   - if $Q_H < Q_{thr}$ then a steady state is observed where $\Delta I = 0$;
   - if $Q_H > Q_{thr}$ then spontaneous oscillations occur at a well-defined frequency, and ΔI increases with increasing Q_H ;

2. between oscillating and steady states, the system selects the configuration which minimizes the value of ΔS, i.e. maximizes the value of $\oint \frac{\mathbf{q}}{T} \cdot d\mathbf{a}$ (see Sect. 6.2.13 for further discussion). Just like in the fluid discussed in the large-Ra problem of [86] in Sect. 5.6.7, the system the heat flux $\mathbf{q}$ comes from (i.e., the two heat exchangers) has negligible entropy production in the bulk. If oscillations occur, these quantities are replaced by their time averages over many oscillations. In other words, the maximization of the net amount $\overline{\oint \frac{\mathbf{q}}{T} \cdot d\mathbf{a}}$ of entropy which comes out of the system of heat exchangers per unit time acts as a selection rule between a stable steady state and a stable, oscillating state;
3. even if different modes of oscillation at different frequencies are possible, the system selects always the mode which minimizes ΔS.

This maximization above seems to allow a thermodynamic description of the onset of spontaneous, periodic oscillations. We have found such oscillations—and the time-averaging on many periods of oscillation—when dealing e.g. with seasonal variations in [96]. As such, Biwa et al.'s result deserves further discussion.

5.8.3 Meija et al.'s Experiment

In Biwa et al.'s experiment [116] of Sect. 5.8.2, $\frac{d_i S}{dt} = 0$ in the system of the two heat exchangers, and the selection rule involves the amount of entropy coming out of this system. Nothing is said about the fluid. In particular, $\frac{d_i S}{dt}$ may be > 0 inside the fluid. We are therefore allowed to imagine that $\frac{d_i S}{dt}$ is due to some exothermal chemical reaction inside the fluid. Moreover, nothing is said about the physical origin of Q_H. Thus, we may modify the layout of Biwa et al.'s experiment as follows.

Let a flame be in contact with a wall, and let the system made of the flame and the wall be initially in steady state. 'Premixed' combustion[91] consists of the chemical reactions within a flow of a premixture of reacting species. In steady state, the amount of heat released by combustion inside the flame per unit time is constant and is exactly equal to the amount of heat that goes from the fluid into the wall. We may easily keep this amount of heat at a constant value by controlling the mass flows of fuel and

---

[91] Familiar, e.g. to designers of gas turbine burners.

oxidizer. Moreover, we may keep the temperature of the wall at a constant value by a suitable cooling system. For example, we may think of a system of pipes behind the wall, which carry cold water which in turn cools the wall; we control the temperature of the wall by controlling the mass flow of the cold water in the pipes: the larger this flow, the larger the amount of heat carried away by the water, the cooler the wall.

Experiments [117] show that if we raise the mass flow of cold water above a threshold while keeping the mass flows of fuel and oxidizer at a constant value, then the flame starts producing sound in the gaseous mixture of oxidizer and fuel. The acoustic oscillations are due to oscillations in the heat release of the flame (the wall stands still). Under an oscillating thermal load, a compressible fluid undergoes oscillations of pressure, which propagate as sound. Not surprisingly, this sound is dubbed 'thermoacoustic' in the literature.

This result seems to confirm Biwa et al.'s findings [116]. Here Q_H is the amount of heat released by combustion per unit time. Since its physical origin is not relevant, the flame itself plays the role of H in Biwa et al's experiment, while the wall plays the role of C, and its temperature plays the role of T_C. Now, however, it is T_C which plays the role of the control parameter, rather than Q_H. When we raise the mass flow of the cool water in the pipes, we lower T_C. By doing so, however, we may reasonably assume that we leave both Q_H and T_H unchanged; the former and the latter, because the mass flows of fuel and oxidizer are kept constant, and because the complicated details of combustion inside the flame depend only weakly on the cold water in the pipes, respectively. In steady state $\Delta I = 0$; but as we—i.e., the external world—lower T_C we raise ΔS, and as a response the system tries to compensate the impact of the external world (by Le Châtelier's principle) and to minimize ΔS by raising ΔI above zero, i.e. by triggering thermoacoustic oscillations.

Alternatively, we may say that by lowering the temperature of the wall while keeping the heat released by combustion per unit time constant we push a larger amount of entropy per unit time from the fluid into the wall, as the amount of heat entering the wall per unit time is exactly equal to the heat released by combustion per unit time. Now, the wall acts as the cool heat exchanger in Biwa et al's experiment, and $\frac{d_i S}{dt}$ vanishes inside the wall (at least if it is so thin that $\nabla T = 0$ inside the wall, so that the contribution of heat conduction to $\frac{d_i S}{dt}$ is negligible). Then, according to Biwa et al's results, the system tends to maximize the amount of entropy coming out from the wall into the fluid per unit time, hence hindering the growth of the amount of entropy coming from the fluid into the wall per unit time imposed by the external world. The system does so by triggering thermoacoustic oscillations, i.e. by raising ΔI above zero.[92]

---

[92] See Sect. 6.2.13 for further discussion.

5.8.4   Hong et al.'s Experiment

The lesson of Biwa et al.'s and Meija et al.'s experiments in Sects. 5.8.2 and 5.8.3, respectively, is that a selection rule exists—for systems with vanishing $\frac{d_iS}{dt}$ at least—which is related to the amount of entropy coming out from the system per unit time: the system seems to select the state (steady or oscillating) which maximizes this amount. But then there is no need to raise the flow of cool water in Meija et al.'s experiment: if such selection rule actually works, in order to trigger thermoacoustic oscillations starting from a steady state it is enough to replace the material the wall is made of with another material with larger thermal conductivity. This way, indeed, we change nothing but the flow of heat from the fluid into the wall, thus obtaining the same effect observed in Meija et al.'s experiment when raising the mass flow of cool water.

Experiments [118] confirm this prediction, although in a somehow reversed way: thermoacoustic oscillations can be prevented or significantly delayed by using a material with low thermal conductivity (ceramics) rather than a material with high thermal conductivity (steel) at the flame anchoring region.[93]

5.8.5   Flame Quenching

A non-trivial reinterpretation of the results discussed in Sect. 5.8.4 is available. Let us focus our attention on a laminar (i.e., no turbulent) flame in premixed combustion which is in thermal contact with a wall. This contact tends to cool the flame; conduction carries heat towards the wall as far as we neglect radiation and convection. We assume that Fourier's law (Sect. 5.2.1) holds. As for $\frac{d_iS}{dt}$, it vanishes inside the wall and differs from zero in the flame. We have suggested in Sect. 5.8.4 that the system made of the flame and the wall selects the state (steady or oscillating) which maximizes the amount of entropy coming out per unit time from the $\frac{d_iS}{dt} = 0$ region, i.e. from the wall towards the flame. This is equivalent to say that the system minimizes the amount of entropy coming out per unit time from the flame towards the wall. Now, let us start with a stable, steady state, and let us try to destabilize it. For this purpose, we may act on the wall, either by cooling it (as in Ref. [117]) or by replacing it with materials with higher thermal conductivity (as in Ref. [118]). But then, let us leave the wall unaffected, and raise rather the thermal conductivity χ of the fluid. In all cases we raise the heat flux from the flame towards the wall, thus driving away the system from its initial, stable state. Sooner or later, if we keep on raising χ then the system shall become unstable. The thermal conductivity of the fluid affects the physics of the flame-wall interaction through the dimensionless 'Péclet number' $Pe \equiv \frac{S_L D}{\alpha}$, where S_L, $\alpha \equiv \frac{\chi}{\rho c_p}$ and D are the laminar flame speed,[94] the thermal

---

[93] See Sect. 6.2.13 for further discussion.

[94] I.e. the propagation velocity of the normal flame front relative to the unburnt mixture, which depends on the stoichiometry, etc.

diffusivity (computed for the unburnt mixture) and the typical distance between the flame and the wall, respectively [119]. Thus, raising χ is equivalent to lower Pe. In other words, stability of the steady state of a system made of a laminar premixed flame and a wall in thermal contact with it requires that Pe is larger than a lower bound, say Pe_{thr}; if $Pe < Pe_{thr}$ then stability is lost. In practice, experimenters lower D in order to modify Pe, while leaving all other things unaffected; the existence of a lower bound Pe_{thr} on Pe implies therefore that a lower bound D_{thr} (the 'quenching distance') such that no stable flame exists for $D < D_{thr}$. Intuitively, if the flame is too near to the wall, then the wall effectively cools the flame and the latter gets quenched. Experiments confirm these predictions [119]. Finally, it can be shown that $\int \mathbf{q} \cdot \nabla \left(\frac{1}{T}\right) d\mathbf{x} \propto \frac{1}{Pe^2}$ [120]; the existence of a lower bound Pe_{thr} means therefore that the maximization of $\int \mathbf{q} \cdot \nabla \left(\frac{1}{T}\right) d\mathbf{x}$ observed in problems of heat conduction in gases—see Sect. 5.6.6—is actually constrained.

5.8.6 Holyst et al.'s Simulations

When dealing with the transition from conduction to convection in a slab of an externally heated perfect gas, Holyst et al. have shown [85] that the system spontaneously selects the configuration which minimizes the content of internal energy in the gas for given heating power. This result generalizes the outcome of similar investigations on conduction (Sect. 5.6.6) to convection; in the latter case, of course, the time average of internal energy taken on many typical rotation periods of the convective cells is to be taken.

5.8.7 Rauschenbach's Hypothesis

Once again, let us invoke the energy balance of a fluid (Sect. 4.2.8):

$$\frac{\partial}{\partial t} \left(\rho \frac{|\mathbf{v}|^2}{2} + \rho u + \frac{1}{2\mu_0} |\mathbf{B}|^2 + \frac{1}{2} \varepsilon_0 |\mathbf{E}|^2 + \rho \phi_g \right) +$$

$$+ \nabla \cdot \left[\rho \mathbf{v} \left(\frac{|\mathbf{v}|^2}{2} + h \right) - \mathbf{v} \cdot \sigma' + \frac{\mathbf{E} \wedge \mathbf{B}}{\mu_0} + \mathbf{q} \right] = 0$$

and apply it to the fluid in Biwa et al's experiment [116] of Sect. 5.8.2. We recall that this energy balance holds regardless of the actual value and the detailed structure of σ in the fluid. We take the time average of both sides so that the contribution of the time derivative on the L.H.S. vanishes. We also neglect both viscosity and electromagnetic fields, so that the second and the third term in the argument of the divergence vanish identically. We perform a volume integration of both sides on the volume embedding the fluid. Gauss' theorem of divergence gives

$$\Delta I = \oint d\mathbf{a} \cdot \overline{\left[\rho \mathbf{v} \left(\frac{|\mathbf{v}|^2}{2} + h \right) \right]}$$

We take $\mathbf{v}_0 = 0$ in the frame of reference of the heat exchangers. In steady state, both sides vanish. If oscillations occur, then the R.H.S. is equal to $\oint d\mathbf{a} \cdot \overline{\left[p_1 \mathbf{v}_1 + (\rho u)_1 \, \mathbf{v}_1 \right]}$ (where we have invoked the relationship $h = u + \frac{p}{\rho}$ and we have neglected higher-order terms). If the fluid is a perfect gas, then $(\rho u)_1 \propto p_1$, so that $\Delta I \propto \oint d\mathbf{a} \cdot \overline{p_1 \mathbf{v}_1}$. We have seen that the system tends to maximize ΔI. Then, the relaxed state tends to maximize $\oint d\mathbf{a} \cdot \overline{p_1 \mathbf{v}_1}$.

When dealing with thermoacoustic oscillations related to combustion, this result is in agreement with the following hypothesis, formulated as a rule of thumb by a preeminent Soviet rocket engineer in the Sixties, Rauschenbach (see Chap. 9, Sect. 45 of Ref. [121]): *the development of vibrations in an oscillating combustion system evolves towards those relationships involving amplitudes and phases which maximize the amount of acoustic energy irradiated from the region where combustion occurs.* As we are going to show below, $p_1 \mathbf{v}_1$ is precisely the energy flux of acoustic waves propagating across a fluid at rest.[95]

5.8.8 Rayleigh's Criterion of Thermoacoustics

Let us assume that the amplitude of the acoustic waves is not too large, so that a linearized description of these waves is suitable. We find a new way to write down Rauschenbach's hypothesis as well as a new interpretation of its physical meaning provided that we invoke the linearized energy balance of acoustic waves (Eq. 8.128 of [122]) in an inviscid, non-gravitating, compressible mixture of perfect gases with the same specific heat per unit volume and the same molecular weight, with $\mathbf{v}_0 = 0$, $\nabla s_0 = 0$ and a source of heat which provides the fluid with an amount P_h of heat per unit time and volume:

$$\frac{\partial}{\partial t} \left(\frac{1}{2} \rho_0 |\mathbf{v}_1|^2 + \frac{1}{2} \frac{p_1^2}{\rho_0 c_s^2} \right) + \nabla \cdot (p_1 \mathbf{v}_1) = \left(\frac{\gamma - 1}{\gamma p_0} \right) p_1 P_{h1}$$

whereby c_s we mean the speed of sound in the unperturbed fluid, $\gamma \equiv \frac{c_p}{c_v}$, c_p and c_v are the specific heat per unit mass at constant pressure and constant volume, respectively. The bracketed term under the time derivative on the L.H.S. is the energy density of the acoustic wave. The term acted upon by the divergence operator on the L.H.S. is the amount of mechanical work done per unit volume and time by the acoustic wave. The R.H.S. is the source term; as stated above, any oscillation in the heating process (regardless of the detailed nature of the latter) induces a corresponding oscillation of the mass density of the fluid, hence of its pressure, which in turn propagates.

---

[95] See Sect. 6.2.13 for further discussion.

Now, let us perform a volume integration of both sides on the volume embedding the fluid. Then, let us take the time average. We are left with [122–125]:

$$\oint d\mathbf{a} \cdot \overline{p_1 \mathbf{v}_1} = \left(\frac{\gamma - 1}{\gamma p_0} \right) \int d\mathbf{x} \overline{p_1 P_{h1}}$$

so that Rauschenbach's hypothesis reduces to maximization of $\int d\mathbf{x} \overline{p_1 P_{h1}}$ in a relaxed, oscillating state.[96] Such maximization requires that the value of the so-called 'Rayleigh index' $\overline{p_1 P_{h1}}$ is as large as possible everywhere. If P_{h1} oscillates in time with a given frequency, however, p_1 does the same everywhere with the same frequency, as we assumed the acoustics to be linear. Then, Rayleigh's index gets maximized only if the *phases* of P_{h1} and p_1 are equal. This requirement is trivially satisfied in steady state, where the amplitudes of all oscillations vanish identically, but is far from trivial if oscillations occur. In other words, if sound is spontaneously produced in a fluid at the expense of some source of heat then P_{h1} must be maximum when p_1 is maximum.[97]

If we recall that sound is but an ordered form of energy, while heat is a form of disordered energy, then we find a simple analogy between this result and the familiar Brayton cycle of thermodynamics. In Brayton cycle, indeed, there are two isentropic transformations and two isobaric transformations, say at pressure p_A and $p_B > p_A$. The system achieves maximum efficiency of transformation of heat into mechanical energy when heat is supplied to the system during the isobaric transformation at the highest pressure $p = p_B$ and is subtracted from the system during the isobaric transformation at the lowest pressure $p = p_A$. If a sound wave propagates across an inviscid fluid and we may neglect the dissipation related to this propagation, then we may describe it as a cyclic transformation from a (possibly quite short) state at maximum compression of the fluid to a state at minimum compression of the fluid, and all the way around again et again. Then, the Brayton cycle teaches us that the most efficient way to transform heat into mechanical energy (in this case, the energy of the wave itself) is to provide heat at the time of maximum pressure and to subtract it at the time of minimum pressure.

Back in the XIX century, Rayleigh has summarized this fact in a seminal lecture [126] describing an experiment of thermoacoustics with air as the working fluid. The work contains the by now well-known words, now referred to 'Rayleigh's criterion of thermoacoustics' in the literature *If heat be given to the air at the moment of greatest condensation, or be taken from it at the moment of greatest rarefaction, the vibration is encouraged. On the other hand, if heat be given at the moment of greatest rarefaction, or abstracted at the moment of greatest condensation, the vibration is discouraged.*

The origin of the heat is definitely not relevant. Our comparison above between Biwa et al.'s experiment [116] and Meija et al.'s experiment [117] of Sects. 5.8.2 and

---

[96] Everything vanishes in steady state as $P_{h1} \equiv 0$.

[97] The quantity P_{h1} is usually dubbed q_1 in engineering textbooks.

5.8.3, respectively, shows that the heat may come either from combustion or from the wall (i.e. the external world), indifferently. This is to say, P_h is just the density of Q_H. This is not just a formal relationship; we are going to invoke it in the following.

In order to clarify this point, we recall that the requirement that $\oint d\mathbf{a} \cdot \overline{p_1 \mathbf{v}_1}$ is maximized in relaxed state, i.e. Rauschenbach's hypothesis [121], is something new. It follows from some form of MEPP concerning the entropy flow across the boundaries; it is also equivalent to require that if the relaxed state exists among all possible solutions of the thermoacoustics in our fluid, then it has either no oscillations at all or vanishing phase between the oscillation of pressure and the oscillation of heat. The fact that the well-known results on the Brayton cycle confirm this requirement strongly suggests that some form of MEPP concerning the entropy flow across the boundaries actually holds.

5.8.9 Rijke's Tube

Historically, the first application of Rayleigh's criterion was Rijke's tube. In 1859, Rijke discovered a way of using heat to sustain a sound in a cylindrical tube open at both ends [127]. He used a glass tube, about 0.8 m long and 3.5 cm in diameter. Inside it, about 20 cm from one end, he placed a disc of wire gauze; see Fig. 5.8. Gauze friction with the walls of the tube is sufficient to keep the gauze in position. With the tube vertical and the gauze in the lower half, Rijke heated the gauze with a flame until it was glowing red hot. Upon removing the flame, he obtained a loud sound from the tube which lasted until the gauze cooled down (about 10 s).

Instead of heating the gauze with a flame, Rijke also tried electrical heating. Making the gauze with electrical resistance wire causes it to glow red when a sufficiently large current is passed. With the heat being continuously supplied, the sound is also continuous and rather loud. Rijke seems to have received complaints from his university colleagues because he reports that the sound could be easily heard three rooms away from his laboratory. The electrical power required to achieve this is about 1 KW.

Later, Rayleigh reproduced Rijke's experiments [126]. He made use of two layers of gauze made from iron wire inserted about a quarter of the way up the tube. The extra gauze is to retain more heat, which makes the sound longer lasting. The sound comes from a standing wave whose wavelength is about twice the length of the tube, giving the fundamental frequency. The flow of air past the gauze is a combination of two motions. At the initial time, only upwards convection occurs. If a pressure perturbation occurs such that a pressure peak occurs right at the centre, then it pushes away the fluid from the centre. Accordingly, the central pressure peak lowers the upwards, convection-driven vertical motion of the fluid across the lower half of the tube where the gauze mesh is located. As a result, a longer time is available for heat exchange between mesh and air, the latter gets heated better and the pressure peak is enforced. As time goes by, the pressure peak flattens more and more, and even less fresh air gets heated.

Fig. 5.8 A simple
construction of the Rijke
tube with a gauze in the
lower half of a vertical metal
pipe. Here a wire mesh
replaces the gauze. The tube
is suspended over a Bunsen
burner. The latter heats the
wire mesh

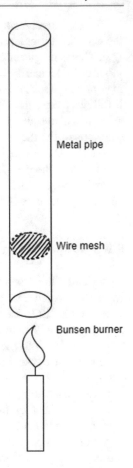

Metal pipe

Wire mesh

Bunsen burner

Alternatively, things may as well go all the other way around: there is a pressure drop on the centre, air gets sucked, there is less time available for heat exchange and air heating is less effective, so that less heat is added by the mesh to the air, and the central pressure is further depressed. Heat exchange between air and gauze plays a crucial role. There is a uniform upwards motion of the air due to a convection current resulting from the gauze heating up the air. Superimposed on this is the motion due to the sound wave. For half the vibration cycle, the air flows into the tube from both ends until the pressure reaches a maximum. During the other half cycle, the flow of air is outwards until the minimum pressure is reached. All air flowing past the gauze is heated to the temperature of the gauze and any transfer of heat to the air will increase its temperature and its pressure, according to the gas law.

Admittedly, as the air flows upwards past the gauze most of it will already be hot because it has just come downwards past the gauze during the previous half cycle. However, just before the pressure maximum, any small quantity of cool air which comes into contact with the gauze gets heated, and its pressure is increased. This increases the pressure maximum, so reinforcing the vibration.

During the other half cycle, when the pressure is decreasing, the air above the gauze is forced downwards past the gauze again. Since it is already hot, no pressure change due to the gauze takes place, since there is no transfer of heat. The sound wave is therefore reinforced once every vibration cycle and it quickly builds up to a large amplitude. This explains why there is no sound when the flame is heating the gauze. All air flowing through the tube is heated by the flame, so when it reaches the gauze, it is already hot and no pressure increase takes place.

When the gauze is in the upper half of the tube, there is no sound. In this case, the cool air brought in from the bottom by the convection current reaches the gauze towards the end of the outward vibration movement. This is immediately before the pressure minimum, so a sudden increase in pressure due to the heat transfer tends to cancel out the sound wave instead of reinforcing it.

The position of the gauze in the tube is not critical as long as it is in the lower half. To work out its best position, there are two things to consider. Most heat will be transferred to a small mass element of air where the acceleration of this small mass element along the tube is a minimum, as heat exchange may occur in this case before the small mass element gets shifted significantly away from the pressure of the wave (for given initial position and velocity of the small mass element). In turn, this occurs where the pressure gradient of the wave is a minimum, i.e. near the end of the tube. However, the effect of increasing the pressure (in thermodynamic jargon: the efficiency of conversion of heat into mechanical work) is greatest where there is the strongest heating, i.e. in the middle of the tube. There is therefore no perfect position for the gauze. A good compromise is obtained by placing the gauze midway between these two positions, i.e. one quarter of the way in from the bottom end.

Implicitly, we have postulated that the chemical nature of the fluid is actually not relevant to the experiment. Indeed, experiments by Chladni and Faraday successfully replicated Rijke's results with gases of different chemical compositions, including hydrogen.

We stress the point that the spontaneous production of sound depends on the exact value of the oscillation period only weakly. In fact, two Rijke's tubes with the same geometry and basically the same fundamental acoustic eigenfrequency may produce dramatically different amounts of acoustic power if the gauze is located just below or above the half-length of the tube. Accordingly, the problem of computing the acoustic spectrum and the problem of predicting the actual production of sound are quite different.

In spite of its apparent simplicity, Rijke's tube provides us with a lot of useful information. First of all, general consensus underlines the crucial role played by the relative phase of oscillations of pressure and heat release in Rayleigh's criterion: it is this phase that explains why sound is heard only if the gauze is located at a suitable position inside the tube [126,128]. It turns out that both the occurrence of acoustic oscillations (sometimes referred to as 'humming' in the literature) and the value of $Re\{\omega\}$ ($\omega = 2\pi$ times the oscillation frequency) depend on this phase [129]. As we are going to see, however, this relative phase is far from being the only quantity which is relevant to the triggering of the oscillations.

Secondly, the original Rijke's observation is that sound is spontaneously produced while the gauze cools down. In modern language, Rijke' tube is an example of the so-called dissipative structures[98] which arise from relaxation (the cooling of the gauze), i.e. from entropy growth. And where entropy is at stake, thermodynamics has a say.[99]

Thirdly, Rayleigh's original words *If heat be given [...] or be taken* quoted above [126] refer to exchanged heat, not to heat release -let alone combustion. This point is stressed again and again in many passages of Rayleigh's work. For example, when reviewing an experiment, the author writes [126]:

> *in order that the whole effect of heat be on the side of encouragement, it is necessary that previous to condensation the air should pass not merely towards a hotter part of the tube, but towards a part of the tube which is hotter than the air will be when it arrives there*

Here Rayleigh's emphasis on the words *hotter than* suggests that it was the difference in temperature, hence the heat flow, which drew his attention. Rather than the detailed mechanism of gauze heating, it turns out that the heat exchange between the gauze and the surrounding environment is crucial to the onset of humming in Rijke's tube. For instance, sound is heard even when the gauze is cooler than the environment [130]. More recently, spontaneous growth of acoustic (*Taconis'*) oscillations in pipes that are partially filled with liquid helium has been reported. See, e.g. Ref. [131].

Moreover, a louder and longer sound is heard when many gauze discs are inserted inside the tube beyond the heated gauze disc, thus delaying air motion and leaving more time for heat transfer to occur. In Rijke's own words [127][100] :

> *when, instead of a single disc, several were placed in the tube, the sound lasted longer [...] that is because the presence of a greater number of discs, by diminishing the rapidity of the air current diminishes also the rapidity of cooling of the first disc*

This *rapidity of the air current* acts therefore as a further crucial parameter for sound generation, beyond the relative phase quoted above. This *rapidity* is just the relative velocity of the air and the hot gauze, as the hot gauze is at rest in the laboratory reference system. This velocity rules the heat exchange between the fluid and the external world, which plays a fundamental role here too.

Finally, we have seen that Rijke heard the sound fading away after a while in his first experiments. In order to obtain a sustained sound, Rijke heated the gauze with DC current. Sound generation starts, provided that the electric current is large enough (i.e. that the Joule power dissipated in the gauze is not too small). Moreover, with the heat being continuously supplied, the sound is also continuous. Historically, it was precisely the self-sustaining character of the acoustic oscillations heard by Rijke

---

[98] See Sect. 6.1.3.

[99] see Sect. 6.2.13.

[100] In his original German paper [132], Rijke refers to the *rapidity* with the term *Schnelligkeit* only when it comes to the *air current*. *Schnelligkeit* means *velocity*.

that triggered the attention of both Rayleigh and modern manufacturers of humming-affected combustors.[101] This fact, regardless of the detailed origin of the heat flowing either from the external world to the fluid or all the other way around, suggests that both the relative phase of pressure and heat release oscillations, the *rapidity of the air current* and this heat flow play a crucial role when it comes to sustaining the generation of sound.

5.8.10 Sondhauss' Tube

Rijke's tube acts as a 'half-wave resonator', i.e. its length is equal to one half of the wavelength of the pressure perturbation. Rijke's tube operates with both ends open. However, a tube with one end closed will also generate sound from heat, if the closed end is very hot. Such a device is called a *Sondhauss' tube* [133] after Sondhauss who described it in 1850—see Fig. 5.9.

The phenomenon was first observed by glassblowers. Sondhauss' tube operates in a way that is basically similar to the Rijke's tube. Initially, air moves towards the hot, closed end of the tube, where it is heated, so that the pressure at that end increases. The hot, higher-pressure air then flows from the closed end towards the cooler, open end of the tube. The air transfers its heat to the tube and cools. The air surges slightly beyond the open end of the tube, briefly compressing the atmosphere. The compression propagates through the atmosphere as a sound wave. The atmosphere then pushes the air back into the tube, and the cycle repeats. Sondhauss' tube acts as a 'quarter-wave resonator', i.e. its length is equal to one fourth of the wavelength of the pressure perturbation.

Unlike Rijke's tube, Sondhauss's tube does not require a steady flow of air through it. All the same, we anticipate here for the sake of future reference that heat exchange plays a crucial role in both Rijke's tube and Sondhauss' tube. As for the latter, it was discovered that placing a porous heater -as well as a *stack*, i.e. a porous plug- in the tube greatly increases the oscillation amplitude in both Rijke and Sondhauss' devices.

---

[101] In Rayleigh's words [126]

> *When a piece of fine metallic gauze, stretching across the lower part of a tube open at both ends and held vertically, is heated by a gas flame placed under it, a sound of considerable power, and lasting for several seconds, is observed almost immediately after the removal of the flame. [...] the generation of sound was found by Rijke to be closely connected with the formation of a through draught, which impinges upon the heated gauze. In this form of the experiment the heat is soon abstracted, and then the sound ceases; but by keeping the gauze hot by the current from a powerful galvanic battery, Rijke was able to obtain the prolongation of the sound for an indefinite period. In any case from the point of view of the lecture the sound is to be regarded as a maintained sound.*

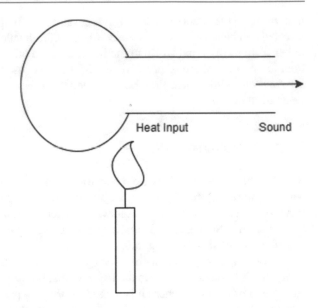

Fig. 5.9 Sondhauss'
tube—from Ref. [133]

These stacks play the same role as discs in the Rijke's tube discussed above. Today, stacks are suitably designed in order to raise heat exchange between the working fluid and the material walls in modern thermoacoustic devices [133].[102]

5.8.11 Welander's Loop

We have discussed the isentropic case so far. In the general $\nabla s_0 \neq 0$ case, the following generalization of the expression for $\oint d\mathbf{a} \cdot \overline{p_1 \mathbf{v}_1}$ holds [134]:

$$\oint d\mathbf{a} \cdot \overline{p_1 \mathbf{v}_1} = \int d\mathbf{x} \overline{\left(\frac{T_1 P_{h1}}{T_0} - s_1 \mathbf{v}_1 \cdot \frac{p_0}{r c_p} \nabla s_0 \right)} \quad ; \quad r = c_p \frac{\gamma - 1}{\gamma} \quad ; \quad \gamma = \frac{c_p}{c_v}$$

(where c_v is the specific heat at constant volume per unit mass) so that Rauschenbach's hypothesis reduces to maximization of $\int d\mathbf{x} \overline{p_1 \mathbf{v}_1}$ in a relaxed, oscillating state (in steady state everything vanishes as $P_{h1} \equiv 0$). Such maximization requires that the value of $\overline{\frac{T_1 P_{h1}}{T_0} - s_1 \mathbf{v}_1 \cdot \frac{p_0}{r c_p} \nabla s_0}$ is as large as possible everywhere.

This requirement undergoes dramatic simplification in the limit of subsonic motion. In this case, indeed, we may assume the speed of sound c_s to be infinitely large, i.e. the system to be basically incompressible, as the larger the adiabatic compressibility $\left(\frac{\partial p}{\partial v} \right)_{s, c_1, \dots}$, the lower c_s. But then, we recall that LTE holds

---

[102] See Sect. 6.2.13 for further discussion.

everywhere at all times and invoke[103] the thermodynamic identity $c_p = c_v - T\left[\left(\frac{\partial v}{\partial T}\right)_{p,c_1,...}\right]^2 \left[\left(\frac{\partial v}{\partial p}\right)_{T,c_1,...}\right]^{-1}$ where $0 < \left[\left(\frac{\partial v}{\partial p}\right)_{T,c_1,...}\right]^{-1} = \left(\frac{\partial p}{\partial v}\right)_{T,c_1,...} < \left(\frac{\partial p}{\partial v}\right)_{s,c_1,...}$ because of Le Châtelier's principle. If $c_s \to \infty$, then $\left(\frac{\partial p}{\partial v}\right)_{s,c_1,...} \to 0$, $\left(\frac{\partial p}{\partial v}\right)_{T,c_1,...} \to 0, c_p \to c_v, \gamma \to 1$ and $\frac{p_0}{r c_p} = \frac{p_0 r}{r^2 c_p} = \frac{p_0}{r^2}\frac{\gamma-1}{\gamma} \to 0$. Thus, Rauschenbach's hypothesis requires that $\frac{T_1 P_{h1}}{T_0}$ is as large as possible everywhere—i.e. that T_1 and P_{h1} are in phase—whenever an oscillation is sustained in the system.

Given the link of P_h and Q_H discussed above, even if there is no internal source of heat in the system, Rauschenbach's hypothesis still requires that the perturbation of the amount of heat supplied by the external world to a unit volume of the system oscillates in phase with the oscillation of temperature, if an oscillation is to be sustained.[104]

Our result is in agreement with the outcome of Welander's discussion of stability of a differentially heated fluid loop [135]. This discussion is relevant e.g. to the problem of 'thermohaline circulation', i.e. the part of the large-scale ocean circulation that is driven by density gradients.[105] Welander solves the equations of motion of an incompressible fluid in an external, uniform gravitational field and with given thermal conductivity and given viscosity (which possibly stands for the contributions to the friction of both the fluid bulk and the wall). The fluid flows in a closed loop (Fig. 5.10).

The loop is made of two vertical pipes with two short horizontal connections at the top and the bottom. The vertical pipes are thermally insulated. In contrast, the wall of the bottom and the top connection is kept at a constant temperature T_c and $T_h > T_c$, respectively. Heat transfer occurs by conduction between the wall and the fluid at the top and the bottom horizontal connections. The larger the speed of the fluid, the less effective this heat transfer. (In Welander's loop, any region between two cross sections of the pipes may be utilized for writing down Rauschenbach's hypothesis).

If $T_h - T_c$ is large enough, then $Ra > Ra_{thr}$ and convection is triggered. In this case, the fluid moves at constant speed v_0 upwards along one pipe, then horizontally

---

[103] Here we refer to the results of Sect. 3.3 with $c_p = \left(\frac{\partial h}{\partial T}\right)_{p,c_1,...}$ and $c_v = \left(\frac{\partial u}{\partial T}\right)_{v,c_1,...}$.

[104] Admittedly, this result is only an order-of-estimate assessment. Strictly speaking, indeed, Rayleigh's treatment generalized in [134] holds for perfect gases only, where γ is a constant quantity depending on the microscopic nature of the gas, and does not depend on c_s. All the same, we stick to it as to a necessary condition for the stability of the relaxed oscillating state, because the neglected term describes in [134] the $\overline{p_1 v_1}$-lowering effect—in agreement with Le Châtelier's principle—of oscillations carrying a time-averaged amount $\propto \overline{v_1 s_1}$ of entropy per unit time towards (away from) regions of increasing (decreasing) entropy content, i.e. in the direction of positive (negative) ∇s_0. Now, should no internal heat of source (external or internal) and no heat flow be present, this neglected term would be able to sustain no oscillation indefinitely, as the system would be thermally isolated and ∇s_0 would eventually relax to zero.

[105] Which in turn may be driven by inhomogeneities in temperature, salinity, etc.

Fig. 5.10 Welander's loop.
The short horizontal
connections at the top and at
the bottom are kept at
constant temperature T_c and
$T_h > T_c$ respectively. The
vertical pipes are thermally
insulated—from Ref. [135]

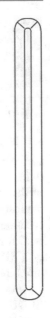

from this pipe to the other one, then downwards along the other pipe, then again
horizontally towards the first pipe, in a closed loop. With no loss of generality, we
may suppose that v_0 is uniform in each pipe (this is, e.g. required by the mass balance
in steady state if the cross section of the pipes is the same everywhere). In this case,
when we investigate the balance of entropy in each pipe we may get rid of v_0 (locally
at least) with the help of Galileian transformation of coordinates. Then, we may
apply the results of [134] and conclude that if this steady-state convection is unstable
against perturbations, and the system relaxes towards a new, stable, oscillating state,
then in the latter, the oscillation of temperature is exactly in phase with the oscillations
of the amount of heat supplied to the fluid by the external world through the top and
the bottom horizontal connections.

At a first glance, in a steady motion, viscous dissipation and thermal conductivity
oppose any change in velocity. Any increase (decrease) in velocity, for example,
would cause an increase (a decrease) in friction and a decrease (an increase) in total
buoyancy, because the heating per unit length of fluid column is weaker (stronger)
as the transfer of heat from the wall is less (more) effective due to shorter (longer)
time available for transfer.

One may therefore think that a steady motion is always stable. However, these two
restraining effects may not be in phase and an overshooting can then occur, eventually
producing growing oscillations. For example, if a small perturbation slightly raises T
in a small element of fluid as the latter passes the bottom connection then buoyancy
is built up and the small mass of fluid accelerates, so that when this small mass
reaches the top connection (where the wall temperature is lowest) the velocity is
maximum and the cooling effect due to the heat exchange with the top connection
wall is minimized. In contrast, when the small mass comes back to the bottom the

velocity achieves a minimum and the heating effect due to the heat exchange with the bottom connection wall is maximized.

Welander shows that no oscillation is triggered if either viscosity or thermal conductivity is too large. The result concerning viscosity is far from surprising, as viscosity raises dissipation of mechanical energy and damps the oscillation. As for the thermal conductivity, too large a value means that the amount of heat provided by the walls to the fluid in steady state is already quite large, and the triggering of oscillation is, therefore, more difficult. Remarkably, the smaller the viscosity the larger Ra, and the same is true when the thermal conductivity gets smaller. Thus, if $Ra - Ra_{thr}$ is not only positive but also larger than a given threshold, then the relaxed state achieved by convection ceases to be a steady state and becomes an oscillating state.

Welander's analytical treatment is a linear one. When the amplitude grows enough, no linear treatment holds anymore. The amplification of the perturbation stops as the above described mechanism ruling the amplification leads the relative phase between the oscillation in temperature and the oscillation in the amount of heat released per unit time by the external world through the walls towards the fluid down to its minimum allowable value, namely zero. The balance between such saturated amplification and the damping due to viscosity sets the amplitude of the oscillation in the relaxed state, in agreement with Rauschenberg's hypothesis of Sect. 5.8.7. In contrast with the problems of Meija et al. and Hong et al. in Sects. 5.8.3 and 5.8.4, respectively, however, here the oscillation affects the exchange of heat with the external world (through the modulation of the fluid velocity), rather than the production of heat inside the system. This is in agreement with the connection between P_h and Q_H established in Sect. 5.8.8.[106]

5.8.12 Eddington and the Cepheids

There is a striking example of the generality of the result of Biwa et al.'s experiment [116] of Sect. 5.8.2. This example allows us to identify a relaxed state with a maximum of $\oint \frac{\mathbf{q}}{T} \cdot d\mathbf{a}$. Accordingly, if we perturb a relaxed state ($\mathbf{q} \to \mathbf{q}_0 + \mathbf{q}_1$, $\left(\frac{1}{T}\right) \to \left(\frac{1}{T}\right)_0 + \left(\frac{1}{T}\right)_1$) then we lower the value of $\oint \frac{\mathbf{q}}{T} \cdot d\mathbf{a}$, i.e. we obtain $0 > \oint \frac{\mathbf{q}}{T} \cdot d\mathbf{a} - \left(\oint \frac{\mathbf{q}}{T} \cdot d\mathbf{a}\right)_0 = \oint \overline{\left(\frac{1}{T}\right)_1 \mathbf{q}_1} \cdot d\mathbf{a}$ as first-order terms vanish after time-averaging and higher-order terms are negligible if the amplitude of the perturbation is not too large. We have rewritten the selection rule as a necessary condition of stability as:.

$$\oint \overline{\left(\frac{1}{T}\right)_1 \mathbf{q}_1} \cdot d\mathbf{a} < 0$$

Whenever $p_1 \left(\frac{1}{T}\right)_1 < 0$ (as, e.g. in perfect gases) this condition is equivalent to $\oint \overline{p_1 \mathbf{q}_1} \cdot d\mathbf{a} > 0$. The physical meaning is explained by the following words of

---

[106] See Sect. 6.2.13 for further discussion.

Eddington [136]: *Consider the mode in which thermal dissipation acts in the case of a sound wave. The air is hottest at a point of maximum compression. If this heat leaks away, the compressed fluid loses some of its spring, and the expansion which follows has diminished energy—consequently, the waves decay. If, on the other hand, the air could be persuaded to lose heat at points where it was ratified and coolest, the ensuing compression would be assisted and the waves reinforced [...] Intermediately, a loss or gain of heat at a point of normal density neither dissipates nor increases the energy [...] the material relatively loses heat as it passes through its normal density expanding, and gains heat at the same stage contracting; but neither gains nor loses when most compressed or most ratified.* These three cases correspond to the occurrence of the operator $>$, $<$ and $=$ in the condition above, respectively. The third case (with $=$) is also well described by Rayleigh's words [126]: *if the air is at its normal density at the moment where the transfer of heat takes place, the vibration is neither encouraged nor discouraged.*[107]

Remarkably, however, Eddington was not concerned with the onset of spontaneous acoustic oscillations in a Rijke's tube. Rather, he investigated spontaneous oscillations of a Cepheid star. A Cepheid is a member of a class of pulsating variable stars. Typical Cepheids pulsate for periods of a few days to months, and their radii change by several million Km (30%) in the process—see e.g. Table III of [137]. Helium is the gas thought to be most active in Cepheid pulsation. The more helium is heated, the more ionized it becomes. But He^{++} is more opaque than He^+. At the dimmest part of a Cepheid's cycle, the ionized gas in the outer layers (envelope) of the star is opaque, and so is heated by the star's radiation, and therefore, it begins to expand. As it expands, it cools, and so becomes less ionized, and therefore, more transparent, allowing the radiation to escape. Then the expansion stops and reverses due to the star's own gravitational attraction. The process then repeats [138]. The mechanics of a Cepheid oscillation as a heat engine was proposed by Eddington in 1917, while Rayleigh worked with Rijke's tube in 1876. However, to the author's knowledge, Eddington was not aware of Rayleigh's observations.

As for the assumptions underlying our condition of stability, no heating (nuclear) process occurs in the envelope, and heat is mainly transferred across the layers from the nucleus of the star towards the external world, Moreover, it can be safely assumed that the time-averaged speed vanishes on the boundaries, provided at least that the outflow of matter has a negligible effect on the star mass balance during one oscillation. Indeed, radiation plays a crucial role in the oscillation, in contrast e.g. to oscillations involving flames. Just like in Welander's loop, the mechanism underlying the oscillation relies on the modulation of the heat exchange with the external world. Despite the huge differences between the two spontaneously oscillating systems, Rijke's tube and a Cepheid, Rayleigh and Eddington's arguments are quite similar. This is even more surprising when we recall that stars are ruled by gravity, which has been neglected when discussing Rayleigh's criterion. However, gravity

---

[107] See Sect. 6.2.13 for further discussion.

affects the energy balance (Sect. 4.2.7), not the entropy balance.[108] Now, entropy is a thermodynamic quantity. Thus, the common ground between Rijke's tube and the Cepheids is thermodynamics.

References

1. Richardson, I.W.: Biophys. J . **9**(2), 265–267 (1969)
2. Gage, D.H., Schiffer, M., Kline, S.J., Reynolds, W.C.: The non-existence of a general thermokinetic variational principle. In: Donnelly, R.J. (ed.), Non-Equilibrium Thermodynamics: Variational Techniques and Stability. University of Chicago Press, Chicago (1966)
3. Lavenda, B.: Thermodynamics of Irreversible Processes. McMillan, London (1979)
4. Andersson, N., Comer, G.L.: Living Rev. Relat. **24**, 3 (2021)
5. Kuznetsov, Y.A.: Elements of Applied Bifurcation Theory. Springer, New York (2004)
6. DeGroot, S.R., Mazur, P.: Non-Equilibrium Thermodynamics. North Holland, Amsterdam (1962)
7. Kondraputi, D., Prigogine, I.: Modern Thermodynamics. Wiley, New York (1998)
8. Jaynes, E.T.: Ann. Rev. Phys. Chem. **31**, 579 (1980)
9. Hinton, F.L., Hazeltine, R.D.: Rev. Mod. Phys. **48**, 239–308 (1976)
10. Strutt, J.W.: (Lord Rayleigh) Proceedings of the London mathematical society **s1–4**(1), 357–368 (1871)
11. Balescu, R.: Phys. Fluids B **3**, 564 (1991). https://doi.org/10.1063/1.859855
12. Di Vita, A.: J. Plasma Phys. **46**(3), 423–436 (1991)
13. Brusati, M., Di Vita, A.: J. Plasma Phys. **50**(2), 201–230 (1993)
14. Barbera, E.: Continuum Mech. Thermodyn. **11**, 327–330 (1999)
15. Gyarmati, I.: Non-equilibrium Thermodynamics. Springer, Berlin (1970)
16. Landau, L.D., Lifshitz, E.: Fluid Mechanics. Pergamon, Oxford (1960)
17. Landau, L.D., Lifshitz, E.: Electrodynamics of Continuous Media. Pergamon, Oxford (1960)
18. Hermann, F.: Eur. J. Phys. **7**, 130 (1986)
19. Montgomery, D., Phillips, L.: Phys. Rev. A **38**, 2953–2964 (1988)
20. Compton, K.T., Morse, P.M.: Phys. Rev. **30**, 305 (1927)
21. Steenbeck, M.: Z. Phys. **72**, 505 (1931). (In German)
22. Steenbeck, M.: Wissenschaftlichen Veroeffentlichungen aus den Siemens Werke **1**, 59 (1940). (in German)
23. Peters, T.: Z. Phys. **144**, 612–631 (1956). (in German)
24. Frost, S.L., Liebermann, R.W.: Proc. IEEE **59**, 474 (1971)
25. Lamb, H.: Hydrodynamics. Cambridge University Press, Cambridge (1906)
26. Rebhan, E.: Phys. Rev. A **32**, 581 (1985)
27. von Helmholtz, H., Wissenschaftliche Abhandlungen, Bd. 1, (1882) (in German). http://echo.mpiwg-berlin.mpg.de/ECHOdocuViewfull?url=/mpiwg/online/permanent/einstein_exhibition/sources/QWH2FNX8/index.meta&start=231&viewMode=images&pn=237&mode=texttool
28. Korteweg, D.J.: London, Edinburgh and Dublin Philosophical. J. Sci. **16**(98), 112–118 (1883)
29. Bertola, V., Cafaro, E.: Int. J. Heat Mass Transf. **51**, 1907–1912 (2008)
30. Thompson, D.W.: On Growth and Form. Cambridge, New York (1945)
31. Murray, C.D.: Proc. Natl. Acad. Sci. (Physiol.) **12**, 207–214 (1926)
32. Sun, T., Meakin, P., Jossang, T.: PRE **49**(6), 4865–4872 (1994)

---

[108] It is implicitly assumed that the internal pressure of the star is able to prevent gravitational collapse (Sect. 3.4.1).

33. Lorenz, R.D.: Earth system dynamics discussions 1–13 (2019). https://doi.org/10.5194/esd-2019-73
34. Liu, H.H.: Fluid flow in the subsurface. Springer Int. Publ. (2017). https://doi.org/10.1007/978-3-319-43449-0
35. Liu, H.H.: Chin. Sci. Bull. **59**(16), 1880–1884 (2014)
36. Botré, C., Lucarini, C., Memoli, A., D'Ascenzo, E.: Bioelectrochemistry and bioenergetics **8**, 201–212 (1981). (a section of J. Electroanal. Chem., Constit. **128** (1981))
37. Horne, W.C., Karamcheti, K.: Extrema Principles of Entropy Production and Energy Dissipation in Fluid Mechanics NASA Technical Memorandum 100992 (1998)
38. Xu, M.: Entransy dissipation theory and its application in heat transfer. In: Developments in Heat Transfer, (Edited by Marco Aurélio dos Santos Bernardes) InTechOpen (2011). (https://doi.org/10.5772/19573)
39. Chandrasekhar, S.: Hydrodynamic and Hydromagnetic Stability. Oxford University Press, New York (1961)
40. Busse, F.H.: J. Fluid Mech. **30**(4), 625–649 (1967)
41. Malkus, W.V.R.: J. Fluid Mech. **1**, 521 (1956)
42. Elsgolts, I.V.: Differential Equations and Variational Calculus. Mir, Moscow, URSS (1981)
43. Di Vita, A.: Phys. Rev. E **81**, 041137 (2010)
44. Reynolds, W.C., Tiederman, W.G.: J. Fluid Mech. **27**(2), 253–272 (1967)
45. Bejan, A.: J. Adv. Transp. **30**(2), 85–107 (1996)
46. Bejan, A.: J. Heat Mass Trans. **40**(4), 799–810 (1997)
47. Bejan, A., Lorente, S.: J. Appl. Phys. **100**, 041301 (2006)
48. Bejan, A., Ikegami, Y., Ledezda, G.A.: J. Heat Mass Trans. **41**(13), 1945–1954 (1998)
49. Niemeyer, L., Pietranero, L., Wiesmann, H.J.: PRL **52**(12), 1033–1036 (1984)
50. Cheng, X.T.: Critique of Constructal Theory. Cambridge Scholars Publishing, Cambridge (2019). (ISBN-13: 978-1-5275-3839-9)
51. Jones, C.: Non-designer design. https://www.skeptic.com/eskeptic/13-03-27/, 2013 the 27th
52. Evolution News.: Good Grief. No, Airplanes Do Not "Evolve" by Natural Law. 2014 July the 25th https://evolutionnews.org/2014/07/good_grief_no_a/
53. Razavi, M.S., Shirani, E., Salimpour, M.R., Kassab, G.S.: PLoS ONE **9**(12), e116260 (2014). https://doi.org/10.1371/journal.pone.0116260
54. Ferrero, G.: L'Inertie Mentale et la Loi du Moindre Effort. Revue Philosophique de la France et de l'Etranger **37**, 169 (1894). (in French)
55. Zipf, G.K.: Human Behavior and the Principle of Least Effort. Addison-Wesley, Cambridge (1949)
56. Neuman, Y.: How Small Social Systems Work: From Soccer Teams to Jazz Trios and Families. Springer Nature Switzerland AG (2021)
57. Altmann, G.: Zipfian linguistics Glottometrics **3**, 19–26 (2002)
58. Visser, M.: Zipf's law, power laws and maximum entropy. J. Phys. **15**, 043021 (2013)
59. Zhu, Y., Zhang, B., Wang, Q.A., Li, W., Cai, X.: J. Phys: Conf. Ser. **1113**, 012007 (2018)
60. Zipf, G.K.: Am. Sociol. Rev. **11**(6), 677–686 (1946)
61. Wang, Q.A.: Chaos. Solitons Fractals **153**, 111489 (2021)
62. Hrebicek, L.: Zipf's law and text Glottometrics **3**, 27–38 (2002)
63. Pareto, V.: Cours d'economie politique, Éd. Rouge, Lausanne, (1897), vol. II p. 305 (in French)
64. Sun, Q., Wang, S., Zhang, K., Ma, F., Guo, X., Li, T.: Math. Prob. Eng. 6509726 (2019). https://doi.org/10.1155/2019/6509726
65. Wilson, A.: Geograph. Anal. **42**, 364–394 (2010)
66. Haynes, K.E., Phillips, F.Y., Mohrfeld, J.W.: Socio-Econ. Plan. Sci. **14**, 137–145 (1980)
67. Senior, M.L.: Progress Human Geogr. **3**, 175–210 (1979)
68. Shreider, Y.A.: Problemy Peredachi Informatiki **3**, 1, 57–63 (1967) (in Russian)
69. Martyushev, L.M., Seleznev, V.D.: Phys. Rep. **426**, 1–45 (2006)
70. Martyushev, L.M.: Phil. Trans. R. Soc. B **365**, 1333–1334 (2010)
71. Reis, A.H.: Ann. Phys. **346**, 22–27 (2014)

72. Bruers, S.: Energy and ecology. On entropy production and the analogy between fluid, climate and ecosystems Ph.D. Thesis, University of Leuven, Belgium (2007)
73. Bruers, S.: Classification and discussion of macroscopic entropy production principles. arXiv:cond-mat/0604482v2 [cond-mat.stat-mech]
74. Endres, R.G.: Nat. Sci. Rep. **7**, 14437 (2017)
75. Glimm, J., Lazarev, D., Chen, G.Q.G.: SN Appl. Sci. **2**(12), 1–9 (2020)
76. Kleidon, A., Malhi, Y., Cox, P.M.: Phil. Trans. R. Soc. B **365**, 1297–1302 (2010)
77. Struchtrup, H., Weiss, W.: PRL **80**(23), 5048–5051 (1998)
78. Kawazura, Y., Yoshida, Z.: PRE **82**, 066403 (2010)
79. Wesson, J.: Tokamaks. Oxford University Press, Oxford (2011)
80. Grad, H.: Commun. Pure Appl. Math. **14**, 234–240 (1961)
81. Sawada, Y.: Progr. Theor. Phys. **66**, 68 (1981)
82. Rebhan, E.: Phys. Rev. A **42**, 781 (1990)
83. Christen, T.: J. Phys. D Appl. Phys. **39**, 4497–4503 (2006)
84. Zupanovic, P., Juretic, D., Botric, S.: PRE **70**, 056108 (2004)
85. Holyst, R., Maciołek, A., Zhang, Y., Litniewski, M., Knychała, P., Kasprzak, M., Banaszak, M.: PRE **99**, 042118 (2019)
86. Ozawa, H., Shimokawa, S., Sakuma, H.: Phys. Rev. E **64**, 026303 (2001)
87. Malkus, W.V.R., Veronis, G.: J. Fluid Mech. **4**, 225 (1958)
88. Yoshida, Z., Mahajan, S.M.: Phys. Plasmas **15**, 032307 (2008)
89. Rubin, D.M., Ikeda, H.: Sedimentology **37**, 673–684 (1990)
90. Haase, R.: Zeitschrift für Naturforschung A **6**(10), 522–540 (1951)(in German)
91. Fomin, N.: J. Eng. Phys. Thermophys. **90**, 3 (2017)
92. Martyushev, L.M., Seleznev, V.D., Kuznetsova, I.E., Eksp, Zh.: Theor. Fiz. **118**, 149 (2000)
93. Kirkaldy, J.S.: Metall. Trans. A **16A**(10), 1781 (1985)
94. Hill, A.: Nature **348**, 426 (1990)
95. Sekhar, J.A.: J. Mater. Sci. **46**, 6172 (2011)
96. Paltridge, G.W.: Nature **279**, 630 (1979)
97. Ozawa, H., Ohmura, A., Lorenz, R.D., Pujol, T.: Rev. Geophys. **41**(4), 1018 (2003)
98. Pascale1, S., Gregory, J.M., Ambaum, M.H.P., Tailleux, R., Lucarini, V.: Earth Syst. Dynam. **3**, 19–32 (2012)
99. Kleidon, A., Lorenz, R.D.: Entropy production by earth system processes. In: Kleidon, A., Lorenz, R.D. (eds.), Non-equilibrium Thermodynamics and the Production of Entropy: Life, Earth, and Beyond. Springer, Heidelberg (2004). ISBN: 3-540-22495-5
100. Labarre, V., Paillard, D., Dubrulle, B.: Earth Syst. Dyn. **10**, 365–378 (2019)
101. Marston, J.B.: Ann. Rev. Cond. Matter Phys. **3**, 285–310 (2012)
102. Shutts, G.J.: Quart. J. R. Met. Soc. **107**, 503–520 (1981)
103. Nicolis, C., Meteorol, Q.J.R.: Soc. **125**, 1859–1878 (1999)
104. Bartlett, S., Virgo, N.: Entropy **18**, 431 (2016)
105. Lotka, A.J.: Proc. Natl. Acad. Sci. USA **8** (6), 151–154 (1922)
106. DeLong, J.P.: Oikos **117**, 1329–1336 (2008)
107. Lotka, A.J.: Proc. Natl. Acad. Sci. **8**(6), 147–151 (1922)
108. Odum, H.T.: Ecol. Model. **20**, 71–82 (1983)
109. Cai, T.T., Montague, C.L., Davis, J.S.: Ecol. Model. **190**, 317–335 (2006)
110. Heinberg, R.: Power: Limits And Prospects For Human Survival. New Society Publishers, Gabriola Is (2021)
111. Hall, Ch.A.S.: Ecol. Model. **178**, 107–113 (2004)
112. Curzon, F.L., Ahlborn, B.: Am. J. Phys. **43**, 22 (1975)
113. Kangas, P.: Ecol. Model. **178**, 101–106 (2004)
114. Tykodi, R.J.: Thermodynamics of Steady States. Macmillan, New York (1967)
115. Labarre, V., Paillard, D., Dubrulle, B.: Entropy **22**, 966 (2020)
116. Biwa, T., Ueda, Y., Yazaki, T., Mizutani, U.: EPL **60**, 363 (2002)
117. Meija, D., Selle, L., Bazile, R., Poinsot, T.: Proc. Comb. Inst. **35**, 3201–3208 (2014)

118. Hong, S., Shanbhogue, S.J., Ghoniem, A.F.: Impact of the flameholder heat conductivity on combustion instability characteristics. In: GT2012-70057 Proceedings of ASME Turbo Expo 2012 June 11–15, 2012. Copenhagen, Denmark (2012)
119. Ferguson, C.R., Keck, J.C.: Combust. Flame **28**, 197–205 (1977)
120. Arpaci, V.S., Selamet, A.: Prog. Energy Combust. Sci. **18**, 429–445 (1992)
121. Rauschenbach, B.V.: Vibrational Combustion State Editions of Physico-Mathematical Literature. URSS, Moscow (1961) (in Russian)
122. Poinsot, T.J., Veynante, D.: Theoretical and Numerical Combustion. R. T. Edwards (2001)
123. Putnam, A.A., Dennis, W.R.: J. Acoust. Soc. Am. **26**(5), 716–725 (1954)
124. Akamatsu, S., Dowling, A.P.: Three dimensional thermo-acoustic oscillation in a premixed combustor. In: Proceedings of ASME TURBO EXPO 2001 June 4–7, 2001, New Orleans, Louisiana 2001-GT-0034 (2001)
125. Dowling, A.P.: J. Sound Vibr. **180**(4), 557–581 (1995)
126. Rayleigh, J.W.S.: Nature **18**, 319–321 (1878)
127. Rijke, P.L.: Phil. Mag. **17**, 419–422 (1859)
128. Lieuwen, T.C., Yang, V. (eds.): Combustion Instabilities in Gas Turbine Engines: Operational Experience, Fundamental Mechanisms, and Modeling Prog. Astronaut. Aeronaut. 210. Am. Inst. Aeronaut. Astronaut, Reston (2005)
129. Lieuwen, T.C.: Unsteady Combustor Physics Cambridge University Press, Cambridge (2012)
130. Riess, P.: Ann. Phys. **185**, 145–147 (1860). (in German)
131. Kramers, H.A.: Physica **15**(11–12), 971–984 (1949)
132. Rijke, P.L.: Ann. Phys. **183**, 339–343 (1859). (in German)
133. Girgin, I., Turker, M.: J. Naval Sci. Eng. **8**(1), 14–32 (2012)
134. Chu, B.T.: Acta Mech. **1**(3), 215–234 (1965)
135. Welander, P.: J. Fluid Mech. **29**(1), 17–30 (1967)
136. Eddington, A.S.: The pulsation theory of Cepheid variables. Observatory **40**, 290–293 (1917)
137. Rodgers, A.W.: Mon. Not. R. Astron. Soc. **117**, 84–94 (1956)
138. Zhevakin, S.A.: Ann. Rev. Astron. Astrophys. **1**, 367–400 (1963)

A Room, a Heater and a Window

6

Abstract

Firstly, we discuss the performances of the available approaches to non-equilibrium thermodynamics when it comes to describe the steady, stable configurations of a deceitfully simple system made of a room, a heater and a window: the linear non-equilibrium thermodynamics, Kirchhoff's, Korteweg-Helmholtz', Chandrasekhar's, 'maximum economy' and 'maximal entropy' principle, Ziegler's orthogonality principle, the constructal law, the 'excess entropy production' (with the related notion of 'dissipative structure'), the 'selective decay' (which include Taylor's and Turner's principle in magnetohydrodynamics and extended magnetohydrodynamics, respectively), the 'extended irreversible thermodynamics', the 'steepest ascent', the 'second entropy', the 'information thermodynamics' (and the related 'MaxEnt'), the 'quasi-thermodynamic approach' and the 'entropy generation' (related to Gouy-Stodola's theorem). Secondly, we derive two necessary conditions of stability in systems at local thermodynamic equilibrium from Le Châtelier's principle. Finally, we start from these two necessary conditions and retrieve all results listed above concerning stable steady states outside linear non-equilibrium thermodynamics as particular cases; moreover, we also retrieve Kohler's principle for gases described by Boltzmann's kinetic equation and the extremum properties of entropy production in both a radiation field, a radiating body, Liesegang rings in supersaturated solutions and gelation of polymers.

6.1 When Principles Collide

6.1.1 The Problem

We can compare most of the variational principles of non-equilibrium thermodynamics with each other while analysing a simple, familiar system of everyday life: a room with a heater and a window. A room has a window and contains a heater. Initially, the window is closed. After a while, the distribution of temperature across the room relaxes to a stable configuration $T(\mathbf{x})$, under the competing effect of the

© The Author(s), under exclusive license to Springer Nature Switzerland AG 2022   157
A. Di Vita, *Non-equilibrium Thermodynamics*, Lecture Notes in Physics 1007,
https://doi.org/10.1007/978-3-031-12221-7_6

heating and the cooling provided by the heater and the radiation and the conduction of heat across the window towards the external world. Both the heater and the
window keep our room in a steady state far from thermodynamic equilibrium. Then,
we open the window. Convection across the window carries energy outwards, and
the temperature profile eventually relaxes towards a new steady state, which is also
far from thermodynamic equilibrium. We ask ourselves how different approaches to
non-equilibrium thermodynamics describe our room.

6.1.2   Insufficient Approaches

We have seen that LNET fails whenever heat conduction is involved. Its predictions
are indisputably wrong even in the simplest monodimensional problem of heat conduction. For example, for uniform thermal conductivity, LNET predicts that $\Delta \frac{1}{T} = 0$,
in contrast with Fourier's heat equation [1]—see also Ref. [2] for further counterexamples. And with convection too, the relevance of LNET is dubious, to say the least
[3]. *A fortiori*, LNET is of no use in our room. In particular, Onsager and Machlup's
principle of Sect. 4.1.5 follows from LNET and is, therefore, useless here.

Beyond LNET, no MinEP discussed so far is also relevant to our room. The result
of Kirchhoff in Sect. 5.3.1 holds for Joule heating and vanishing ∇T only. Korteweg-
Helmholtz' principle of Sect. 5.3.7 holds for viscous heating and vanishing ∇T only.
Chandrasekhar's result of Sect. 5.3.11 holds for small ∇T and vanishing flow at the
boundary (i.e. closed window) only. The 'maximum economy' principle of Sect. 5.3.8
holds for viscous heating only. In our room, in contrast, the nature of the heating
process in the heater is not specified, the temperature gradient is not necessarily
small, and we may open the window.

As for the constructal law of Sect. 5.4, we need an unambiguous definition of
'flow'. Here, there are at least two things that flow, i.e. heat and air. Moreover, heat
flows in three distinct ways, namely conduction, convection and radiation. In order to
describe our room with the help of the constructal law, an unambiguous prescription
on the construction of a suitable Lyapunov function relevant to the problem is needed.

6.1.3   Excess Entropy Production and Dissipative Structures

When discussing the consequences of the only indisputed maximization principle in
thermodynamics, namely the Second Principle of thermodynamics, Glansdorff and
Prigogine present a general and rigorous treatment in Refs. [4,5]. Let us start from the
global form (Sect. 4.2.9) of entropy production $\frac{dS}{dt} = P + \frac{d_e S}{dt}$ (with the notation $P \equiv$
$\frac{d_i S}{dt}$) and expand S near thermodynamic equilibrium: $S = S_0 + \delta S(t) + \frac{1}{2}\delta^2 S(t)$.
Being $P = X_i J_i$ and $X_i = 0$, $J_i = 0$ and $\frac{d_e S}{dt} = 0$ at thermodynamic equilibrium,
the contributions to the first and the second order to the entropy balance are $\frac{d\delta S}{dt} =$
$\delta \frac{d_e S}{dt}$ and $\frac{1}{2}\frac{d\delta^2 S}{dt} = \delta^2 P + \delta^2 \frac{d_e S}{dt}$, respectively, where $\delta P = 0$ and $\delta^2 P = \delta X_i \delta J_i$
('excess entropy production') is the first non-vanishing contribution to P near

thermodynamic equilibrium; basically, it is a perturbation of the entropy production in the bulk. All perturbations which leave $\frac{d_e S}{dt}$ unaffected[1] satisfy therefore the condition $\frac{1}{2}\frac{d\delta^2 S}{dt} = \delta^2 P$. Now, $S =$ max at thermodynamic equilibrium, then $-\delta^2 S > 0$ outside equilibrium; accordingly, if $\delta^2 P > 0$ too outside equilibrium, then $-d^2 S$ is a Lyapunov function and thermodynamic equilibrium is stable.[2] Positiveness of $\delta^2 P$ is, therefore, a sufficient condition for stability.[3]

Of course, we knew from the beginning that thermodynamic equilibrium is stable; indeed, in its neighbourhood, the excess entropy production is just equal to P (up to higher orders in the perturbation) and $P = \int \sigma d\mathbf{x} > 0$ because $\sigma > 0$. But the strength of Glansdorff and Prigogine's argument is that it holds also for perturbations near LTE, not just near full thermodynamic equilibrium, because we are free to limit ourselves to a perturbation occurring within a small mass element of the system. Since everything in a small mass element at LTE runs exactly as in thermodynamic equilibrium, we may say that LTE locally corresponds to a maximum of S and we may repeat the argument step-by-step provided that we are interested in perturbations occurring locally.

The validity of this analysis has been checked many times, e.g. when LNET holds, when diffusion and chemical reactions occur, and when the centre-of-mass of the small mass element moves at velocity $\mathbf{v}$.[4] It has been confirmed also in problems of nonlinear diffusion [6] and synchronization of oscillators [7]. While $\delta^2 P > 0$ near full thermodynamic equilibrium, it may become negative if the system is somehow sufficient 'far' from it. As $\delta^2 P$ crosses zero spontaneous self-organization, a behaviour unknown to full thermodynamic equilibrium becomes possible; in Prigogine's poetic language, *dissipative structures* arise [8]. In a dissipative structure order (e.g. temporal order, i.e. an oscillation with well-defined frequency) arises spontaneously as an outcome of relaxation. Dissipative structures are usually found in open thermodynamic systems, i.e. in systems where exchanges of energy and

---

[1] Again, we focus on the entropy produced inside the system.

[2] Regardless of Lyapunov, we recall that the stability of thermodynamic equilibrium is related to the fact that the probability of observing the system in a configuration different from thermodynamic equilibrium is $\propto \exp\left(\frac{\delta^2 S}{k_B}\right)$ according to Einstein's formula (Sect. 4.1.1); this probability achieves a maximum as $\delta^2 S \to 0$. This is precisely the starting point of the analysis in Ref. [5].

[3] The following link with GEC (Sect. 3.6) exists in a system with diffusion of particles of reacting species (Sect. 4.3.9) in a system at constant temperature with negligible heat flux and where no LNET is assumed to hold (so that $\frac{d_X P}{dt}$ may differ from $\frac{d_Y P}{dt}$). Being $\frac{d_e S}{dt} = 0$, the mass current densities of all chemical species are constant (just like in Sect. 4.3.9), and all reaction rates are constant at constant T. Thus, $\frac{d_Y P}{dt} = 0$, hence $\frac{dP}{dt} = \frac{d_X P}{dt}$ in Sect. 4.3.8. But $\frac{d_X P}{dt} \leq 0$ because of GEC (Sect. 4.3.9) with the operator $<$ being replaced by $=$ in steady state only, hence $\frac{dP}{dt} \leq 0$. Since $\frac{dP}{dt} = \frac{d\delta^2 P}{dt}$ as $\delta P = 0$, $\frac{d\delta^2 P}{dt} \leq 0$ so that $\delta^2 P > 0$ is a sufficient condition for stability. The fact that it is a sufficient condition of stability follows from Lyapunov theory of stability, hinted at in Sect. 4.3.10.

[4] In this case, the addition of a term proportional to $|\mathbf{v}|^2$ allows suitable generalization of $\delta^2 P$ while leaving all other things unchanged, just like in the note of Sect. 3.6.

matter with the external world occur.[5] The Belousov-Zhabotinsky chemical reactions provide a well-known example of dissipative structures [9].

In Eigen's model [10] of the evolution of the macromolecules which the origin of life stems from, the occurrence of a mutant 'species' (i.e. a new variant of a pre-existing self-replicating entity) exhibiting a selective advantage corresponds to a negative fluctuation of entropy production in the bulk, i.e. to a negative value of $\delta X_i \delta J_i$. This causes an instability, i.e. a breakdown of the steady state. In the final state, the mutant grows to a dominant level, and the older species become extinct. As for the rate of entropy production, it relaxes back to its previous value; but the total amount of entropy of the biosphere at any time remains lower than the value it would have achieved without the mutation. In other words, the final effect of the mutation is to increase the order and the overall level of organization and of the biosphere.[6]

The onset of Bénard convection cells[7] can be explained the same way [14] as Eigen's line of reasoning applies step-by-step to the onset of Bénard cells [15].[8] The approach of excess entropy production has indirectly inspired the analytical results of Ref. [16] on Taylor's stability problem for a viscous fluid contained between rotating coaxial cylinders with and without a radial temperature gradient and the discussions in Refs. [17, 18][9] concerning non-Newtonian fluids, as well as the experiments reported in Ref. [19] in electrochemistry.[10]

The effort aimed at writing down a Lyapunov function usually leads to look for a quantity that is a minimum with respect to small perturbations of the steady state. Computations routinely keep some quantity related to the unperturbed state at a constant value. In turn, this leads to a numerical algorithm for stability analysis. The price to be paid is that some *a priori* knowledge of the relaxed state is required, and iterations may fail to converge due to the local nature of the postulated minima. The approach of excess entropy production has inspired an intriguing result [20] concerning the construction of a Lyapunov function for particle diffusion with non-constant diffusion coefficients. Unfortunately, this work postulates that the steady state whose stability is investigated undergoes only small perturbations. In particular, the 'potential' minimized in Eq. (4.6) of Ref. [20] in order to pinpoint the stable state explicitly depends on the stable state itself. The resulting iteration scheme usually converges if the deviation from the stable steady state is not too large. This fact allows only 'local' analysis of stability against perturbations.

---

[5] And where the interpretation of stability as a matter involving just the entropy production P in the bulk with no role of $\frac{d_e S}{dt}$ is debatable, to say the least.

[6] Section 3.6 of Ref. [11] includes a discussion of the excess entropy approach in biophysics, as well as references to further bibliography.

[7] Which are formally described in analogy with a biological system in Ref. [12] with the help of the well-known Lorentz's dynamical system originally developed for the physics of atmosphere. We recall that MEPP does not apply to the latter system [13]; see Sect. 5.6.13.

[8] See notes in Sects. 6.1.9 and 6.2.19.

[9] See Sect. 6.2.9.

[10] See Sect. 6.2.8.

The trouble with excess entropy production is that its positiveness is a sufficient, not a necessary condition for stability. *Per se*, the violation of a sufficient condition of stability like $\delta^2 P > 0$ is no sufficient condition for instability. Stable, far-from-equilibrium configurations with $\delta^2 P < 0$ are, therefore, possible. Generally speaking, once a reasonable guess about the stable state is available, the inequality $\delta^2 P > 0$ provides the logical ground for algorithms aiming at checking local stability against small perturbations. However, with a few exceptions (including full thermodynamic equilibrium, of course), Glansdorff and Prigogine's result turns out to be scarcely useful when it comes to locate stable states from scratch, because stable, relaxed states may exist which violate this criterion of stability. There are many problems where it fails to locate relaxed states, even assuming that an explicit expression for $\delta^2 P$ is available (a far from trivial task in itself).

In their critical discussion of Ref. [21], Keizer and Fox investigate a chemical reaction where steady, stable states exist with $\delta^2 P < 0$ and write: *By considering the stability of the equilibrium state, we conclude that the second differential of the entropy, which is at the heart of the Glansdorff-Prigogine criterion, is likely to be relevant for stability questions close to equilibrium only. [...] The Lyapunov-stability conditions are known to be sufficient conditions [...], and while this point appears in Glansdorff and Prigogine's theory, it is not in our opinion sufficiently stressed to prevent inadvertent misapplication.* In his review of the subject, De Sobrino [22] concludes that *it is extremely difficult to delimit with precision the strength of a stability criterion as general as the Glansdorff-Prigogine criterion. [...] One cannot expect to have a strong criterion not based on a Lyapunov function whose construction makes full use of the coefficients of the linearized equations.* The last words hint at the fact that full knowledge of the unperturbed state is required in order to write down the Lyapunov equation, which sounds good news for the numerical approaches described above but not-so-good when it comes to find and describe previously unknown relaxed states. Finally, the excess entropy production approach has also been rejected by Lavenda [23] who has shown that even if the sign of excess entropy production is known, in the general case (e.g. in our room), stability depends on the actual eigenvalues of the linearized equations of motion. Eventually, evaluating the sign of excess entropy production is equivalent to evaluating the sign of the imaginary part of the eigenfrequencies in a linearized description of the system under investigation near its relaxed state, i.e. the same task of familiar modal analysis.

6.1.4   Selective Decay

This rather frustrating series of failures may suggest that the problem lies in the common assumption of all the proposals discussed so far: LTE. For example, the equations of motion in many physical systems show that different quantities evolve with different time-scales and spatial scales; this result does not rely on LTE (explicitly, at least). Accordingly, it is at least conceivable that relaxation satisfies the 'selective decay' scenario.

The selective decay hypothesis is characterized by the following. If one considers the 'ideal invariants' of the system (namely, the quantities which would remain exactly constant during the evolution of the system should no dissipation occur), once dissipation has been introduced these quantities do not remain constant but start decaying, unless the external world somehow maintains their value constant. It is often found that one of these quantities is somehow 'better conserved' or 'more rugged' than others, i.e. that its typical decay time is much longer than the decay time of other quantities.[11] If one minimizes the expression for the poorly conserved invariant subject to the constraint that the rugged invariant is conserved using the technique of Lagrange multipliers (Sect. A.3), an Euler-Lagrange equation for the field variables in the relaxed state results. The Lagrange multiplier is the ratio of the poorly conserved invariant to the ruggedly conserved one.

Typically, selective decay applies to problems in two-dimensional and three-dimensional magnetohydrodynamics ('MHD'), where the couples 'rugged invariant versus poorly conserved invariant' are 'energy versus mean square vector potential' and 'energy versus magnetic helicity', respectively. In MHD, for example, a well-known example of relaxed state is described by Taylor's principle of minimum magnetic energy $\propto \int |\mathbf{B}|^2 d\mathbf{x}$ with fixed magnetic helicity $\int (\mathbf{A} \cdot \mathbf{B}) d\mathbf{x}$ (where $\mathbf{B} = \nabla \wedge \mathbf{A}$ and $\mathbf{A}$ is the vector potential) [25]. In Hall MHD, i.e. a macroscopic description of magnetized plasmas (made of two species, electrons and ions with ion mass m_{ion} and ion electric charge q_{ion}) where electrons are effectively decoupled from ions, the couple 'rugged invariant versus poorly conserved invariant' is 'total (magnetic + kinetic) energy' versus 'magnetic helicity and generalized ion helicity $\int (\mathbf{V} \cdot \mathbf{\Omega}) d\mathbf{x}$', with $\mathbf{V} \equiv \mathbf{v} + \frac{q_{ion}}{m_{ion}} \mathbf{A}$ and $\mathbf{\Omega} \equiv \nabla \wedge \mathbf{V}$. Taylor's principle is replaced by Turner's principle [26] of minimization of total energy with two constraints: fixed magnetic helicity and fixed generalized helicity. Remarkably, and in qualitative agreement with Kirchhoff's principle of Sect. 5.3.1, in order to describe plasmas in the solar corona it has been postulated [27] to replace Turner's principle with the constrained minimization of Joule heating power; fixed generalized ion helicity and its electron counterpart are the constraints.[12]

However, our room is a thermodynamically open system. The room exchanges either heat or both heat and mass with the external world. Strictly speaking, however, selective decay applies only to isolated systems. It is for this reason that researchers

---

[11] In order to fix the ideas, we follow the discussion of Sect. 4.4 of Ref. [24] and suppose that a physical system has two distinct ideal invariants $\alpha \propto O(|\nabla^m u|)$ and $\beta \propto O(|\nabla^n u|)$ where $u = u(\mathbf{x})$, m and $n < m$ are quantity depending on space and two integer numbers, respectively, while ∇^m and ∇^n denote the generic partial derivative on spatial coordinates of order m and n, respectively. (Should we rather assume that $\alpha \propto O(|\nabla u|^m)$ and $\beta \propto O(|\nabla u|^n)$, nothing would change in the following). Let us apply a dissipative process to the system, which acts on a vanishingly small spatial scale l. As a result, $u \to u + \delta u$, $\alpha \to \alpha + \delta \alpha$ and $\beta \to \beta + \delta \beta$ with $|\delta \alpha| \approx l^{-m} |\delta u| \gg |\delta \beta| \approx l^{-n} |\delta u|$. As a rule of thumb, therefore, it is β, i.e. the quantity which contains the lower-order derivatives, which plays the role of rugged invariant.

[12] If we identify $u(\mathbf{x})$ with any component of $\mathbf{A}$ and recall that $\mathbf{B} = \nabla \wedge \mathbf{A}$ and $\mathbf{j} \propto \nabla \wedge \mathbf{B}$, then the total energy and Joule heating power scale as $|\mathbf{B}|^2$ (i.e. $m = 2$) and $|\mathbf{j}|^2$ (i.e. $m = 4$) while the rugged invariants scale as $\mathbf{A} \cdot \mathbf{B} = \mathbf{A} \cdot \nabla \wedge \mathbf{A}$ (i.e. $n = 1$). For further discussion, see Sect. 6.2.8.

are used to focus their attention on systems that are freely decaying or are otherwise disconnected from external energy sources. (The situation is far less clear for driven systems, though many of the features found in isolated systems carry over). As for selective decay, in the relaxed state, such state dissipation is supposed to be compensated indefinitely, as the value of the rugged invariant is at a constant value in spite of continuous dissipation. Of course, heat and matter flow from the heater and across the window to sustain the relaxed state in our room: but the problem if such support is actually enough to ensure stability is precisely what no approach focussed on dissipation in isolated systems may successfully solve. In a nutshell, stability is either to be proven for each problem or to be observed empirically.

6.1.5 Maximal Entropy

The principle of 'maximal entropy'[13] dictates that the air in a room initially distributed in clumps moves towards smooth uniformity; thermodynamic equilibrium does not admit large-scale structures. However, for a system with a constrained phase space, maximal entropy can generate large-scale structures as a long-lived intermediate state. Remarkably, no LTE is explicitly invoked. To apply the principle of maximal entropy, one needs to consider a discrete or quantized version of the field variables. If we have N such quanta of the field,[14] we consider the number of ways these N quanta can be arranged in a given state (like spins up or down). The most probable state is the one with the most permutations or the highest entropy subject to other constraints (such as conservation of energy and particle number); here entropy is defined as the logarithm of the number of permutations times Boltzmann constant. The description of the system is perfectly analogous to the familiar description of the 2D spin system in statistical mechanics of thermodynamic equilibrium [32]. The maximal entropy perspective addresses the question: are these observed large-scale, self-organized structures in some sense statistically more probable than other less simple ones?

Again, our room is a thermodynamically open system. The room exchanges either heat (across the closed window) or both heat and mass (across the open window) with the external world. In a relaxed state, the long-lived, large-scale structures are supposed to live not just for a long time, but indefinitely. The relevance of these approaches to the relaxed state of our room is, therefore, yet to be proven.

---

[13] For an excellent review of both maximal entropy approach and selective decay in Sect. 6.1.4, see [28].

[14] E.g. vortices in 2D fluid dynamics [29] and bundles of magnetic flux or electric current filaments in MHD [30, 31].

6.1.6   Extended Irreversible Thermodynamics

Does a description of the relaxed state exist which is as general and problem-independent as thermodynamics and not related to LTE? According to the Extended Irreversible Thermodynamics ('EIT'), the answer is yes—see Ref. [33] for a review. EIT drops the LTE assumption and postulates that entropy depends locally not only on internal energy, particle density, etc. but also on the heat flux. Inclusion of the latter in the list of fundamental thermodynamic quantities allows the derivation of a heat transport equation and of familiar equilibrium relationships, e.g. those concerning specific heat on an equal footing; moreover, the EIT equation of heat transport allows straightforward generalization to a relativistically invariant version (in contrast with Fourier's law of Sect. 5.2.1) and is, therefore, of great interest when it comes to describe relativistic systems [34]. Together with Einstein's formula for the probability of fluctuations, EIT leads to predictions that agree with well-known results of a kinetic theory near thermodynamic equilibrium. EIT looks also to have inspired (indirectly at least) Sienutycz et al.'s work in Ref. [35].[15]

In spite of the alleged independence of EIT from LTE, however, the latter remains somehow involved. For example, the contribution of heat conduction to EIT entropy is basically the product of a relaxation time and the corresponding term in the entropy production rate at LTE. Much worse, the temperature is no more the multiplying factor of the differential of entropy in the First Principle of thermodynamics; the familiar formulation of energy balance, e.g. in fluid mechanics of Sect. 4.2.8 [36] is, therefore, at stake, as well as the relevance of EIT to the physically meaningful description of our room.

6.1.7   Steepest Ascent

What if entropy, rather than a statistical, information-theoretic, macroscopic or phenomenological concept, were an intrinsic property of matter in the same sense as energy is universally understood to be an intrinsic property of matter? What if irreversibility were an intrinsic feature of the fundamental dynamical laws obeyed by all physical objects, macroscopic and microscopic, complex and simple, large and small? What if the second law of thermodynamics, in the hierarchy of physical laws, was at the same level as the fundamental laws of mechanics, such as the great conservation principles? [37] According to the claims of Refs. [38–40], the answer to these questions, originally put forward by Corbage and Prigogine (see Ref. [41] and Refs. therein), stems from the fact that irreversibility in Nature is a fundamental microscopic dynamical feature and as such it must be built into the fundamental laws of time evolution. MEPP follows, therefore, from a fundamental extension of known physics at the quantum level. In the words of Ref. [38], *every trajectory unfolds along a path of steepest entropy ascent compatible with the constraints [...]. For an isolated*

---

[15] See note in Sect. 4.3.2.

system, the constraints represent constants of the motion. [...] the qualifying and unifying feature of this dynamical principle is the direction of maximal entropy increase. [...] The challenge with this approach is to ascertain if the intrinsic irreversibility it implies at the single particle (local, microscopic) level is experimentally verifiable, or else its mathematics must only be considered yet another phenomenological tool. When it comes to classical physics, the constraints acting on the relaxation identify a hyper-surface in the phase space of the system. In the words of [40], *on this surface we can identify contour curves of constant entropy, generated by intersecting it with the constant-entropy surfaces. Every trajectory [...] lies* on this surface *and is at each point orthogonal to the constant-entropy contour passing through that point. In this sense, the trajectory follows a path of steepest entropy ascent compatible with the constraints.*[16] However, this conclusion relies on the postulate that relaxation obeys a particular law: we refer to Eq.(12) of [39] and Eq.(25) of [40]. Such equation contains a characteristic time τ *which is a function of the system state and is the relaxation time which describes the speed at which the state evolves in state space in the direction of steepest entropy ascent* [39]. No recipe for the computation of τ is provided for a given problem like our room's evolution. Then, the relevance of this approach is to be checked.

6.1.8 Second Entropy

Attard's proposal [42] of a 'second entropy', a function of the same quantities involved in LNET but with an additional, explicit dependence on a typical time-scale τ of the relaxing system which is to be maximized in order to describe the relaxed state, works fine in problems with convection cells. However, no explicit recipe is provided for the computation of τ in the general case, and the relevance to the description of our room remains therefore to be seen.[17]

6.1.9 Information Thermodynamics and MaxEnt

Dewar [44] and Niven [45] claim to provide formally rigorous, LTE-independent justification of the maximization of entropy production in the relaxed state. They start from the results ('information thermodynamics') contained in two papers published by Jaynes [46,47], where the author emphasizes a natural correspondence

---

[16] For a review of models which are allegedly compatible with the present 'steepest ascent' approach, we refer to [37].

[17] Now we may appreciate Sawada's original approach [43] discussed in Sect.5.6.3, a proposal which neither invokes LTE nor is limited to isolated systems (unlike both selective decay and maximal entropy.) nor relies on the constraint of fixed thermodynamic forces (unlike the orthogonality principle) nor modifies the definition of entropy (unlike EIT and second entropy). It puts in evidence the crucial role played by energy transport across the boundaries of the system in any MEPP—if any MEPP ever holds.

between statistical mechanics (which provides us with the foundation of thermo-dynamics) and information theory. In particular, Jaynes argues that the entropy of statistical mechanics [32] and Shannon's entropy of information theory [48] behave basically the same way. Consequently, statistical mechanics should be seen just as a particular application of a general tool of logical inference and information the-ory.[18] In most practical cases, the testable information is given by a set of conserved quantities (average values of some moment functions), or, equivalently, by a set of symmetries of the probability distribution. Given testable information, the maximum entropy ('MaxEnt') procedure consists of seeking the probability distribution which maximizes information entropy, subject to the constraints related to testable infor-mation.[19] Then, the principle of maximum entropy can be seen as a generalization of the familiar principle of insufficient reason.[20]

MaxEnt has met considerable success outside physics, e.g. in image reconstruction software, natural language processing, etc. [51]. To say the least, it is a brilliant and thought-provoking analysis of our information-gathering processes, like a set of rules for identikit-makers. It is uniquely useful in data analysis—for an excellent review of both MaxEnt approach and its applications in biology, see Ref [52]. Reference [53] provides us with a review of MEPP which is focussed on MaxEnt. Moreover, MaxEnt has been applied to atmospheric physics [54–56] and to problems with convection [54,57]. We have discussed Paltridge's and Ozawa et al.'s arguments in Sects. 5.6.13 and 5.6.7, respectively. A popular application of MEPP to the Earth as a whole is discussed by Kleidon in [58], where the planet is described as a single system that is strongly shaped by life—at least from the point of view of geochemistry. Many researchers in ecology have invoked MEPP as a heuristic tool—for a review, we refer to Refs. [12,59].[21] There is some connection between the work of Ref. [58] and the so-called 'Gaia hypothesis' [62,63]. The Gaia hypothesis posits that the Earth is a self-regulating complex system involving the biosphere, the atmosphere, the hydrospheres and the soil, tightly coupled as an evolving system. The hypothesis

---

[18] Landauer's principle [49] states that the erasure of a bit corresponds to an entropy increase. In other words, if an observer loses information about a system, then the observer's ability to extract useful work from that system gets diminished. Both the formation of Bénard cells in convection [15] and the mutations in biology [10] are examples of processes which decrease $\frac{d_i S}{dt}$ and which raise the information content of the system. Jaynes' argument, however, is focussed on entropy, rather than on entropy production.

[19] Entropy maximization with no testable information takes place under a single constraint: the sum of the probabilities must be equal to 1.

[20] Also dubbed 'principle of indifference' in the literature, this principle states that in the absence of any relevant evidence, agents should distribute their credence (or 'degrees of belief') equally among all the possible outcomes under consideration [50].

[21] We have to be careful of the—sometimes far-fetched—application of thermodynamic ideas. As for ecology, for example, even the words 'far from equilibrium' stand for different things in different authors—for a review, see [60]. In some particular cases, this application has a precise meaning; for example, in a bird flock, the time it takes for one bird to realign the orientation of its flight along the locally averaged orientation of its neigbours' flights is so short that the notion of LTE still makes sense [61].

contends that this system as a whole, called Gaia,[22] seeks a physical and chemical environment optimal for life. Just like Bejan's constructal law, Gaia hypothesis is far from meeting general consensus [64]. Here, we stress the point that—as usual with MEPP—$\frac{d_i S}{dt}$ is neglected: according to Ref. [58], indeed, should MEPP hold then *complex systems would actually evolve too and maintain such a steady state just like an engineer would work towards achieving the Carnot limit when designing an engine*,[23] and the amount of entropy produced per unit time in the bulk of an engine operating near the Carnot limit is vanishingly small.[24]

---

[22] From the Greek name of Mother Earth goddess.

[23] Remarkably, in a description of Earth's atmosphere where the mechanical power related to wind and waves is computed as the outcome of a thermal machine driven by the Sun, it has been claimed that this power achieves a maximum in an analysis [65] which explicitly takes into account Earth's rotation, in contrast with Paltridge's one [55]. This maximization is equivalent to the maximization of the efficiency of Earth's climate seen as a machine that transforms part of the power received by the Sun into mechanical power. If this maximization is a feature of a stable configuration of the climate, then it follows that a loss of stability corresponds to a drop in efficiency. This conclusion is independently confirmed by the results of Ref. [66], which show that Earth's climate is a bistable system with two well distinct configurations, the 'snowball Earth' and the 'warm Earth'. Snowball Earth displays global glaciation of water and extremely dry atmosphere; heat transport and viscous dissipation rule entropy production because the planet is almost entirely dry. Warm Earth displays relatively small sea-ice cover; the main contribution to entropy production comes from latent heat due to large-scale and convective precipitation. Mathematically, the transition between the two states is a bifurcation [67]. Physically, *a general property which has been found is that, in both regimes, the efficiency increases when we get closer to the bifurcation point and at the bifurcation point the transition to the newly realized stationary state is accompanied by a large decrease in the efficiency* [66]. Given the boundary conditions—i.e. at given $\frac{d_e S}{dt}$—we expect that a drop in efficiency of a thermal machine corresponds to an increasing role of dissipative processes inside the machine, i.e. to a larger $\frac{d_i S}{dt}$. Conversely, if loss of stability corresponds to an increase in P, then stable configurations correspond to minimum values of $\frac{d_i S}{dt}$, in agreement with the conclusions of Sect. 6.1.3. This agreement is all the more relevant, as the onset, e.g. of Bénard cells discussed in Sect. 6.1.3 is precisely a bifurcation.

[24] Endres puts forward a clever argument in favour of MaxEnt by showing its success in describing the well-known behaviour of a simple, analytically solvable, one-dimensional bistable chemical system [68]. Starting from microscopic dynamics, he shows that the system relaxes to a steady state where fluctuations still occur but follows a probability distibution that does not depend on time. He writes down an explicit expression of this probability distribution and shows that the configuration with maximum probability (for given fluxes of particles of the species involved at the boundary) is the configuration that maximizes the production of entropy in the bulk, i.e. $\frac{d_i S}{dt}$. Furthermore, he shows that his results agree with Jaynes' results. Endres' work is remarkable because the maximized quantity is $\frac{d_i S}{dt}$, rather than $\frac{d_e S}{dt}$ as in most works about MEPP. However, in the steady, relaxed state described in [68] $0 = \frac{dS}{dt} = \frac{d_i S}{dt} + \frac{d_e S}{dt} = \frac{d_i S}{dt} - \int_{boundary} \mathbf{j}_s \cdot d\mathbf{a}$ (after suitable time-averaging on the fluctuation time-scale at least) so that Endres' maximization is just another example of maximization of the entropy outflow $\int_{boundary} \mathbf{j}_s \cdot d\mathbf{a}$. It remains to be seen if this proves that MEPP holds even beyond this particular chemical reaction. Finally, it turns out that the configuration of maximum probability may shift from a MEPP state to another, MinEP state; this swap (mathematically, a bifurcation [67]) occurs depending on the actual boundary and initial conditions. Reference [68] provides us, therefore, with a further example of the coexistence of different extremum conditions, as suggested in Sect. 5.6.2.

In spite of their past popularity, however, arguments supporting MEPP which rely on information thermodynamics meet no general consensus. Fallacies in the arguments of Ref. [44] have been found in Refs. [69,70]; the proof in [45] has been challenged in [71]. MEPP tenets have been put in doubt in Refs. [72]. As for our room, the fact that the evidence supporting MEPP is related to phenomena where $\frac{d_i S}{dt}$ is neglected suggests that MEPP does not properly describe the irreversible heating due to the heater in our room. Moreover, in contrast with physical intuition (which requires that the nearer the heater, the higher the temperature), maximization of entropy production—which is an increasing function of the ratio of the heating power and the temperature—would *minimize* the temperature near the heater.

6.1.10 Orthogonality Principle

Ziegler [73] invokes LTE explicitly and postulates what he calls the 'orthogonality principle' which is basically equivalent to maximization of entropy production at fixed thermodynamic forces. According to Ref. [72], which provides us with a review of MEPP focussed on Ziegler's work, Ziegler's principle includes Onsager and Machlup's principle of Sect. 4.1.5 as a particular case. As a matter of principle, this fact allows us to reconcile MEPP and MinEP—a long-standing problem of non-equilibrium thermodynamics. The argument runs as follows and starts from the assumption that different time-scales rule the evolution of the system. On the shortest time-scales, the system maximizes entropy production at preset fixed thermodynamic forces at a given moment. On the slower time-scales, thermodynamic forces are allowed to undergo slight modifications, which as such can be described in a linearized theory like LNET, and the system changes free thermodynamic forces so as to decrease entropy production. According to Ref. [74], Ziegler's MEPP can be considered just as a working hypothesis for nonlinear non-equilibrium thermodynamics. According to the discussions in Refs. [72,75], indeed, Ziegler's principle has its statistical substantiation only if the deviation from equilibrium is small, which is definitely not the case of our room.

In order to overcome this difficulty, Martyushev has published a proof of Ziegler's principle [76]. In a nutshell, this proof runs as follows. Let us start from our original expression $\frac{dS}{dt} = XY$ for the time derivative of the entropy, with X and Y thermodynamic force and thermodynamic flow, respectively; we assume that there is just one degree of freedom and that $X > 0$, with no loss of generality. The Second Principle of thermodynamics requires that $\frac{dS}{dt} > 0$ whenever irreversible phenomena occur. The orthogonality principle says that if X is fixed, then the relaxed state of the system corresponds to a maximum of $\frac{dS}{dt}$. Since $X > 0$, this means that the system selects the value of Y as large as possible, say $Y = Y_{max}$. Now, let us suppose that when performing this maximization the system may choose among several different flows, say $Y_j, j = 1, 2, \ldots \leq Y_{max}$. Martyushev argues that the actual value of the Y_j's depends on the observer's choice of spatial and time coordinates. Thus, a suitable rescaling of such coordinates makes it always possible to set $Y_{max} = 0$, so that all Y_j's $\neq Y_{max}$ are ≤ 0. In the author's own words, *the maximum flow is taken as a zero flow [...] in practice, this can be realized, e.g. by time/space scaling.* Now, should Ziegler be wrong, the system would select a value of Y different from $Y = Y_{max}$ in the relaxed state, then $\frac{dS}{dt} = XY < 0$ in contrast with the Second Principle of thermodynamics, which is absurd.

As explicitly stated in [76], however, a tenet of the proof is the validity of the Second Principle of thermodynamics even after the transformation of coordinates which sets $Y_{max} = 0$. This is obvious, as the Second Principle is a universal law of Nature. As a consequence, the orthogonality principle seems to be far from being a universal law of Nature, as it follows from the Second Principle in the particular set of spatial and time coordinates where $Y_{max} = 0$ *only*, rather than being valid regardless of the coordinates (like the Second Principle). In other words, the orthogonality principle holds provided that $Y_{max} = 0$, so that no entropy is produced in the relaxed state ($\frac{dS}{dt} = XY_{max} = 0$), i.e. provided that the relaxed state is just the thermodynamic equilibrium—in agreement with the theorem of Sect. A.4 and with Ref. [77]. Moreover, once the system has achieved thermodynamic equilibrium, no change in coordinates may push it away and assign to $\frac{dS}{dt}$ a non-zero value. Accordingly, the orthogonality principle seems to be scarcely relevant to our room.

6.1.11 Quasi-Thermodynamic Approach

The roots of Lavenda's 'quasi-thermodynamic approach' ('QTA') [23] may be found in Ref. [78]. The starting point is the replacement the familiar form of the First Principle of thermodynamics $dU = TdS - pdV + \mu_k dN_k$ (the sum being extended to all chemical species in the system) for isothermal ($T = $ const.) and isochoric ($V = $ const.) systems with $d\left(\frac{dU}{dt}\right) = Td\left(\frac{dS}{dt}\right) + \mu_k d\left(\frac{dN_k}{dt}\right)$. Then, *not too large deviations from the stationary state* are taken into account. Finally, phenomenological coefficients are assumed to be constant. As a result, a local balance of energy is obtained ('power equation') which links the second-order differential of the entropy production density, Rayleigh's dissipation function (Sect. 4.1.3) and the second-order differential of the rate at which work is done on the system by the external forces. It is claimed that the last quantity is never negative, in agreement with Le Châtelier's principle; this fact leads to many different variational principles, depending on the particular problem. Basically, Lavenda's claim is the generalization of the reinterpretation of Korteweg-Helmholtz' principle of Sect. 5.3.7 outside the domain of $Re \ll 1$ fluids. The interpretation of stability is mechanical in nature.

QTA applies to a system that is ruled by the equations of motion of a generalized, forced, linear harmonic oscillator with constant coefficients [23]. The particular case of constant mass density is discussed in Ref. [79]. Many well-known principles of plasma physics can be rewritten with the help of Lavenda's power equation [80]. QTA applies also successfully to a particular nonlinear problem—the limit cycle of a Van der Pol oscillator [81]—provided that the motion behaves as if it were periodic during any single period, whereas the effects of dissipation are only noticeable over the longer space-time of evolution. Unfortunately, our room satisfies none of these assumptions.

6.1.12 Gouy-Stodola's Theorem and Entropy Generation

An approach involving both QTA [23] and Gouy-Stodola's theorem [82] is developed by Lucia [83]. It has been applied to problems in biophysics [84] and hydrology [85].

As for Gouy-Stodola's theorem, a well-established result of thermodynamics, imagine a body in thermal contact with an environment with entropy S and S_0, respectively, just like in Sect. 2.2. If $P_h = 0$ everywhere at all times in the environment and the temperature is constant and uniform (say, $= T_0$) in the environment, then the local form of the environment entropy balance of Sect. 4.2.8 implies $T_0 \frac{dS_0}{dt} = -\int \mathbf{q} \cdot d\mathbf{a}$ after multiplication by T_0.[25] The R.H.S. is computed on the interface between the environment and the body and $d\mathbf{a}$ is directed from the environment towards the body, so that $\int \mathbf{q} \cdot d\mathbf{a}$ is (-1) times the total amount of heat flowing per unit time from the body towards the environment. In turn, the latter amount is equal to the total amount P_{TOT} of heat produced per unit time in the body provided that the body is in steady state or if it undergoes a cyclic transformation.[26] Then, $T_0 \frac{dS_0}{dt} = P_{TOT}$. After integration of both sides on a time interval Δt, we obtain Gouy-Stodola's theorem $T_0 \Delta S_0 = W_{diss}$ where W_{diss} is the amount of work lost into heat and can be regarded as the difference between the maximum reversible useful work (sometimes referred to as 'exergy'[27]) that could be produced by the system and the actual work produced, during the same time interval Δt. Being S equal to its initial value as the body is in steady state or under cyclic transformation, ΔS_0 is the increment of the entropy of the Universe; it is referred to as S_g ('entropy generation'). As for W_{diss}, it can be also interpreted as the difference between the actual amount R of work the environment must do on the body in order to induce in it a given transformation and the minimum amount R_{min} that would be necessary if no dissipations were present; according to Sect. 2.2, if the final and the initial values of S are equal, then $W_{diss} = T_0 \Delta S_0$.

Lucia [83] invokes QTA and Gouy-Stodola's theorem and claims that $W_{diss} =$ extremum in steady state; in particular, the 'entropy generation rate' $\frac{dS_g}{dt}$ is maximum in [85]. QTA (Sect. 6.1.11) claims that the second-order differential of the rate at which work is done *on the system by external forces* is ≥ 0, so that this amount of work may only be a minimum per unit time. Then,[28] the amount of work done *by the system on the external world* per unit time may only be a maximum.[29] Now, in each unit of time, this maximum amount of work raises the internal energy of

---

[25] Here, we have invoked both the fact that $\frac{dS}{dt} = \frac{d}{dt} \int \rho s d\mathbf{x} = \int \rho \frac{ds}{dt} d\mathbf{x}$ and Gauss' theorem of divergence.

[26] As in both cases the net time-averaged amount of heat delivered to the body is zero. We refer to the discussion of Kohler's principle in Sect. 6.2.5.

[27] For a heat engine, the exergy can be simply defined as the energy input times the Carnot efficiency. Since many systems can be modelled as a heat engine, this definition can be useful for many applications.

[28] Because of Newton's Third Law of dynamics.

[29] There is an analogy with the maximization of the amount of heat flowing across the boundaries of the system towards the external world in Sect. 5.7.

the environment while leaving its volume and temperature unaffected and raises therefore its entropy; maximization of $\frac{dS_g}{dt}$ follows.[30] Being rooted in Lavenda's results [23], Lucia's approach [83] seems to share their limitations.[31]

6.1.13   Much Ado for Nothing?

We conclude that, in spite of decades of efforts, *none* of the theories of non-equilibrium thermodynamics discussed above seem to be able to provide us with a self-consistent, unambiguous (and therefore reliable) description of the stability of the various relaxed states in our simple system made of a room, a heater and a window. The reason is that our room, although deceptively simple, is a thermodynamically open system where irreversible phenomena occur both in the system's bulk (the heater) and across the system boundary (the window). By the way, this is just what happens in most natural phenomena. In contrast, available theories:

- either need some unambiguous definition of 'flow' while there are at least three distinct flows of energy—by radiation, convection and conduction—and one flow of matter through the open window (constructal law, Sect. 5.4);
- or rely on too restrictive assumptions (Kirchhoff's principle, Sect. 5.3.1; Korteweg-Helmholtz' principle, Sect. 5.3.7; Chandrasekhar's principle, Sect. 5.3.11; LNET, Sect. 6.1.2; QTA, Sect. 6.1.11; entropy generation, Sect. 6.1.12);
- or may fail to locate relaxed states (excess entropy production, Sect. 6.1.3);
- or are limited to selected dissipation processes (Kirchhoff's principle Sect. 5.3.1; Korteweg-Helmholtz' principle, Sect. 5.3.7; maximum economy, Sect. 5.3.8);
- or apply to isolated systems only (selective decay, Sect. 6.1.4; maximal entropy, Sect. 6.1.5);
- or require *a priori* knowledge of a unique characteristic time τ for all processes occurring in the room (steepest ascent, Sect. 6.1.7; second entropy, Sect. 6.1.8);
- or rely on just one unjustified, postulated variational principle for all irreversible phenomena, both in the bulk and across the boundaries (EIT, Sect. 6.1.6; MaxEnt, Sect. 6.1.9).

---

[30] Two examples. When applied to the flooding of a river, this result implies that the water makes the maximum mess possible in the environment surrounding the river—see also Sect. 6.2.18. When applied to Bénard convection, the maximized $\frac{dS_g}{dt}$ coincides [86] with the maximized quantity in Ref. [57] in Sect. 6.1.9 after suitable averaging—see equations 5-5' of [57] and Eq. 14 of Ref. [86].
[31] The smaller the rate at which work is done on the system by external forces per unit time, the smaller the drop of exergy per unit time due to dissipation inside the body. According to Eq. 9 of [87], minimization of exergy dissipation in a viscous fluid with constant kinematic viscosity in the limit of vanishing Reynolds number is equivalent to $\frac{dK}{dt} + P_{ext} - \int \sigma'_{ik} \frac{\partial v_i}{\partial x_k} d\mathbf{x} = \min$ (in the language of Sect. 5.3.10), and the mechanical energy balance $\frac{dK}{dt} = P_{ext} - \int \sigma'_{ik} \frac{\partial v_i}{\partial x_k} d\mathbf{x}$ gives therefore $P_{ext} = \int \sigma'_{ik} \frac{\partial v_i}{\partial x_k} d\mathbf{x}+$ a minimized quantity. Minimization of the work P_{ext} done by the external world on the fluid is, therefore, equivalent to Korteweg-Helmholtz' principle. See Ref. [88] for further discussion about the connection between the results of [83], [87] and Sect. 5.7.

6.2   One Principle to Bind Them All?

6.2.1   A 1st Necessary Condition for Stability

Rather than looking for a general-purpose, universal answer, in the following, we invoke Le Châtelier's principle and the assumption of LTE everywhere at all times (Sect. 3.2), in order to obtain information concering the relaxed states of our room.[32] Glansdorff and Prigogine's original intuition [14] provides us with the rationale of our choice: the very fact that any arbitrary small mass element of the system that satisfies LTE at all times puts a constraint on the evolution of the whole system. Correspondingly, if relaxation occurs, then it is conceivable that LTE alone may provide us with useful information on the outcome of this relaxation, the relaxed state. Firstly, we start with a closed window.[33] Then, we discuss the case of an open window.[34]

LTE differs from full thermodynamic equilibrium because of the non-vanishing value of some thermodynamic force (Sect. 3.2). In continuous systems, thermodynamic forces are gradients of some function of position (Sect. 4.3.2). In this case, Le Châtelier's principle at LTE leads to the so-called 'restated Second Law' postulated by Schneider and Kay [90]: *the thermodynamic principle which governs the behaviour of systems is that as they are moved away from equilibrium they will utilize all avenues available to counter the applied gradients. As the applied gradients increase, so does the system's ability to oppose further movement from equilibrium.* We stress the following point: once the formal requirement that the thermodynamic forces are gradients is satisfied, no loss of generality affects our discussion, as the selection of a given set of thermodynamic forces is to be given no particular meaning (Sect. 2.5).

Let us recall the definition of the heating power density $P_h \equiv \nabla \cdot \mathbf{q} + \rho T \frac{ds}{dt}$ in Sect. 4.2.8. In most cases discussed above, e.g. whenever the only heating mechanisms inside the system are Joule and viscous heating as well as exothermal (chemical, nuclear) inter-particle reactions heat the system, P_h is just the total amount of heat released per unit time and volume by all heating processes and is, therefore, negative nowhere. We limit ourselves to these cases and drop the dependence on space and time in the following for simplicity. In our room, P_h vanishes outside the heater. However, we may easily remove this assumption. In fact, nothing essential changes if we assume P_h to vanish nowhere: think, e.g. of a great number of microscopic heaters scattered all across the room. Below we suppose the value of the total

---

[32] For our purposes, we shall refer to the couple made of Le Châtelier's principle and the assumption of LTE everywhere at all times indifferently with either the wordings 'Le Châtelier's principle' or 'LTE' here and in the following.

[33] We stress the point that this is just a thought experiment. Any realistic description of rooms with heaters and windows should include the effects of wall permeability, leaks and the like—we refer to Ref. [89] and to the bibliography therein.

[34] Admittedly, our arguments are qualitative. See Ref. [71] for rigorous proofs which stem from another consequence of the same assumption, namely GEC—see Sect. 3.6.

heating power $\int P_h d\mathbf{x}$ to attain a known value P_{TOT} in the relaxed state (the domain of integration is the room volume).

To start with, we require that our relaxed state exists and is stable against perturbations. Now, let the external world lead to a perturbation which slightly raises T inside a small volume $d\mathbf{x}$ inside the room, say centred at $\mathbf{x}$ at a given time t while leaving the rest of the room unaffected. The decomposition $a = a_0 + a_1$ described above for the generic quantity a allows us to write $T_1 > 0$. Since $\frac{1}{T}$ decreases with increasing T, this corresponds to $\left(\frac{1}{T}\right)_1 < 0$. Le Châtelier's principle dictates that the system counteracts the impact of such perturbation. In particular, it follows that $P_{h1} d\mathbf{x} \leq 0$; should $P_{h1} d\mathbf{x}$ be > 0, indeed, a small growth of temperature would raise the heating power density locally, thus heating the room further and triggering a positive feedback which in turn would drive the system far from the initial state.[35] Accordingly, $\left(\frac{1}{T}\right)_1 P_{h1} d\mathbf{x} \geq 0$.[36] But then the decomposition $a = a_0 + a_1$ allows us to write $\frac{P_h}{T} d\mathbf{x} = \left(\frac{P_h}{T}\right)_0 d\mathbf{x} + P_{h1} \left(\frac{1}{T}\right)_0 d\mathbf{x} + P_{h0} \left(\frac{1}{T}\right)_1 d\mathbf{x} + P_{h1} \left(\frac{1}{T}\right)_1 d\mathbf{x} \geq \left(\frac{P_h}{T}\right)_0 d\mathbf{x} + P_{h1} \left(\frac{1}{T}\right)_0 d\mathbf{x} + P_{h0} \left(\frac{1}{T}\right)_1 d\mathbf{x}$. After time-averaging and volume integration on the room volume, we obtain $\overline{\int d\mathbf{x} \frac{P_h}{T}} \geq \left(\overline{\int d\mathbf{x} \frac{P_h}{T}}\right)_0$ where we have taken into account that $\overline{a_1} = 0$. This means that, in order to be stable against perturbations, the relaxed state corresponds to a (constrained) minimum of $\overline{\int d\mathbf{x} \frac{P_h}{T}}$. For the same reason, we may rewrite the constraint as $\overline{\int P_h d\mathbf{x}} = P_{TOT}$. Thus, a relaxed state of the room satisfies the condition:

$$\overline{\int d\mathbf{x} \frac{P_h}{T}} = \min \quad \text{with the constraint} \quad \overline{\int P_h d\mathbf{x}} = P_{TOT}$$

Intuitively, in the relaxed state and on average at least, we expect that T is maximum near the heater and that P_h attains a maximum in the region of space occupied by the heater. Accordingly, we expect that T is maximum where P_h is maximum. Such maximization is equivalent to reduce $\frac{P_h}{T}$ everywhere.

So far, we have said nothing about the frequency spectrum of a_1. Our results hold even if $a_1 \propto \exp(i\omega t)$ with period $\frac{2\pi}{\omega} \ll$ the characteristic relaxation time of out system made of the room, the heater and the window. We may imagine that there is a fan in our room, whose blades keep on rotating at high frequency even after relaxation has occurred. In this case, the relaxed configuration is a periodically oscillating state.

Our discussion invokes no detailed description of the physical mechanism underlying heating. Moreover, nothing is said about the actual origin of the power supplied to the heater: it may be either internal to the room (e.g. burning coal) or external (a

---

[35] Le Châtelier's principle is concerned with perturbations of thermodynamic forces (Sect. 3.1). By definition, P_h is the gradient of no physical quantity, but this raises no problem as we are not in LNET (Sect. 4.3.2) and the freedom in our choice of the relevant thermodynamic forces (Sects. 2.5 and 3.2) allows us to apply Châtelier's principle to P_{h1} anyway.

[36] Had we started from a perturbation which *cools* the room locally, i.e. with $T_1 < 0$, Le Châtelier's principle would have implied the same result.

plug connecting an electric heater to the grid). Furthermore, our result links provide us with a link between the distributions of temperature and heating power density in a stable, steady state even if we know nothing about the exact mechanism of heat transport throughout the room (including, e.g. the values of thermal conductivity, radiation opacity or heat exchange coefficients). Finally, our variational principle provides us with no complete description of the relaxed configuration of our room. In fact, such a description requires further information, e.g. the equation of state of the air in the room, the boundary values of temperature at both the heater, the walls and the window, etc. Different stable profiles of temperature and heating power density may exist, depending on the information listed above. The actual description of the room is provided by the balance equations of mass, momentum and energy, including turbulence, radiative effect, etc. Stable steady solutions of these equations are those steady solutions that satisfy also the constrained minimization above. Such equations act as further constraints on the minimization. Minimization may, therefore, be useful when bringing forward the—often cumbersome—analysis of the stability of the solutions of the balance equations.

The minimized quantity $\overline{\int \frac{P_h}{T} d\mathbf{x}}$ has the dimension of entropy/time and may be considered as the amount of entropy produced per unit time by the heating processes inside the room. Remarkably, it differs from the total amount of entropy produced per unit time by all irreversible processes. To fix the ideas, let a small, helium-filled balloon be inside the room at the initial time, and let a very small hole be on the balloon, so that the helium may leak slowly. After a while, the helium diffuses freely across the room. This is definitely an irreversible process that raises entropy. However, the corresponding entropy production is related in no way to the heating process in the heater and leaves, therefore, our discussion unaffected. As we shall see below, different contributions to the entropy balance enjoy different properties in a stable state.[37]

6.2.2   Convection at Moderate Ra, Retrieved

When dealing with the problem of convection in a fluid with Joule heating and viscous heating at values of Ra just above threshold, we invoke the definition of P_h and take

---

[37] This point is usually overlooked in the literature. For example, minimization of $\int d\mathbf{x} \frac{P_h}{T}$ is invoked [91,92] in order to compute the laminar flame velocity (Sect. 5.8.5). The discussion includes both heat conduction and exothermal combustion reactions in a simplified way. Basically, however, the theory relies on LNET, which is scarcely satisfied in laminar flames; for instance, the validity of Onsager's relationships requires that the heat conductivity is uniform across the flame, which is definitely not the case. Moreover, it is not clear how to reconcile a MinEP with the large values of $\int \mathbf{q} \cdot \nabla \left(\frac{1}{T} \right) d\mathbf{x}$ at flame quenching. Admittedly, it is still possible to derive the relevant balance equations in a simplified form in the framework of LNET; but the price to be paid is the introduction of the flame velocity itself into the explicit expression of both the thermodynamic fluxes and the phenomenological coefficients (see, e.g. Eq. (2.23) of [92]), a price which casts further doubt on the validity of the underlying assumption of Onsager's symmetry.

the time-average of $\int \frac{P_h}{T} d\mathbf{x}$ on many periods of rotation of the convective cells. We obtain $\int \frac{P_h}{T} d\mathbf{x} = \int \frac{\nabla \cdot \mathbf{q}}{T} d\mathbf{x} + \int \rho \frac{ds}{dt} d\mathbf{x}$, where[38] $\int \rho \frac{ds}{dt} d\mathbf{x} = \frac{d}{dt} \int \rho s d\mathbf{x} = \frac{dS}{dt} = 0$.[39] Since convection is the physical process which is mainly responsible for the transport of heat and since is triggered by the temperature gradient, it is only reasonable to assume that $\int \frac{\nabla \cdot \mathbf{q}}{T} d\mathbf{x}$ increases with increasing $|\nabla T|$, and that the two quantities are proportional to each other for small temperature gradient (i.e. not so large Ra) at least. Then, minimization of $\int \frac{P_h}{T} d\mathbf{x}$ in Sect. 6.2.1 corresponds to minimization of Ra.

As for the constraint, conservation of energy requires that the time-average over many periods of rotation of the convection cells of the power dissipated (both via Joule heating and viscous heating) is equal to the time-averaged value of the power provided by the buoyancy force. We have, therefore, retrieved Chandrasekhar's result of Sect. 5.3.11.

Minimization of Ra implies also minimization of $\int_0^{\epsilon 0} Ra(\epsilon) d\epsilon$, like in Ref. [93]. Moreover, Busse's constraint of given convective heat flux follows from the fact that (just above threshold at least) this heat flux increases with increasing typical velocity of the fluid in the convection cell, which is in turn an increasing function of ε_g, and Chandrasekhar's constraint is, therefore, retrieved.

6.2.3   Detonation Waves, Retrieved

Detonation waves (Sect. 5.6.11) provide us with an example of the application of the necessary condition of stability. The contribution of the heat flux to the entropy balance is negligible in detonation waves, hence the local form of the entropy balance of Sect. 4.2.8 reduces to $\rho \frac{ds}{dt} = \frac{P_h}{T}$. Volume integration[40] on the volume V_S occupied by a wavefront of given area A_S for $\frac{\partial}{\partial t} = 0$ leads to $\int_V d\mathbf{x} \frac{P_h}{T} = \int_V d\mathbf{x} \nabla \cdot (\rho s \mathbf{u}) = \oint d\mathbf{a} \cdot \rho s \mathbf{v} = A_S [\rho s v_n]$.[41] Minimization of the L.H.S. (constrained by a constant heat release per unit time) reduces, therefore, to the minimization of $[\rho s v_n]$, and the result of Sect. 5.6.11 is retrieved.

---

[38] By the identity $\frac{d}{dt} \int \rho a d\mathbf{x} = \int \rho \frac{da}{dt} d\mathbf{x}$ with $a = s$.

[39] We invoke the identity $\frac{da}{dt} = 0$ for the generic quantity $a = a(t)$ which satisfies $|a(t)| < \infty$ at all times here, with $a = s$.

[40] The identity $\frac{d}{dt} \int_V d\mathbf{x} a = \int_V d\mathbf{x} \frac{\partial a}{\partial t} + \int_V d\mathbf{x} \nabla \cdot (a\mathbf{u})$ holds, where $\mathbf{u}$ is the speed of the generic point on the boundary surface of the integration volume V. If V is the fluid volume, then $\mathbf{u} = \mathbf{v}$ on the boundary of the fluid.

[41] Again, we have invoked Gauss' theorem of divergence.

6.2.4   Two Applications of Bejan's Constructal Law

When it comes to the description of the network of roads which minimizes the time required to reach a given destination M inside a region Ω (Sect. 5.4), Bejan [94] stresses the point that this optimal network is the outcome of an iterative algorithm. Each step requires time, then the solution of the variational problem (where the constraint is the number $\dot{N}$ of uniformly distributed persons reaching M per unit time) is the result of an evolutionary process. It follows that—once this optimum configuration has been achieved—any perturbation evolves back to it, i.e. that the relaxed state is stable against perturbations. If e.g. an obstacle (say, a landslide) suddenly obstructs a road, travellers will change their path: as the obstacle is removed, however, people will keep on following the old road provided that such road minimizes the time required to reach M and provided that the traffic $\dot{N}$ remains unchanged.[42] This constrained minimization is, therefore, a necessary condition of stability. Moreover, it describes a relaxed state which does not oscillate; then, we may drop the time-average and write the necessary condition of stability in Sect. 6.2.1 as $\int \frac{P_h}{T} d\mathbf{x} = \min$ with the constraint $\int P_h d\mathbf{x} = P_{TOT}$. Let us investigate the consequences of this condition.

The total number of persons travelling in a time Δt towards M is $\Delta t \cdot \dot{N}$. In a small section of the trip towards the final destination, each traveller needs a positive-definite amount dW of mechanical power, which delivers an amount $\frac{dW}{T}$ of entropy per unit time[43] inside Ω. The total amount of entropy released by one traveller is $\int \frac{dW}{T}$, the integral being the traveller's path. No traveller is privileged. Then, the total amount of entropy produced in the bulk of Ω (which we referred to as to $\int \frac{P_h}{T} d\mathbf{x}$) is the product of the total number $\Delta t \cdot \dot{N}$ of traveller and of the averaged amount $\langle \int \frac{dW}{T} \rangle$ of entropy produced per unit time by one traveller, where the average is taken over all travellers. Formally, we may, therefore, write $\int \frac{P_h}{T} d\mathbf{x} = \Delta t \cdot \dot{N} \cdot \langle \int \frac{dW}{T} \rangle$, and since all factors on the R.H.S. are positive, the minimization of the L.H.S. implies minimization of Δt.

As for the constraint, no heat flows (so that $\mathbf{q} = 0$). Then, the definition of P_h allows us to write $\int P_h d\mathbf{x} = \int \rho T \frac{ds}{dt}$ where $\rho = nm$ and n, m, T and s are the density of travellers, their mass (supposed to be the same for all travellers for simplicity), their temperature (equal for all travellers, nobody has a fever) and the entropy per unit mass of traveller, respectively. Since T is uniform, we write $\int \rho T \frac{ds}{dt} d\mathbf{x} = T \int \rho \frac{ds}{dt} d\mathbf{x}$ and the definition of $\frac{d}{dt}$ allows us to write $\int \rho \frac{ds}{dt} d\mathbf{x} = \int d\mathbf{x} \rho \left(\frac{\partial s}{\partial t} + \mathbf{v} \cdot \nabla s \right) =$

---

[42] Here, we describe traffic as a (possibly non-Newtonian) fluid.

[43] For example, if a person with 70 Kg weight climbs vertically upwards 100 m, then the required amount of energy is 68600 J, to be supplied by food. If the temperature of this person is 36.5 °C, then the person's entropy increases by 221.5 J · K^{-1} through irreversible processes like the metabolism, etc. In a steady state, the person's entropy is constant; hence, the same amount of entropy is supplied by the person to the surrounding environment while climbing (by radiating energy away, by sweating, by the friction of the boots, etc.). If the climbing takes one hour, the amount of entropy $\frac{dW}{T}$ delivered per unit time inside Ω in this long section of the trip is 0.062 J · K^{-1} · s^{-1}.

$\int d\mathbf{x} \left[\frac{\partial(\rho s)}{\partial t} + \nabla \cdot (\rho s \mathbf{v}) \right] - \int d\mathbf{x} s \left[\frac{\partial \rho}{\partial t} + \nabla \cdot (\rho \mathbf{v}) \right]$. The first term in the R.H.S. is equal to $\frac{d}{dt} \int \rho s d\mathbf{x} = \frac{dS}{dt} = 0$ because the total entropy S of the travellers is constant in steady state. After dividing and multiplying by m, the second term is equal to $-(ms) \cdot \left(\frac{1}{m} \right) \cdot \int d\mathbf{x} \left[\frac{\partial \rho}{\partial t} + \nabla \cdot (\rho \mathbf{v}) \right]$ as the entropy ms of each traveller does not change during the trip; finally, the fact that a constant mass $m\dot{N}$ of persons per unit time leaves Ω after reaching M means that M acts as a sink of mass with $\int d\mathbf{x} \left[\frac{\partial \rho}{\partial t} + \nabla \cdot (\rho \mathbf{v}) \right] = -m\dot{N}$. Thus, the constraint of fixed $\int P_h d\mathbf{x}$ is equivalent to a constraint of fixed $T (ms) \dot{N}$, i.e. of fixed $\dot{N}$ as both the temperature and the entropy content of each traveller are given.

For the roads' problem at least, we conclude that Bejan' variational principle (minimize Δt at fixed $\dot{N}$) follows from the necessary condition of constrained minimization of $\frac{P_h}{T} d\mathbf{x}$ for the stability of relaxed states against perturbations, a condition which in turn follows from Le Châtelier's principle. The latter provides the constructal law with a sound physical ground, as it allows stability of the optimum configuration which the system tends to according to the constructal law.

A similar treatment is relevant to the problem [95] of locating and connecting the heat conductors in a region Ω filled with electronic circuits. The constraint $\int d\mathbf{x} P_h = P_{TOT}$ is obviously just the constraint of a given amount of Joule heating power, provided that P_h is the Joule heating power density. If there is no heating (i.e. the circuitry is switched off), then $\nabla T = 0$ and $\int \frac{P_h}{T} d\mathbf{x} = 0$. As far as $|\nabla T|$ remains small, the non-negative quantity $\int \frac{P_h}{T} d\mathbf{x}$ is an increasing function of it (see our discussion on Chandrasekhar's principle in Sect. 5.3.11). If there is heating and the configuration is optimized, then $|\nabla T|$ is small (but $\neq 0$) and $\int \frac{P_h}{T} d\mathbf{x} \propto O(|\nabla T|)$; minimization of $|\nabla T|$ corresponds, therefore, to minimization of $\int \frac{P_h}{T} d\mathbf{x}$, as required in [95]. This seemingly trivial result deserves further attention.

We have seen that Le Châtelier provides us with a necessary condition of stability against perturbations. The exact meaning of the wordings 'stability against perturbations' remains to be discussed in the problem of the heat conductors. Of course, once the optimal solution has been found, any deviation from it is far from triggering a spontaneous evolution of the network of heat conductors back to the optimal state, as the heat conductors do not move—in contrast with the travellers in the problem above—and do not change their own location spontaneously. We are not allowed to speak of 'spontaneous relaxation' here: intelligent design, not spontaneous evolution, leads to better and better approximations to the optimal solution—see, e.g. the thought-provoking discussion in [96].

But be that as it may, it remains true that if an external disturbance—say, a grain of dust—obstructs the optimal heat flow, then an intelligent operator has to restore the original, optimized configuration in order to prevent excessive, undesired heating—either by removing the grain or by replacing the whole motherboard altogether with a new one. In other words, once the system made of both the circuitry and the operator has achieved an optimum state (as far as Joule heating is concerned, at least) and the external world has pushed this system away from this optimum, then the system answers in order to counteract this disturbance.

This pattern of action is formally equivalent to the restoring processes which occur in a stable system when perturbed and described by Le Châtelier's principle—regardless of the actual nature of the restoring factor, be it intelligent design or mindless, spontaneous relaxation.

The examples above suggest that thermodynamics (in particular, the LTE assumption everywhere at all times) provides us with the same necessary condition for stability of the steady state against perturbations—namely, Le Châtelier's principle and its consequences—no matter how complicated its structure and regardless of its origin (intelligent design, mindless evolution...). This may explain the striking similarity between some artificial structures and some fractal-like structures in Nature.

6.2.5   Kohler's Principle

Our discussion of Chandrasekhar and Bejan's problems in Sects. 5.3.11 and 5.4, respectively, has focussed on the minimized quantity in Sect. 6.2.1, the entropy produced per unit time in the bulk of the system. Now, let us discuss the constraint in more detail. After multiplication by temperature, integration on the volume of the system and suitable time-averaging of both sides, the local balance of entropy of Sect. 4.2.8 leads to the following normalization condition:

$$\overline{\int P_h d\mathbf{x}} = \overline{\int \mathbf{q} \cdot d\mathbf{a}}$$

provided that $\overline{\int \rho T \frac{ds}{dt} d\mathbf{x}} = 0$, i.e. no net time-averaged amount of heat is delivered into the system. This may occur, e.g. either in a steady state or in cyclic transformations. The relaxed state satisfies, therefore, minimization of the entropy produced by heating inside the system with the constraint of a given net flux of heat across the boundaries. We stress the point that this is just a necessary condition for stability, to be satisfied by those solutions of the equations of motion which are also stable. In particular, depending on the problem of interest, further constraints may be applied. This result agrees with the result of Kohler [97], who has rigorously shown that the entropy produced per unit time and volume by inter-particle collisions in a gas that is subject to a given amount of heat flow and friction stress and is described by Boltzmann's kinetic equation is a minimum in steady state. Viscosity rules heating in a gas; however, Kohler's analysis extends to the transport of electrons in a metal, and in this case, the result applies also to Joule heating. Kohler's minimization is the fundamental tool for the computation of transport coefficients in systems at LTE even in the quantum mechanical domain [98]. Kohler assumed both vanishing magnetic field and small perturbances of thermodynamic equilibrium; both assumptions have been later removed [99].[44]

---

[44] Not surprisingly, the relaxed state obtained in Kohler's analysis satisfies the condition that the amount of entropy produced per unit time and by molecular collisions is equal to the macroscopic

6.2.6   Entropy Production in a Radiation Field

Just like Le Châtelier's principle, the normalization constraint discussed in Sect. 6.2.5 holds regardless of the detailed nature of the heat transport mechanism. For example, radiative transfer carries energy through a succession of many events—including the emission, the transmission, the scattering and the absorption of photons. As a whole, however, the radiative transfer is an irreversible process, and it is perfectly meaningful to speak of the entropy density of radiation—for a thorough discussion, see Ref. [101].[45] In particular, when considering a set of atoms exchanging photons with each other, we may focus our attention on the set of photons with one particular frequency. With respect to this particular system, the 'external world' is made of both the atoms and all other photons. For the photons of our system, sending energy to the external world means just being absorbed by some atom, which will then interact with other atoms through other photons, etc. Analogously, our system receives energy from the external world when an atom emits a photon with the frequency of the photons of our system. The net energy flow to and from the external world depends on the state of the atoms, as the latter can emit and absorb photons.[46] If the latter state is given, then our constrained minimization in Sect. 6.2.1 dictates that our system of photons has minimum $\overline{\int d\mathbf{x} \frac{P_h}{T}}$.

The physical meaning of this conclusion is clear if we recall that emission, scattering and absorption are the only physical processes occurring in our system, and that the only contribution to P_h of these processes is of electromagnetic origin. After radiative transfer of heat has flattened all gradients of temperature while the photons reaching equilibrium with all the matter interacting with them, the only surviving contribution to the entropy production density σ in the local entropy balance of Sect. 4.2.8 is the electromagnetic one, which describes precisely the interaction of the electromagnetic field with matter.[47] Thus, we conclude that the amount of entropy produced per unit time in our system of photons by emission and absorption tends to a minimum as the system relaxes to a stable, steady state. Our result is confirmed by the quantum mechanical analysis of [103].[48]

Remarkably, this result is not subject to the restrictive conditions (linear phenomenological relations, constant phenomenological coefficients) of LNET; it is not

---

entropy production, calculated as the product of the heat flux and the corresponding thermodynamic force. This is a quite natural condition indeed, as far as the small perturbances of thermodynamic equilibrium are assumed to be small. If this result is taken as a constraint, replacing the constraint of fixed heat flow, then the relaxed state corresponds to a maximum of entropy production, according to [100] and in agreement with the reciprocity principle of Sect. 5.3.12.

[45] For example, a homogeneous monochromatic ray of light or heat is, from a phenomenological standpoint, necessarily endowed with entropy. This is due to the fact that the natural electromagnetic waves it is composed of are subject to random fluctuations in phase, amplitude and direction. See Ref. [102] and Refs. therein.

[46] Here, we neglect photon-photon interactions.

[47] In quantum mechanical language, it takes into account both emission and absorption of photons.

[48] A particular case is retrieved in the model of steady-state atmosphere of Ref. [102]—see Sect. 6.2.12.

even required that the total number of photons remains constant during the relaxation (in contrast, e.g. with the number of electrons in Kohler's treatment of electron transport across a metal—see Sect. 6.2.5).

Even the electromagnetic nature of the radiation is not a necessary requirement for the validity of Le Châtelier's principle. Irreversible transfer of energy may as well occur through many emission, scattering and absorption events of particles different from photons. *A fortiori*, the analogy holds when dealing with particles whose total number (like photons) does not remain necessarily constant and have extremely low rest mass (photons have exactly zero rest mass). Accordingly, we expect that nothing changes if we replace photons, e.g. with neutrinos. This result is confirmed by the quantum mechanical analysis of Ref. [104].

As for gravitation, the role of entropy is still the subject of a hot debate. As discussed in Sect. 3.4.1, it is difficult to achieve LTE with gravity but in selected cases.[49] Remarkably, however, the strongest argument in favour of the physical existence of gravitational waves, which some physicists (including Eddington and Einstein) were tempted to consider just like mathematical artifacts of General Relativity, was put forward by Feynman well before their detection in 2015 and involves entropy [106]. Let us focus our attention on a bead sliding freely (but with a small amount of friction) on a rigid stick, and let a gravitational wave impinge on the system made of the stick and the bead. If the stick is oriented transversely to the direction of propagation of the wave, then General Relativity predicts that the wave causes the bead to slide back and forth. As a consequence, friction produces heating of the bead and the stick and raises entropy. Of course, the physics of gravity does not depend on the detailed description of the friction between the bead and the stick. On the other hand, entropy growth is no mathematical artifact; the wave carries energy from the source (say, a celestial body far away) and this energy is dissipated by viscosity in the stick+bead system. Just like the entropy growth of this system, gravitational waves are, therefore, no mathematical artifact, but a solid physical reality.

6.2.7   Uniform Temperature: A Reciprocal Problem,...

The limit of the vanishing temperature gradient deserves further discussion. This limit is justified whenever heating processes are not able to provide any significant temperature gradient across the system, due to competing energy transport. We limit ourselves to steady relaxed states for simplicity. The constrained minimization in Sect. 6.2.1 reduces to $\frac{1}{T} \int P_h d\mathbf{x} = \min$ with the constraint $\int P_h d\mathbf{x} = P_{TOT}$, which is equivalent to $T = \max$ with the constraint $\int P_h d\mathbf{x} = P_{TOT}$. Let the actual value of T be T_0. Then, the reciprocity principle of Sect. 5.3.12 ensures that the solution

---

[49] One could argue that one of the most topical problems in gravitational physics involves two stars (or black holes) in a binary system, which lose angular momentum through the emission of gravitational waves. The emission of gravitational waves allows irreversible collapse and is ruled by General Relativity, whose fundamental equations satisfy a variational principle [105]. But then, the collapsing system is scarcely in far-from-equilibrium, steady relaxed states until the end.

of the latter variational problem is also the solution of the 'reciprocal' variational problem:

$$\int P_h d\mathbf{x} = \min \quad ; \quad \text{with the constraint} \quad T = T_0$$

6.2.8   ...Joule Heating,...

In an electrical conductor, if $P_h = \frac{|\mathbf{j}_{el}|^2}{\sigma_\Omega}$ with $\nabla \cdot \mathbf{j}_{el} = 0, \sigma_\Omega = \sigma_\Omega\,(T)$ and $\nabla T = 0$, then we retrieve Kirchhoff's principle of Sect. 5.3.1 from the variational principle of Sect. 6.2.7. Our discussion invokes neither any explicit form of Ohm's law nor the assumption of fixed resistivity, unlike Hameiri et Bhattacharjee [107]. Unlike Kohler [97], we require neither $\mathbf{B} = 0$ nor small applied electric fields.[50] Unlike Peters [108], we assume no Onsager symmetry relationship. Finally, we do not assume $\nabla \wedge \nabla \wedge \mathbf{j}_{el} = 0$, unlike Montgomery et Phillips [109]. Beyond the range of validity of linear Ohm's law, minimization of Joule dissipated power at a given electric current is independently retrieved in Ref. [110].

Being the total Joule power just equal to the product of electric current and applied voltage, Kirchhoff's principle and the reciprocity principle of Sect. 5.3.12 lead to the conclusion that if we replace the constraint of given electric current with the constraint of given applied voltage, then the stable steady state corresponds to a maximum of current, in agreement with the reciprocity principle of Sect. 5.3.12—for a discussion, see [111]. In this case, the stability criterion takes the form of maximization of the entropy production [112] (Sect. 5.6.5) and provides us with predictions in agreement with Steenbeck's principle of Sect. 5.3.2 [113] with no need of invoking MEPP. Such maximization underlies many attempts at proving MEPP—for a review, see [114]. A spectacular confirmation of the spontaneous maximization of Joule heating in a system of carbon nanotubes at constant voltage may be found in Ref. [115].

In the relaxed states of plasmas which are both magnetized, electrically conducting, viscous and turbulent, $\nabla T = 0$ due to competing energy transport. Correspondingly, constrained minimization of the sum P_h of Joule and viscous heating power with the constraints provided by the balance equations of mass and momentum, Maxwell's equations of electromagnetism and the proper expression of Ohm's law plasmas [116] leads to the same Euler-Lagrange equations (in the Lagrangian coordinates $\mathbf{B}$ and $\mathbf{v}$) of Taylor's principle [25] and Turner's principle [26] of Sect. 6.1.4 in the domain of validity of MHD and Hall MHD, respectively. The Euler-Lagrange equations of Turner's principle are formally similar to the Euler-Lagrange equations of the constrained minimization of Joule heating power of Ref. [27] (see Sect. 6.1.4) and describe plasmoids in space physics [117].[51]

---

[50] As for the corresponding generalization of the kinetic treatment of Ref. [97], see Ref. [99] and the bibliography therein.

[51] Remarkably, tokamak plasmas (Sect. 5.6.8) in steady state satisfy neither Taylor's nor Turner's principles. This is far from surprising, as $\nabla T \neq 0$ almost everywhere in tokamak plasmas.

6.2.9   ...Viscous Heating,...

In a viscous fluid, if $P_h = \eta |\nabla \wedge \mathbf{v}|^2$ with $\eta = \eta(T)$ and $\nabla \cdot \mathbf{v} = 0$, then we retrieve Korteweg-Helmholtz' principle of Sect. 5.3.7 from the variational principle of Sect. 6.2.7. As discussed above, the requirement $\nabla T = 0$ throughout the fluid implies that the heat produced by viscosity is small; for given viscosity, this implies that the absolute values of the gradients of the components of $\mathbf{v}$ are also small. Then, we expect Korteweg-Helmholtz to hold for $Re \ll 1$ only.

Our discussion of Sect. 6.2.1, however, does *not* rely on the assumption that the fluid is Newtonian. Extension of the minimization of viscous power to $Re \ll 1$ flows of non-Newtonian fluids follows. This generalization has been rigorously proven by Schechter in [17] for non-Newtonian fluids of the Reiner-Rivlin type in the $Re \ll 1$ limit. In a loose analogy with Helmholtz' approach (Sect. 5.3.10), Schechter looks for an extremum of the sum of P_{ext} and of a quantity which increases with increasing $\frac{dK}{dt}$, namely the volume integral of $\frac{1}{2} \left(\frac{\partial v_i}{\partial x_j} + \frac{\partial v_i}{\partial x_j} \right) \Pi_{ij}$ where $i, j = 1, 2, 3$ and Π_{ij} is the momentum flux density tensor of the fluid [36] (with sligthly modified viscosity coefficients); given values of stresses and velocity on the boundary are the constraint. This is a true variational principle, i.e. its Euler-Lagrange equations are the conservation equations for the fluid. A tenet of the proof is that all phenomenological coefficients which play the role of viscosities are assumed to be constant and uniform quantities, just like η is constant and uniform in Korteweg-Helmholtz' principle. In the case of Reiner-Rivlin fluids, however, this assumption is not trivial [118] and requires separate confirmation on a case-by-case basis—see Ref. [119] for an example. Moreover (and this is the worse limitation), even some components of the momentum flux density tensor are assumed to be in the unperturbed state and are, therefore, not subject to variation: this limits the domain of applicability of Schechter's principle to small perturbations and suggests that MinEP holds only locally in this case. Korteweg-Helmholtz' result can be generalized also to other types of non-Newtonian fluids—see Ref. [120] and the bibliography therein.[52]

Generalization to periodic motions is straightforward, provided that we replace the minimized quantity with its time-averaged value, the average being taken over many periods of oscillation as in Sect. 5.8.1. When applied to the circulation of blood in the human body (where $\nabla T \approx 0$), the principle of maximum economy of Sect. 5.3.8 follows, as the human blood can be described as a Reiner-Rivlin non-Newtonian fluid with constant and uniform coefficients [121]. Analogously, generalization to the $\nabla T = 0$, $Re \gg 1$ problems of the formation processes of river basins and of yardangs (which are affected by many cyclic phenomena both on the circadian and

---

[52] A different variational principle for Reiner-Rivlin fluids has been proposed which differs from Korteweg-Helmholtz and which holds even if Re is not $\ll 1$, but which postulates that the steady flow whose stability is investigated undergoes only small perturbations, thus allowing only local stability analysis against such perturbations [18]. This work is but the extension to viscous fluids of previous work on diffusion [20]—see Sect. 6.1.3.

the seasonal time-scale) allow us to retrieve the results of [122,123] of Sect. 5.3.8, respectively.

6.2.10 ... And Porous Media

In the porous system of Sect. 5.3.9, we focus our attention on water. In particular, we may safely assume with no loss of generality that the motion of water is so slow that (a) ρ is constant and uniform; (b) the contribution of kinetic energy to the energy balance of Sect. 4.2.8 is negligible; (c) there is enough time for flattening of ∇T everywhere, so that a relaxed state satisfies $\nabla T = 0$ and corresponds to some suitably constrained extremum of $\int P_h d\mathbf{x}$ where $P_h = \rho T \frac{ds}{dt}$ as both heat conduction and radiation are negligible for $\nabla T = 0$ and s is the entropy per unit mass. Here, we keep on invoking the pressure p; should we replace p with the capillary pressure p_c (as in the unsaturated case), no result would formally be affected in the following. There is no electromagnetic field. Accordingly, the energy balance of water in steady state (Sect. 4.2.7, $\frac{\partial}{\partial t} \equiv 0$) reduces to $\nabla \cdot (\rho \mathbf{v} h) = 0$ where $h = u + \frac{p}{\rho} + \phi_g$ is the enthalpy per unit mass (Sect. 3.4.2) and $\phi_g = gz$, g and z are the gravitational potential, the absolute value of the gravitational acceleration and vertical coordinate, respectively, as usual. Moreover, the mass balance in steady state reads $0 = \nabla \cdot (\rho \mathbf{v}) = \rho \nabla \cdot \mathbf{v} = \frac{\rho}{\varphi_p} \nabla \cdot \mathbf{q}_w$ with φ_p porosity of the medium and $\mathbf{q}_w$ volumetric water flow. Together, the energy balance and the mass balance give $0 = \rho \mathbf{v} \cdot \nabla h = \rho \mathbf{v} \cdot \nabla (u + \frac{p}{\rho} + \phi_g)$. Being $d\rho \equiv 0$ and $\frac{\partial}{\partial t} \equiv 0$, we obtain $\rho \mathbf{v} \cdot \nabla u = \rho \frac{du}{dt} = \rho T \frac{ds}{dt} = P_h$ as $du = T ds$ for the First Principle of thermodynamics. On the other hand, being $\nabla \cdot \mathbf{q}_w = 0$ (for mass conservation), $\phi_g = gz$ and $\nabla g = 0$ (for uniform gravitational field across the system), we obtain $\rho \mathbf{v} \cdot \nabla (\frac{p}{\rho} + \phi_g) = \rho g \mathbf{v} \cdot \nabla H_w = \frac{\rho g}{\varphi_p} \mathbf{q}_w \cdot \nabla H_w = \frac{\rho g}{\varphi_p} \Delta E_c$ with H_w hydraulic head and ΔE_c energy expenditure rate per unit volume defined in Ref. [124]. It follows that

$$P_h = -\frac{\rho g}{\varphi_p} \Delta E_c$$

Accordingly, the minimization of $\int \Delta E_c d\mathbf{x}$ postulated in Sect. 5.3.9 corresponds to maximization of $\int P_h d\mathbf{x}$. In order to justify this conclusion, let us investigate further the physical processes underlying P_h. Viscosity, capillarity and gravity rule the system. Unlike viscosity, gravity leaves the entropy balance unaffected. As for capillarity, it is due to the coexistence of water and (say) air. Its contribution to the total time derivative of water is $\Sigma \int (\frac{1}{T_{water}} - \frac{1}{T_{air}}) \mathbf{q} \cdot d\mathbf{a}$, where the sum is extended to all regions of the system where air and water coexist and in each region the integral is computed on the interface surface between air and water, $\mathbf{q}$, T_{water} and T_{air} are the heat flux coming from air into water across this surface, the temperature of water and of air, respectively—see Eq. (7.5) of [125]. Being $\nabla T = 0$ everywhere, $T_{water} = T_{air}$ and the contribution of capillarity to the entropy balance vanishes. Accordingly, the same condition $\nabla T = 0$ which allows us to describe the stability of the relaxed state with the help of $\int P_h d\mathbf{x}$ prevents us from properly taking into account the impact of capillarity. As for non-equilibrium thermodynamics, the problem of water flow in porous media is indistinguishable from the problems discussed above of slow viscous fluids acted upon by a constant drag—here, gravity.

Mass conservation $\nabla \cdot \mathbf{q}_w = 0$ and Darcy's law $\mathbf{q}_w = -K_w \nabla H_w$ give $\Delta E_c = \nabla \cdot (\mathbf{q}_w H_w) = \mathbf{q}_w \cdot \nabla H_w = -K_w |\nabla H_w|^2$.[53] Accordingly, $\frac{\varphi_p}{\rho g} \int K_w |\nabla H_w|^2 dx = -\frac{\varphi_p}{\rho g} \int \Delta E_c dx = \int P_h dx = \int \rho T \frac{ds}{dt} dx = T \int \rho \frac{ds}{dt} dx = T \frac{dS}{dt} = T \int \sigma dx$[54] with S, $\sigma = X_i Y_i$,[55] $X_i = -\frac{\varphi_p}{\rho g T} \frac{\partial H_w}{\partial x_i}$ and $J_i = -K_w \frac{\partial H_w}{\partial x_i} = q_{w_i}$ total entropy of water, entropy production density, thermodynamic force and thermodynamic flux, respectively.[56] Our choice of the X_i's and the J_i's is correct, as the former and the divergence of the latter are the gradient of a physical quantity[57] and zero,[58] respectively—see Sect. 4.3.2.

Let us explain both the choice of the constraint in Sect. 5.3.9 and why $\int P_h dx$ gets maximized rather than minimized. For negligible capillarity $|\nabla H_w| = 1$ and $\int |\nabla H_w|^2 dx = V_{tot}$ (V_{tot} total volume) and the variational principle $\int \Delta E_c dx = $ min with the constraint of fixed $\int |\nabla H_w|^2 dx$ of Sect. 5.3.9 reduces to $\int P_h dx = $ max with the constraint of fixed $\frac{\rho g V_{tot}}{\varphi_p} = -\frac{V_{tot}}{T \cdot X_3}$, i.e. fixed X_3; X_3 is the only non-zero thermodynamic force acting on the system. The corresponding thermodynamic force is $J_3 = q_{w_3}$ and is uniform across the system because of mass conservation.[59] Accordingly, the variational principle of Sect. 5.3.9 becomes $\int P_h dx = T \int \sigma dx = T V_{tot} X_3 q_{w_3} = $ max with fixed X_3, i.e. $q_{w_3} = $ max with fixed X_3.[60] But then, the reciprocity principle of Sect. 5.3.12 ensures that the solution of this variational problem is also the solution of the problem $X_3 = $ min with fixed q_{w_3}. This is equivalent to minimization of $\int P_h dx = T V_{tot} X_3 q_{w_3}$ as both T and V_{tot} are fixed. Being q_{w_3} uniform across the system, moreover, the constraint of fixed q_{w_3} is equivalent to the constraint of fixed total water flow across the system, as required by mass conservation $\nabla \cdot \mathbf{q}_w$ which plays the role of equation of motion. Thus, if capillarity is negligible, then the variational principle of Sect. 5.3.9 is equivalent to the constrained minimization of $\int P_h dx$ with the constraint of fixed T of Sect. 6.2.7 and with the further constraint of mass conservation. This minimization is precisely the requirement for the stability of the relaxed state which follows from Le Châtelier's principle. We

---

[53] Two remarks. Firstly, the fact that $\Delta E_c \leq 0$ agrees with the fact that $\Delta E_c \propto P_h \geq 0$. Secondly, the larger K_w, the lower ΔE_c. Accordingly, all the rest being equal relaxation tends to drive the system towards ever larger K_w, i.e. towards the saturated state, provided that the boundary conditions allow it (i.e. provided enough water is poured upon the system from above). In other words, water tends to occupy all available pores, in agreement with physical intuition.

[54] As we neglect the contribution of heat flux for $\nabla T = 0$.

[55] Where $i = 1, 2, 3$ and we choose to link $i = 3$ with the vertical component, with no loss of generality.

[56] We have invoked Darcy's law in writing J_i.

[57] This physical quantity is $-\frac{\varphi_p H_w}{\rho g T}$.

[58] Which is trivially equal to the partial time derivative of any constant physical quantity.

[59] $\nabla \cdot \mathbf{q}_w = 0$ reduces to $\frac{\partial v_z}{\partial z} = 0$.

[60] As both T and V_{tot} are fixed.

do not require that K_w does not depend on $\mathbf{q}_w$; our results hold also for nonlinear Darcy's law, e.g. for fingering.

6.2.11 A 2nd Necessary Condition for Stability

So far we have dealt with the heater. Now, let us discuss the window. As far as it remains closed, it prevents mass from being exchanged between the room and the external world. Nevertheless, radiation and conduction of energy still occur across the window. The total amount of energy in the room remains constant in a stable, steady state. Accordingly, the power input $\int P_h d\mathbf{x}$ supplied by the heater to the room is to be compensated by a power loss across the window. This loss occurs either through radiation or conduction as convection is forbidden; the corresponding net heat flux $\mathbf{q}$ across the window satisfies the normalization condition $\overline{\int P_h d\mathbf{x}} = P_{TOT}$. The domain of integration in the L.H.S. is the window surface, but generalization to the walls, the floor, etc. is trivial; the 'window' represents the boundary of our volume of interest. Again, let the external world lead to a perturbation that slightly raises T inside a small area with surface vector $d\mathbf{a}$ on the window of the room, while leaving the rest of the window unaffected. Le Châtelier's principle dictates that the system counteracts the impact of such perturbation. In particular, it follows that $\mathbf{q}_1 \cdot d\mathbf{a} \geq 0$; should $\mathbf{q}_1 \cdot d\mathbf{a}$ be < 0, indeed, a small growth of temperature would lower the amount of energy flowing away from the room across the window towards the external world, thus heating the room further and triggering a positive feedback which in turn would drive the system far from the initial state. Accordingly, $\left(\frac{1}{T}\right)_1 \mathbf{q}_1 \cdot d\mathbf{a} \leq 0$.[61] Had we started from a perturbation that cools the room locally, i.e. with $T_1 < 0$, Le Châtelier's principle would have implied the same result. The line of reasoning looks like our argument above concerning $\frac{P_h}{T}$ from here on out.[62] The decomposition $a = a_0 + a_1$ allows us to write $\overline{\int d\mathbf{a} \cdot \frac{\mathbf{q}}{T}} \leq \left(\int d\mathbf{a} \cdot \frac{\mathbf{q}}{T}\right)_0$ after time-averaging and surface integration over the window. This means that the relaxed state corresponds to a (constrained) maximum of $\overline{\int d\mathbf{a} \cdot \frac{\mathbf{q}}{T}}$. Thus, a relaxed state of the room satisfies the condition:

$$\overline{\int \frac{\mathbf{q}}{T} \cdot d\mathbf{a}} = \max \quad \text{with the constraint} \quad \overline{\int P_h d\mathbf{x}} = P_{TOT}$$

Remarkably, the maximization involves no information concerning either the actual position of the heater or the heating process in the heater. Physically, this is far from surprising, as $\frac{\mathbf{q}}{T}$ has the dimension [entropy/(time · surface)] of an entropy flux related to heat radiation and conduction, and these physical processes have nothing to do with the heating process. It follows that the two necessary conditions for stability

---

[61] If the perturbation oscillates periodically, time-averaging and surface integration on the boundary lead to the relationship discussed for Cepheids.

[62] And, again, $\frac{\mathbf{q}}{T}$ is the gradient of no physical quantity. See the note in Sect. 6.2.1.

we have found hold simultaneously, as both are consequences of the same Le Châtelier principle. Moreover, $\int \frac{\mathbf{q}}{T} \cdot d\mathbf{a}$ has the dimension of [entropy/time] and may be considered as the time-averaged total amount of entropy received—through conduction and radiation only—per unit time by the external world from the room. Finally, LTE is the only crucial assumption; the relative weight of conduction and radiation is not relevant.[63] Our discussion invokes no detailed description of the physical mechanism underlying $\mathbf{q}$; it holds regardless, e.g. of the actual values of thermal conductivity, radiation opacity or heat exchange coefficients. Again, it provides us with no complete description of the relaxed configuration of our room. Different stable profiles of temperature and heating power density may exist, depending on the balance equations, the equation of state, etc. Stable steady solutions of these equations are those steady solutions which satisfy also the constrained maximization of $\overline{\int \frac{\mathbf{q}}{T} \cdot d\mathbf{a}}$. This maximization is to be satisfied by stable relaxed states even if more than one window is present; in this case, the domain of integration in the surface integral is the union of the surfaces of all windows.

6.2.12 Entropy Production of a Radiating Body

Generally speaking, the fact that $\mathbf{q}$ is a net flow allows many competing stabilization processes. Let us focus on the case where the transport of energy is due only to electromagnetic radiation. Energy may be radiated in both directions, i.e. either from the window towards the external world or in the opposite direction. A fraction (given by the 'reflection coefficient') of the power impinging on the window will be reflected towards the external world. Stability of the unperturbed state against perturbations is obtained not only by suitably modifying the emission of radiation (as discussed above), but also if any increase (decrease) of T is counteracted by an increase (decrease) of the reflection coefficient, in order to lower (raise) the heating effect due to the external radiation. This kind of feedback mechanism is common, e.g. in Gaia hypothesis [62]. Accordingly, and in agreement with Sect. 6.2.11, a stable state corresponds to a maximum of $\int \frac{\mathbf{q}}{T} \cdot d\mathbf{a}$ which is both constrained by P_{TOT} and allowed by the particular reflection mechanism of the window. In particular, if there is no reflection altogether—i.e. if our window is a 'black body'—then the maximization is limited by reflection no more and $\int \frac{\mathbf{q}}{T} \cdot d\mathbf{a}$ may achieve the maximum value allowed by the remaining constraint involving P_{TOT}. The properties of electromagnetic radiation emitted by a black body at a given temperature are well known [32]. In steady state, the local and the global form of the entropy balance (Sects. 4.2.8 and 4.2.9) when no flow of matter is present imply $0 = \frac{dS}{dt} = \frac{d_i S}{dt} + \frac{d_e S}{dt} = \frac{d_i S}{dt} - \int \frac{\mathbf{q}}{T} \cdot d\mathbf{a}$[64]

---

[63] One could argue that we have tacitly assumed that all the points of the window are at the same temperature. Indeed, this is in agreement with the idea that the window is at LTE, i.e. that all its parts are at equilibrium with each other but not necessarily with the remaining parts of the room and with the external world. Anyway, should we repeat the argument for many small windows, each at a different temperature, nothing would change in our result.

[64] Where we write $\overline{a} = a$ in steady state for $a = \frac{\mathbf{q}}{T}$.

and the constrained maximization of Sect. 6.2.11 dictated by Le Châtelier's principle reduces to $\frac{d_i S}{dt} = \max$ with the constraint of given P_{TOT}[65] : of all bodies which
radiate the same quantity of power, a black body is that which produces the largest
amount of entropy per unit time. This prediction is confirmed by quantum mechanics
[126].

   To avoid confusion, we stress the point that the maximized quantity is not the
entropy produced per unit time during the radiative transfer of energy (see Sect. 6.2.6),
which actually relaxes to a constrained minimum and which is produced per unit time
in a given set of photons [103] (or neutrinos [104]). Rather, it is the amount of entropy
radiated by the black body per unit time. If the radiating body is not exactly black,
then according to our discussion based on Le Châtelier's principle, the radiated
entropy still achieves a constrained maximum, even if its actual value differs from
the value achieved with the black body. This conclusion is in agreement with the
result of the quantum mechanical analysis of Wuerfel and Ruppel [127]. According
to these authors, moreover, Earth is an example of a radiating body that is similar but
not identical to a black body and where the amount of entropy radiated per unit time
towards the external world achieves a maximum, constrained by a given total amount
of energy received by the Sun. This conclusion resembles the arguments of Paltridge
[55]. In a simplified model of the atmosphere in a steady state, the net amount of
entropy carried by electromagnetic radiation across the boundaries of the atmosphere
attains a maximum value and is the sum of the amount of entropy supplied per unit
time to the radiation field due to matter[66] and of the rate of entropy production due to
absorption and emission processes—which achieves a minimum value, in agreement
with our discussion in Sect. 6.2.6—see Eqs. (13), (14) and (20)–(22) as well as Figs. 2
and 3 of Ref. [102].

6.2.13   No Heater, Two Windows

The constrained maximization of $\int \overline{\frac{\mathbf{q}}{T} \cdot d\mathbf{a}}$ in Sect. 6.2.11 holds as a necessary condition of stability even if no heating at all occurs anywhere in the room. In this
case $P_h \equiv 0$, $P_{TOT} = 0$, the constrained minimization of $\int \overline{\frac{P_h}{T} dx}$ leads to an identity. Since the net time-averaged amount of heat delivered to the body is zero, the
normalization condition dictates that $\int \mathbf{q} \cdot d\mathbf{a} = 0$, i.e. the net heat flux across the
boundary is zero; in other words, if a given amount of heat enters the system across
one part of the boundary in a given time interval, then the same amount of heat leaves
the system across another part of the boundary in the same time interval. This is, e.g.
the case of a room with no heater and two windows: where the heat flux coming into

---

[65] Conservation of energy requires that P_{TOT} is equal to the amount of radiated power; the latter
is given by Stefan-Boltzmann's law [32], once the temperature and the area of the surface of the
black body are known. As usual with black bodies, no detailed knowledge of what occurs inside
the black body is relevant here.

[66] More precisely, of $- \int dx \frac{r}{T}$, where r is the energy supply to a material element due to the radiation
field per unit time and volume.

the room through the first window compensates for the heat flux coming from the room outwards through the second window.

A room with no heater and two windows is the basic layout of many problems in non-equilibrium thermodynamics. For example, in fluids at large Re which are flowing between two planes, the entropy production due to viscosity in the bulk of the fluid is vanishingly small. Moreover, maximization of turbulence-enhanced transport of heat corresponds to maximization of $\mathbf{q}$, hence to maximization of $\oint_{boundary} \frac{\mathbf{q}}{T} \cdot d\mathbf{a}$. This is, e.g. the case of both convection cells at large Ra, the plasma edge in a toka-mak (where no nuclear fusion reaction occurs), Earth's oceans and atmosphere, the couple of heat exchangers and the metallic wall in front of a flame and its ceramic counterpart, discussed by Ozawa et al. [54] (Sect. 5.6.7), Yoshida and Mahajan [128] (Sect. 5.6.8),[67] Paltridge [55] (Sect. 5.6.13), Biwa et al. [129] (Sect. 5.8.2), Meija et al. [130] (Sect. 5.8.3) and Hong et al. [131] (Sect. 5.8.4), respectively. The fact that Paltridge's analysis fails to describe convection-related, vertical heat fluxes [132] is not surprising, as this model describes just energy exchanges both between equato-rial and polar regions and between each of the latter and the outer space, and not convection inside these regions: in other words, it deals with irreversible processes occurring at the boundaries of these regions, not in their bulk. Finally, from the point of view of thermodynamics, we have shown in Sects. 5.8.7–5.8.12 that Biwa et al.'s results are equivalent to results in both thermoacoustics (Rauschenbach's hypothesis, Rayleigh's criterion and Rijke and Sondhauss' experiments), thermohaline circula-tion (Sect. 5.8.11) and Eddington's model of Cepheid stars (Sect. 5.8.12).

When applied to the Bénard convection problem at large Ra, maximization of $\oint \frac{\mathbf{q}}{T} \cdot d\mathbf{a}$ has an interesting corollary. This extremum condition is obviously sym-metrical in $\mathbf{q}$ and $\frac{1}{T}$. Moreover, Biwa et al.'s results show that if we raise enough the amount of heat that flows across the system, then we trigger an oscillation with a well-defined frequency. Then, symmetry suggests that if we make $\frac{1}{T}$ (hence, tem-perature) to oscillate at some well-defined frequency, then we raise the amount of heat that flows across the system. Note that $\oint \frac{\mathbf{q}}{T} \cdot d\mathbf{a}$ is a surface integral, its domain of integration being the boundary of the system. In the case of the Bénard convec-tion problem, this boundary is made of the two plates in which the fluid is located between. In order to trigger an oscillation of temperature is, therefore, enough to modulate the temperature of one of the two plates. Indeed, both theory [133] and experiments [134] show that periodic modulation of T on one of the two plates at a suitably chosen frequency leads to a significant heat flux enhancement.

The reason for the wide diffusion of the results concerning a room with no heater with two windows is that a rigorous treatment of the constraint of given P_{TOT} is particularly simple in this case.[68] Being $P_{TOT} = 0$, indeed, the total amount of heat Q coming out of the system per unit time is exactly equal to the total amount coming into the system from the external world. Thus, if the room has two windows, say at

---

[67] See also the note of Sect. 6.2.16.

[68] Here, we discuss the case of non-oscillating systems for simplicity, with no loss of generality.

temperatures T_1 and $T_2 > T_1$, as the heat flows from the hotter to the colder window, the maximization in Sect. 6.2.11 reduces just to maximization of $Q \cdot (\frac{1}{T_1} - \frac{1}{T_2})$, a quite simple expression which is particularly easy to compare with observations.[69]

6.2.14 Resistors, Again

In a network of electric resistors in a steady state, the power supply required depends on the impedances in the system and is equal to the power P_{TOT} dissipated by Joule heating. With no loss of generalities, we may describe each resistance in the system made of $j = 1, \ldots N$ resistances as a current-carrying copper wire. If the jth resistance dissipates the power Q_j (with $\Sigma_j Q_j = P_{TOT}$) and all resistances have the same constant temperature T, we can consider all of them in thermal contact with the same heat reservoir at this temperature and apply therefore the result of Sect. 4.1.6 to each wire $\frac{dS_{j,reservoir}}{dt} = +\frac{Q_j}{T}$ where the L.H.S. is the growth of entropy in the reservoir due to the thermal contact with the jth resistance and $\Sigma_j \frac{dS_{j,reservoir}}{dt} = \frac{dS_{reservoir}}{dt}$ with $S_{reservoir}$ entropy of the reservoir. Regardless of the actual value of P_{TOT}, summation of both sides on all resistances of the network (referred to with the subscript 'ne' below) gives

$$\frac{dS_{reservoir}}{dt} = \frac{P_{TOT}}{T} = \frac{\int_{ne} P_h d\mathbf{x}}{T} = \int_{ne} \frac{P_h}{T} d\mathbf{x} = \int_{ne} \rho \frac{ds}{dt} d\mathbf{x} + \int_{ne} \frac{\nabla \cdot \mathbf{q}}{T} d\mathbf{x} =$$

$$= \frac{dS_{ne}}{dt} + \int_{ne} \frac{\nabla \cdot \mathbf{q}}{T} d\mathbf{x} = \frac{dS_{ne}}{dt} + \int_{ne} \nabla \cdot \left(\frac{\mathbf{q}}{T}\right) d\mathbf{x} = \frac{dS_{ne}}{dt} + \int_{ne} \frac{\mathbf{q}}{T} \cdot d\mathbf{a}$$

where S_{ne} is the entropy of the network, all integrals but the last one are computed on the network volume, the last integral is computed on the boundary surface of the network volume and we have invoked both the definition of P_h, Gauss' theorem of divergence and the fact that T is uniform (i.e. that $\nabla T = 0$). The entropy of the Universe is $S = S_{reservoir} + S_{ne}$, hence $\frac{dS}{dt} = 2\frac{dS_{ne}}{dt} + \int_{ne} \frac{\mathbf{q}}{T} \cdot d\mathbf{a}$. In steady state $S_{ne} = const.$, then

$$\frac{dS}{dt} = \int_{ne} \frac{\mathbf{q}}{T} \cdot d\mathbf{a}$$

When applied to the network, our constrained maximization of $\int \frac{\mathbf{q}}{T} \cdot d\mathbf{a}$ at given P_{TOT} in Sect. 6.2.11 implies that the network maximizes the entropy production at given power, just as predicted by Zupanovic et al. [135] (see Sect. 5.6.5) but with no need of MEPP.

---

[69] As usual by now, further constraints are provided by the conservation equations for each particular problem.

6.2.15 A 2nd Necessary Condition for Stability—General Form

Once the room has relaxed to a configuration described either by the constrained minimization of $\int \frac{P_h}{T} d\mathbf{x}$ or by the constrained maximization of $\int \frac{\mathbf{q}}{T} \cdot d\mathbf{a}$, let us open the window. Convection starts across the window. Generally speaking, convection may be turbulent, and turbulence is intrinsically unsteady. Moreover, we do not expect convection to disappear as time goes by: as far as the temperature gradient across the window is large enough, convective cells never stop rotating. Convection may even play a crucial role in keeping the time-averaged temperature gradient approximately constant.

After a while, the room evolves again towards a new relaxed state, with a steady (on average) distribution of temperature. Since the minimization of $\int \frac{P_h}{T} d\mathbf{x}$ in Sect. 6.2.1 is concerned with heating processes only, it remains valid. However, a new physical phenomenon (convection) acts, correspondingly, we expect that both T and $\mathbf{q}$ are affected and that maximization of $\int \frac{\mathbf{q}}{T} \cdot d\mathbf{a}$ of Sect. 6.2.11 holds no more.

Now, entropy is an additive quantity. Accordingly, convection transports entropy with a flux $\rho s \mathbf{v}$. Again, this quantity has the dimension [entropy/(time·surface)]. The quantity $\int \rho s \mathbf{v} \cdot d\mathbf{a}$ is, therefore, the time-averaged, total amount of entropy received—through convection only—per unit time by the external world from the room. Correspondingly, the quantity $\int \left(\frac{\mathbf{q}}{T} + \rho s \mathbf{v} \right) \cdot d\mathbf{a}$ is the time-averaged, total amount of entropy received—by all means, not just convection—per unit time by the external world from the room.

On one side, indeed, the time-averaged overall content of entropy of the room is constant in a steady state. On the other side, convection is an irreversible phenomenon, i.e. convection-related entropy production is non-negative; it occurs continuously even in the relaxed configuration of our room as convection never disappears and vanishes only if no convection occurs. These conclusions do not contradict each other only if the opening of our window triggers convection processes which in turn add a further, non-negative contribution $\int \rho s \mathbf{v} \cdot d\mathbf{a}$ to the time-averaged amount of entropy lost per unit time towards the external world, in addition to the time-averaged amount $\int \frac{\mathbf{q}}{T} \cdot d\mathbf{a}$ of entropy lost per unit time via conduction and radiation. To put it in other words, the final effect of the window opening on the final, relaxed state is to add a non-negative term to the time-averaged amount of entropy received by the external world per unit time.

We are free to imagine our room inside another, larger room with a window, inside another room with a window... like Russian dolls, indefinitely. Our original room is the only one with a heater, so that P_{TOT} remains unchanged, i.e. equal to the value obtained when the window was closed. With no loss of generality, we may suppose that the external world is always infinitely larger than this system of nested rooms, regardless of the actual number of rooms. Let us open all these windows sequentially, each after another, starting from our original room outwards. On one side, any opening of a window adds a non-negative term to the time-averaged total amount of entropy received by the external world per unit time, and we obtain a succession of larger and larger values of this quantity. On the other side, however, in a relaxed state, the actual value of the time-averaged total amount of entropy received by the external world per unit time, which is the limit of the succession when the number of rooms becomes very large, is both well-defined, finite, >0 and constant. These conclusions do not contradict each other only if the value of $\int \left(\frac{\mathbf{q}}{T} + \rho s \mathbf{v}\right) \cdot d\mathbf{a}$ in the relaxed state attains a maximum. If it differs from the maximum, indeed, we can raise it by adding a new room and by opening a new window. Then, the constrained maximization of $\int \frac{\mathbf{q}}{T} \cdot d\mathbf{a}$ in Sect. 6.2.11 is replaced by its generalization to the case where both convection, conduction and radiation may occur:

$$\overline{\int \left(\frac{\mathbf{q}}{T} + \rho s \mathbf{v}\right) \cdot d\mathbf{a}} = \max \quad \text{with the constraint} \quad \overline{\int P_h d\mathbf{x}} = P_{TOT}$$

This result involves the total amount of entropy received by the external world from the room and holds regardless of the relative weight of convection, conduction and radiation.

6.2.16 Heat Conduction in Gases, Retrieved

Let us rewrite the maximization in Sect. 6.2.15 as follows. To start with, we invoke Gauss' theorem and write $\int \frac{\mathbf{q}}{T} \cdot d\mathbf{a} = \int \nabla \cdot \left(\frac{\mathbf{q}}{T}\right) d\mathbf{x}$. Secondly, we rewrite the integrand as $\nabla \cdot \left(\frac{\mathbf{q}}{T}\right) = \mathbf{q} \cdot \nabla \frac{1}{T} + \frac{\nabla \cdot \mathbf{q}}{T}$. Thirdly, the definition of P_h allows us to write $\frac{\nabla \cdot \mathbf{q}}{T} = \frac{P_h}{T} - \rho \frac{ds}{dt}$. We replace these relationships in the integral, recall that $\int \frac{P_h}{T} d\mathbf{x}$ undergoes constrained minimization and that $\overline{\int \rho \frac{ds}{dt} d\mathbf{x}} = \overline{\frac{dS}{dt}} = 0$[70] and obtain that the constrained maximization of $\int \left(\frac{\mathbf{q}}{T} + \rho s \mathbf{v}\right) \cdot d\mathbf{a}$ reduces to[71]

$$\overline{\int \mathbf{q} \cdot \nabla \frac{1}{T} d\mathbf{x}} + \int \rho s \mathbf{v} \cdot d\mathbf{a} = \max \quad \text{with the constraint} \quad \overline{\int P_h d\mathbf{x}} = P_{TOT}$$

In particular, if no mass flows across the boundaries, then constrained maximization of $\int \mathbf{q} \cdot \nabla \frac{1}{T} d\mathbf{x}$ follows.[72] We stress the point that the maximized quantity is precisely the (time-averaged) amount of entropy produced per unit time by heat transport. For comparison, we recall that LNET requires *minimization*, not maximization, of this quantity in the relaxed state.

---

[70] The time-average of a partial time derivative of a quantity which is finite at all times vanishes, and $\frac{\partial S}{\partial t} = \frac{dS}{dt}$ for a system contained in a fixed volume at all times.

[71] Together with the constrained minimization of $\int d\mathbf{x} \frac{P_h}{T}$, this is the formula derived from GEC (Sect. 3.6) in Ref. [71].

[72] Thus, we have killed two birds with one stone, as we have retrieved also the result of Ref. [128] in Sect. 5.6.8.

Furthermore, if we assume Fourier's law $\mathbf{q} = -\chi \nabla T$ (Sect. 5.2.1) to hold, then constrained maximization of $\int \chi \frac{|\nabla T|^2}{T^2} d\mathbf{x}$ follows. As for the problems in slab geometry investigated by Holyst et al. [136], for example, where a perfect gas with uniform particle density is heated by the external world with a given amount P_{TOT} of radiofrequency power and looses heat towards the walls at $x = 0$ and $x = d$ at fixed boundary temperature $T (x = 0) = T (x = d) = T_{boundary}$ which contain the gas in steady state (Sect. 5.6.6), we expect the temperature of the gas to be peaked at the location x_P where the density $P_h (x)$ of the radiofrequency power deposited in the gas is maximum. Accordingly, our maximization implies that $\chi = \chi (x)$ has a maximum in some location x_χ, which depends on x_P. Since the value of P_{TOT} is the only constraint, we may change x_P as we like, and therefore x_χ. In other words, x_χ is arbitrary, i.e. the relaxed state maximizes χ everywhere, as far as the energy balance $P_{TOT} = Q_{loss}$ described by the normalization condition is satisfied. But the larger χ at given Q_{loss}, the more effective the transport of energy in a system which looses a given amount Q_{loss} of energy per unit time, the lower the amount ΔU of internal energy in excess (with respect to the condition of thermodynamic equilibrium which corresponds to $Q_{loss} = 0$) at given Q_{loss}, the larger Q_{loss} at given ΔU,[73] the lower $\frac{\Delta U}{Q_{loss}}$ at given ΔU, and the result of [136] is retrieved. Time-averaging allows generalization to problems where rotating, convective cells are present—see Sect. 5.8.6.

Finally, in the particular case $x_P = \frac{d}{2}$ where $P_h (x)$ is located on the symmetry axis of the system, then the temperature increases (decreases) monotonically from $x = 0$ to $x = \frac{d}{2}$ (from $x = \frac{d}{2}$ to $x = d$), $T (x) = T \left(\frac{d}{2} - x\right)$ and ΔU increases with increasing $|\nabla T|$. Then, minimization of ΔU at given $Q_{loss} = P_{TOT}$ is equivalent to minimization of $|\nabla T|$ with the constraint of given dissipated power, i.e. to Chandrasekhar's principle of Sect. 5.3.11.

6.2.17 Shock Waves and Dunes, Again

A particular case of the maximization in Sect. 6.2.15 is Rebhan's result [138] for shock waves at large Ma (Sect. 5.6.9). Indeed, the thickness of the shock wave is so thin that the volume V_S of the layer separating the front side and the back side is negligible, and the same holds for P_{TOT} as this quantity is equal to the volume integral of P_h on V_S. It follows that $[q_n] = 0$. And since no heat comes into the shock wave from the front side because of the propagation, $q_n = 0$ on the back side too, and the maximized quantity involves $\rho s \mathbf{v}$ only. Similar arguments hold both for the dunes in Rubin et al.'s experiments [139] (Sect. 5.6.10) and for the analytical results concerning the description of the relaxed state of a system of chemical reactions in [68] (see note in Sect. 6.1.9).

---

[73] Because of the reciprocity principle for isoperimetric problems [137].

6.2.18 Back to Entropy Generation

Let us focus our attention on a body that is in contact with a thermal bath at temperature T_0. The surface integral computed at the interface between the body and the bath is $\int \frac{\mathbf{q}}{T} \cdot d\mathbf{a} = -\frac{\dot{Q}}{T_0}$, where $\dot{Q}$ is the amount of heat per unit time taken > 0 when the heat flows from the bath into the body (while $\mathbf{q} \cdot d\mathbf{a} > 0$ when the heat flows from the body towards the bath). If there are $i = 1, \ldots N$ baths each with its own temperature T_i and with an amount $\dot{Q}_i$ flowing towards the body, then $\int \frac{\mathbf{q}}{T} \cdot d\mathbf{a} = -\Sigma_i \frac{\dot{Q}_i}{T_i}$. Correspondingly, the maximized quantity in Sect. 6.2.15 is $\overline{\int \left(\frac{\mathbf{q}}{T} + \rho s\mathbf{v} \right) \cdot d\mathbf{a}} = \overline{-\Sigma_i \frac{\dot{Q}_i}{T_i} + \int \rho s\mathbf{v} \cdot d\mathbf{a}} = \frac{\partial S}{\partial t} - \Sigma_i \frac{\dot{Q}_i}{T_i} + \int \rho s\mathbf{v} \cdot d\mathbf{a}$.[74] The last quantity is equal to $\overline{\frac{dS_g}{dt}}$,[75] according to Eq. (12) of Ref. [83]. Thus, Le Châtelier's principle leads to maximization of $\frac{dS_g}{dt}$ as in the entropy generation approach of Sect. 6.1.12, even if with the constraint of given time-averaged total amount of heat flowing per unit time from the body towards the environment.

As an example of application, it has been observed that a large flood of a river—interpreted as a loss of stability of the usual river behaviour—corresponds to a gross violation of the rule of maximum growth of entropy generation, i.e. to a sudden drop of S_g [85]. The entropy generation of a river at a time $t > 0$ is $S_g = \int_0^t \frac{\rho g}{T} \dot{V} \frac{dH}{dx} dt$, where x, $\dot{V}$ and H are the spatial coordinate along the direction of the river flow, the volume flow and the elevation above the base level, respectively. Assuming for simplicity that T is the same in the water and in the environment at all times, this result has a simple meaning: $T \frac{dS_g}{dt}$ is the amount of mechanical work done per unit time by the river on the embankment. In a steady, stable state, i.e. if no flood occurs, this amount is maximum, or, in other words, the resistance force offered by the embankment on the water as the river flows is minimum. This prediction is in agreement with the prediction of Korteweg-Helmholtz' principle in the particular case $Re \ll 1$ (Sect. 5.3.10); both predictions, indeed, agree with Le Châtelier's principle. If S_g decreases, then stability is lost and a flood occurs.

The fact that the same stable state of the same physical system (a river that does not overflow) may simultaneously obey two distinct extremization criteria involving two distinct contributions to the entropy balance—namely the minimization of dissipated energy of Ref. [122] in Sect. 5.3.8 and the maximization of the growth rate of environmental entropy of [85] in Sect. 6.1.12—agrees nicely with our derivation of two distinct necessary criteria for stability starting from the same Le Châtelier's principle at LTE and clearly confirms the nonexistence of a general-purpose variational principle outside thermodynamic equilibrium (Sect. 5.1), a fact which is usually overlooked in the literature. A remarkable exception is Ref. [140], where the authors describe a turbulent, viscous fluid by postulating two distinct extremum conditions

---

[74] Where we have taken into account the facts that $\overline{\frac{\partial a}{\partial t}} = 0$ if $|a| < +\infty$ at all times with $a = S$ and that S remains bounded in steady state or under a cyclic transformation.

[75] Which does not vanish as S_g may increase indefinitely with time.

concerning two distinct dissipation-related quantities. We find here another example of the coexistence of different extremum principles far from thermodynamic equilibrium hinted at in Sect. 5.6.2.[76]

Biology provides us with further applications. Let us consider a living cell at rest and with a fixed volume and shape inside a living organism. The organism feeds the cell and plays the role of the environment, with constant temperature T_0; the amount of nutrients available per unit time is constant. A cell in a stable, steady state takes advantage of the nutrients in order to sustain its own metabolism. This metabolism produces heat and waste substances, which are expelled across the cell membrane towards the environment. The resulting $T_0 \frac{dS_g}{dt}$ is a maximum as the cell steady state is stable. Given a constant total mass of the cell, the mass flow of waste across the membrane is also constant.

As the cell starts reproducing itself with a cell division, a fraction of the energy provided by the nutrients per unit time is spent in moving parts of the cell mass in opposite directions. But the amount of energy that is available to the cell per unit time is unchanged, as the amount of available nutrients is fixed. Since part of this available amount of energy is spent on the motion of masses, the actual amount of heat that goes to the environment per unit time is reduced. (The total mass flow of waste towards the environment remains constant, as the sum of the two daughter cells is equal to the mass of the initial cell: reproduction satisfies the requirement of conservation of total mass.) Correspondingly, $T_0 \frac{dS_g}{dt}$ is lowered—and, indeed, the division of the cell may be seen as a loss of stability of the initial cell configuration. The life of the cell may be seen as the outcome of an ongoing competition between two distinct classes of processes: on one side, those which are responsible for the transport of heat and waste towards the external world tend to raise $T_0 \frac{dS_g}{dt}$ and prevent reproduction; on the other side, the processes which tend to lower $T_0 \frac{dS_g}{dt}$ and trigger reproduction.

This picture leads to an interesting corollary. Let us go back to our initial cell. Suppose we are able to provide it with a tiny amount of heat. This is definitely an external perturbation due to the external world. Then, Le Châtelier's principle dictates that the cell reacts by trying to cool itself, i.e. by raising the heat transport towards the external world. As a result, the balance is shifted in favour of the processes which prevent reproduction. In other words, an external, tiny heating of the cell tends to prevent its reproduction. This prediction allows experimental verification. Electromagnetic waves of suitable frequency may selectively heat some classes of living cells; limitation of the reproduction of cancer cells with the help of electromagnetic waves is an active field of research that stems from the thermodynamic discussion above [141, 142].

---

[76] See also Sects. 6.2.19 and 6.2.21.

6.2.19 Convection, Again

As for Malkus and Varonis' requirement [143] for convection between horizontal parallel plates in Sect. 5.6.7, a qualitative argument follows. Both $\mathbf{q}$ and $\rho s \mathbf{v}$ contribute (on average at least) to entropy transport in the same direction from the hotter to the colder plate. Then, in Sect. 6.2.15, we maximize $\int \rho s \mathbf{v} \cdot d\mathbf{a}$ separately. We invoke Gauss' theorem of divergence and write $\int \nabla \cdot (\rho s \mathbf{v}) \, d\mathbf{x} = $ max. For an incompressible fluid ($\nabla \cdot \mathbf{v} = 0$), this reduces to $\int \mathbf{v} \cdot \nabla (\rho s) \, d\mathbf{x} = $ max. By taking $\rho s \propto \ln T$, we obtain max $= \int \mathbf{v} \cdot \nabla (\ln T) \, d\mathbf{x} = \int \mathbf{v} \cdot \left(\frac{\nabla T}{T}\right) d\mathbf{x} = -\beta \int \frac{v_z}{T} d\mathbf{x}$. Any perturbation of the stable state lowers $-\beta \int \frac{v_z}{T} d\mathbf{x}$ from its maximum value, hence $0 \geq$ $\left(-\beta \int \frac{v_z}{T} d\mathbf{x}\right)_1 = -\beta \int \left(\frac{v_z}{T}\right)_1 d\mathbf{x} = \beta \int \frac{v_{z1} T_1}{T^2} d\mathbf{x}^{77} \geq \frac{\beta}{T_{max}^2} \int v_{z1} T_1 d\mathbf{x}.^{78}$ Thus, $0 \geq$ $\beta \int v_{z1} T_1 d\mathbf{x} \propto \beta \{v_{z1} T_1\} = (\beta \{v_z T\})_1$, i.e.[79] $\beta \{v_z T\}$ is a maximum and Malkus and Varonis' requirement is retrieved.

When it comes to convection, it is useful to make a comparison between Holyst et al.'s constrained minimization of the energy content [136] in Sect. 5.6.6, Chandrasekhar's constrained minimization of Ra [144] in Sect. 5.3.11, Welander's vanishing phase between T_1 and Q_{H1} in oscillating relaxed states [145] in Sect. 5.8.11, Ozawa et al.'s maximization of $\oint \frac{\mathbf{q}}{T} \cdot d\mathbf{a}$ [54], Malkus and Varonis' maximization of $\beta \{v_z T\}$ [143] in Sect. 5.6.7 and maximization of heat flow with oscillating T at the boundary in Yang et al. [133] and Urban et al.'s [134] works in Sect. 6.2.13. Holyst et al.'s, Chandrasekhar's and Welander's results hold for $Ra < Ra_{thr}$, $Ra - Ra_{thr} = \varepsilon, 0 < \varepsilon \ll 1$ and $Ra > Ra_{thr}$, respectively. We show in Sect. 6.2.16 that Chandrasekhar's and Holyst et al.'s minimizations agree with each other, in spite of the different geometries. Moreover, Ozawa et al.'s, Malkus and Varonis' and Yang, Urban et al.'s results hold for $Ra \gg Ra_{thr}$. But the story does not end here, for the elegant, analytical Bianciardi and Ulgiati's discussion [15] of the onset of Bénard cells at the transition between conduction and convection shows that the birth of a convection cell in a fluid where no convection cells previously existed is a process with a short, but finite duration and which lowers the entropy production in the bulk.[80] Correspondingly, the amount of energy that is available for transformation into useful work increases: it is the heat transformed into ordered kinetic energy of the going up/coming down fluid molecules in the convective state.

The fact that the same physical phenomenon obeys a minimization principle for some values of a given dimensionless parameter and a maximization principle for other values of the same parameter is a vivid example of the inconsistencies which

---

[77] First-order perturbations disappear after averaging and are not considered here.

[78] Where T_{max} is the upper bound on the temperature of the fluid, e.g. it is equal to the temperature of the hotter plate.

[79] Again, neglecting first-order perturbations.

[80] Remarkably, as far as thermodynamics is concerned, this is the same process Eigen links with a mutation, in its work [10] supporting Glansdorff and Prigogine's excess entropy production approach—see Sect. 6.1.3. Furthermore, Ref. [12] discusses the analogy between the problem of Bénard cells in fluid dynamics and problems in biology.

every attempt at writing a universal, general-purpose, unique variational principle for stable steady states far from thermodynamic equilibrium—in analogy to the Second Principle of thermodynamics for thermodynamic equilibrium—is doomed to stumble upon.[81]

6.2.20 Crystal Growth, Retrieved

When dealing with our room, we have implicitly focussed our attention on a problem with just one species. Generalization to many species is straightforward. The diffusion of each species provides a further contribution to the entropy balance. Crystal growth (Sect. 5.6.12) is an example, where both $\mathbf{q}$ and $\frac{d_i S}{dt}$ vanish. Maximization of the total amount of entropy received by the external world and coming out from the solid across its surface—in agreement with Sect. 6.2.15—implies maximization of the amount $\oint_{boundary} s_k \mathbf{j}_k \cdot d\mathbf{a}$ of entropy exchanged per unit time across the boundary because of diffusion processes, in agreement with observations [72].

6.2.21 Liesegang and Gelation

The coexistence of minimization and maximization criteria for different contributions to entropy production resurfaces many times in our discussion namely in Sects. 5.6.2, 6.2.18 and 6.2.19. It is worthwhile to stress the point that this coexistence follows from the coexistence of different necessary criteria of stability (outlined in Sects. 6.2.1 and 6.2.15) which in turn follow from Le Châtelier's principle.

As a consequence, their validity does not depend on the detailed nature of the irreversible processes occurring in a given system, even if their actual consequences obviously do. The results reported in Ref. [146] provide us with a clear-cut illustration of this point. To start with, the authors investigate the formation of *Liesegang rings*[82]

---

[81] See also Sects. 5.6.2, 6.2.18 and 6.2.21.

[82] Let us consider a chemical system undergoing a precipitation reaction in the absence of convection. 'Precipitation' is the process of transforming a dissolved substance into an insoluble solid from a supersaturated solution. Moreover, let a gel be formed that contains a dilute solution of one of the reactants. If weakly soluble salts are produced from the reaction of two soluble substances, one of which is dissolved in the gel medium, then—under certain conditions of concentration—spatially ordered patterns of concentration arise spontaneously. For example, if a hot solution of agar gel also containing a dilute solution of potassium dichromate is poured into a test tube and (after the gel solidifies) a more concentrated solution of silver nitrate is poured on top of the gel, the silver nitrate will begin to diffuse into the gel. It will then encounter the potassium dichromate and will form a continuous region of precipitate at the top of the tube. After some hours, the continuous region of precipitation is followed by a clear region with no sensible precipitate, followed by a short region of precipitate further down the tube. This process continues down the tube forming several, up to perhaps a couple of dozen, alternating regions of clear gel and precipitate rings. The shape of these ordered patterns is typically band-like in a test tube and ring-like in a Petri dish. In the latter case, the patterns are called 'Liesegang rings'.

and write $\frac{dS}{dt}$ as the sum of the contributions of particle diffusion, heat conduction and chemical reactions to the entropy production in the bulk.[83] In the words of Ref. [146], *the system evolves towards a structure which minimizes the lost work in its formation. The structure thus adopts an architecture such that the work lost in changing the configuration is almost negligible [...] the entropy produced is a minimum* in agreement with the discussion of Sect. 6.2.1.

Reference [146] contains also a description of the entropy production in the process of *gelation*.[84] The authors write $\frac{dS}{dt}$ as the sum of many positive-definite terms, each of them being proportional to a positive-definite time constant (with the dimension [1/s]).[85] All other things being equal—including the boundary conditions and in particular the amount of entropy supplied to the external world by the system where gelation occurs—the larger this sum, the shorter the characteristic time-scales involved in gelation and the faster the transport of entropy across the boundaries of the system towards the external world. Not surprisingly, it turns out that *structures correspond to maximum values of the total entropy produced*, in agreement with the discussion of Sect. 6.2.15.

References

1. Gyarmati, I.: Non-equilibrium Thermodynamics. Springer, Berlin (1970)
2. Jaynes, E.T.: Ann. Rev. Phys. Chem. **31**, 579 (1980)
3. Bertola, V., Cafaro, E.: Int. J. Heat Mass Transf. **51**, 1907–1912 (2008)
4. Glansdorff, P., Prigogine, I.: Thermodynamic Theory of Structure, Stability, and Fluctuations. Wiley-Interscience, New York (1971)
5. Glansdorff, P., Prigogine, I.: Physica **46**, 344–366 (1970)
6. Maes, C., Netočný, K.: J. Stat. Phys. **159**, 1286–1299 (2015)
7. Shimizu, H., Yamaguchi, Y.: Prog. Theor. Phys. Progress Letters **67**, 1 (1982)
8. Prigogine, I.: Time, Structure and Fluctuations. Nobel Lecture (1977). http://www.eoht.info/page/Ilya+Prigogine
9. Prigogine, I., Nicolis, G.: Non-equilibrium Systems. Wiley, New York (1977)
10. Eigen, M.: Naturwissenschaften **58**(10), 465–523 (1971)
11. Demirel, Y., Sandler, S.I.: Biophys. Chem. **97**, 87–111 (2002)
12. Bruers, S.: Energy and ecology. On entropy production and the analogy between fluid, climate and ecosystems. Ph.D. Thesis, University of Leuven, Belgium (2007)
13. Nicolis, C.: Q. J. R. Meteorol. Soc. **125**, 1859–1878 (1999)
14. Glansdorff, P., Prigogine, I.: Physica **30**, 351 (1964)
15. Bianciardi, C., Ulgiati, S.: Ecol. Model. **110**, 255–267 (1998)
16. Butler, H.W., McKee, D.E.: Int. J. Heat Mass Transf. **13**, 43–54 (1970)
17. Schechter, R.S.: Chem. Eng. Sci. **17**, 803–806 (1962)
18. Kern, W., Felderhof, B.U.: Z. Physik B **29**, 293–297 (1978)
19. Botré, C., Lucarini, C., Memoli, A., D'Ascenzo, E.: Bioelectrochem. Bioenerget. **8**, 201–212 (1981) (a section of J. Electroanal. Chem. Const. **128** (1981))

---

[83] See their Eq. (15).

[84] 'Gelation' is the formation of a 'gel' (a fibrous network) from a system with polymers.

[85] See their Eq. (11).

20. Kern, W., Felderhof, B.U.: Z. Phys. **B**(28), 129–134 (1977)
21. Keizer, J., Fox, R.: Proc. Nat. Acad. Sci. USA, **71**(1), 192–196 (1974)
22. De Sobrino, L.: J. Theor. Biol. **54**, 323–333 (1975)
23. Lavenda, B.: Thermodynamics of Irreversible Processes. McMillan, London (1979)
24. Yoshida, Z.: Adv. Phys.: X (2016). https://doi.org/10.1080/23746149.2015.1127773
25. Taylor, J.B.: PRL **33**, 1139–1141 (1974)
26. Turner, L.: IEEE Trans. Plasma Sci. **PS-14**(6), 849–857 (1986)
27. Dasgupta, B., Shaikh, D., Hu, Q., Zank, G.P.: J. Plasma Phys. **75**(2), 273–287 (2009)
28. Brown, M.R.: J. Plasma Physics **57**(1), 203–229 (1997)
29. Welander, P.: Tellus **7**(2), 141–156 (1955)
30. Taylor, J.B., MaNamara, B.: Phys. Fluids **14**, 1492–1500 (1971)
31. Minardi, E., Lampis, G.M.: Plasma Phys. Contr. Fus. **32**(10), 819 (1990)
32. Landau, L.D., Lifshitz, E.: Statistical Physics. Pergamon, Oxford, UK (1960)
33. Jou, D., Casas-Vázquez, J., Lebon, G.: Rep. Prog. Phys. **51**, 1105 (1988)
34. Andersson, N., Comer, G.L.: Living Rev. Relativity **24**, 3 (2021)
35. Sienutycz, S., Berry, R.S.: Phys. Rev. A **46**(10), 6359–6370 (1992)
36. Landau, L.D., Lifshitz, E.: Fluid Mechanics. Pergamon, Oxford, UK (1960)
37. Beretta, G.P.: Steepest entropy ascent model for far-non-equilibrium thermodynamics. unified implementation of the maximum entropy production principle. arXiv:1409.6672v1 [cond-mat.stat-mech]. Accessed 23 Sep 2014
38. Beretta, G.P.: Rep. Math. Phys. **64**(1–2), 139–168 (2009)
39. Li, G.C., von Spakovsky, M.R.: Steepest-entropy-ascent quantum thermodynamic modeling of heat and mass diffusion in a far-from-equilibrium system based on a single particle ensemble. arXiv:1601.01344v2 [cond-at.stat-mech]
40. Beretta, G.P.: Entropy **10**, 160 (2008)
41. Courbage, M., Prigogine, I.: Proc. Natl. Acad. Sci. USA **80**, 2412–2416 (1983)
42. Attard, P.: J. Chem. Phys. **122**, 154101 (2005)
43. Sawada, Y.: Progr. Theor. Phys. **66**, 68 (1981)
44. Dewar, R.D.: J. Phys. A **38**, L371 (2005)
45. Niven, R.K.: PRE **80**, 021113-1-15 (2009)
46. Jaynes, E.T.: Phys. Rev. **106**, 620–630 (1957)
47. Jaynes, E.T.: Phys. Rev. **108**, 171–190 (1957)
48. Shannon, C.E.: Bell Syst. Tech. J. **27**(3), 379–423 (1948)
49. Landauer, R.: IBM J. Res. Dev. **5**(3), 183–191 (1961)
50. Eva, B.: J. Philos. **116**(7), 390–411 (2019)
51. Buck, B., Macaulay, V. (eds.): Maximum Entropy in Action: A Collection of Expository Essays. Oxford University Press, Oxford, UK (1991)
52. De Martino, A., De Martino, D.: Heliyon **4**, e00596 (2018). https://doi.org/10.1016/j.heliyon.2018.e00596
53. Virgo, N.: Entropy **12**(1), 107–126 (2010)
54. Ozawa, H., Shimokawa, S., Sakuma, H.: Phys. Rev. E **64**, 026303 (2001)
55. Paltridge, G.W.: Nature **279**, 630 (1979)
56. Lorenz, R.: Science **299**, 837–838 (2003)
57. Ozawa, H., Ohmura, A., Lorenz, R.D., Pujol, T.: Rev. Geophys. **41**(4), 1018 (2003)
58. Kleidon, A.: Phil. Trans. R. Soc. A **370**, 1012–1040 (2012)
59. Chapman, E.J., Childers, D.L., Vallino, J.J.: BioScience **66**, 27–39 (2016)
60. Ulgiati, S., Bianciardi, C.: Ecol. Model. **96**, 75–89 (1997)
61. Mora, T., Walczak, A.M., Del Castello, L., Ginelli, F., Melillo, S., Parisi, L., Viale, M., Cavagna, A., Giardina, I.: Nat. Phys. **12**(12), 1153–1157 (2016)
62. Lovelock, J.E., Margulis, L.: Tellus **XXVI**, 1–2 (1974)
63. Lovelock, J.E.: Nature **426**, 769–770 (2003)
64. Tyrrell, T.: New Scientist, pp. 30–31 (2013)
65. Pollarolo, G., Sertorio, G.: Il Nuovo Cimento **2C**(3), 335–351 (1979)
66. Boschi, R., Lucarini, V., Pascale, S.: Icarus **226**(2), 1724–1742 (2013)

67. Kuznetsov, Y.A.: Elements of Applied Bifurcation Theory. Springer, New York (2004)
68. Endres, R.G.: Nat. Sci. Rep. **7**, 14437 (2017)
69. Grinstein, G., Linsker, R.J.: Phys. A: Math. Theor. **40**, 9717 (2007)
70. Polettini, M.: Eur. Phys. Lett. **97**, 30003 (2012)
71. Di Vita, A.: Phys. Rev. E **81**, 041137 (2010)
72. Martyushev, L.M., Seleznev, V.D.: Phys. Rep. **426**, 1–45 (2006)
73. Ziegler, H.: ZAMP **34**, 832–844 (1983)
74. Martyushev, L.M., Seleznev, V.D.: Phys. A **233**(410), 17–21 (2014)
75. Polettini, M.: Entropy **15**(7), 2570–2584 (2013)
76. Martyushev, L.M.: Phil. Trans. R. Soc. B **365**, 1333–1334 (2010)
77. Gage, D.H., Schiffer, M., Kline, S.J., Reynolds, W.C.: The non-existence of a general thermokinetic variational principle. In: Donnelly, R.J. (ed.) Non-equilibrium Thermodynamics: Variational Techniques and Stability. University of Chicago Press, Chicago, IL (1966)
78. Lavenda, B.H.: Phys. Rev. A **9**(2), 929–942 (1974)
79. Lucia, U.: Physica A **376**, 289–292 (2007)
80. Sicardi Schifino, A.C., Ferro Fontan, C., Gonzalez, R., Costa, A.: Physica A **334**, 201–221 (2004)
81. Strogatz, S.H.: Nonlinear Dynamics and Chaos. Addison Wesley, Boston (1996)
82. Pal, R.: Int. J. Mech. Eng. Educ. **45**(2), 194–206 (2017)
83. Lucia, U.: Atti della Accademia Peloritana dei Pericolanti **94**(1), A4 (2016)
84. Lucia, U., Maino, G.: EPL **101**, 56002 (2013)
85. Lucia, U., Buzzi, P., Grazzini, G.: Irreversibility in river flow. Intl. J. Heat Technol. **34**, (Special Issue 1), S95–S100 (2016)
86. Lucia, U.: Physica A **391**, 3392–3398 (2012)
87. Sciubba, E.: Int. J. Thermodyn. **7**(3), 115–122 (2004)
88. Lucia, U., Sciubba, E.: Physica A **392**, 3634–3639 (2013)
89. Kreuzer, H.J., Payne, S.H.: Am. J. Phys. **79**(1), 74–77 (2011)
90. Schneider, E.D., Kay, J.J.: Mathl. Comput. Model. **19**(6–8), 25–48 (1994)
91. Kudrin, A.V., Karpov, A.I.: Vestn. Udmurtsk. Univ. Mat. Mekh. Komp. Nauki **4**, 80–85 (2011) (in Russian)
92. Karpov, A.I.: Vestn. Udmurtsk. Univ. Mat. Mekh. Komp. Nauki **3**, 61–68 (2008) (in Russian)
93. Busse, F.H.: J. Fluid Mech. **30**(4), 625–649 (1967)
94. Bejan, A.: J. Adv. Transp. **30**(2), 85–107 (1996)
95. Bejan, A.: J. Heat Mass Transf. **40**(4), 799–810 (1997)
96. Evolution News: Good Grief. No, Airplanes Do Not "Evolve" by Natural Law , 25 July 2014. https://evolutionnews.org/2014/07/good_grief_no_a/
97. Kohler, M.: Z. Phys. **124**, 772–789 (1948) (In German)
98. Christen, T.: Frank Kassubek. J. Phys. D: Appl. Phys. **47**, 363001 (2014)
99. Schlup, W.A.: J. Mathem. Phys. **16**, 1733 (1975)
100. Zupanovic, P., Kuic, D., Juretic, D.: Andrej Dobovisek. Entropy **12**, 926–931 (2010)
101. Oxenius, J.: J. Quant. Spectrosc. Radial. Transfer. **6**, 65–91 (1966)
102. Pelkowski, J.: Entropy **16**, 2291–2308 (2014)
103. Kroell, W.: J. Quant. Spectrosc. Radiat. Transf. **7**, 715–723 (1967)
104. Essex, C., Kennedy, D.C.: J. Stat. Phys. **94**(1–2), 253–267 (1999)
105. Landau, L.D., Lifshitz, E.: The Classical Theory of Fields. Pergamon, Oxford, UK (1971)
106. DeWitt, C.M. (ed.): Conference on the Role of Gravitation in Physics at the University of North Carolina, Chapel Hill, March 1957; WADC Technical Report 57-216 (Wright Air Development Center, Air Research and Development Command, United States Air Force, Wright Patterson Air Force Base, Ohio). https://edition-open-sources.org/sources/5/34/index.html
107. Hameiri, E., Bhattacharjee, A.: Phys. Rev. A **35**, 768–777 (1987)
108. Peters, T.: Z. Phys. **144**, 612–631 (1956) (in German)
109. Montgomery, D., Phillips, L.: Phys. Rev. A **38**, 2953–2964 (1988)
110. Suzuki, M.: Progr. Theor. Phys. Suppl. **195**, 114–119 (2012)
111. Sree Harsha, N.R.: Eur. J. Phys. **40**, 033001 (2019)

112. Christen, T.: Entropy **11**, 1042–1054 (2009)
113. Christen, T.: J. Phys. D: Appl. Phys. **39**, 4497–4503 (2006)
114. Bruers, S.: Classification and discussion of macroscopic entropy production principles. arXiv:cond-mat/0604482v2 [cond-mat.stat-mech]
115. Belkin, A., Hubler, A., Bezryadin, A.: Sci. Rep. **5** (2015). https://doi.org/10.1038/srep08323
116. Di Vita, A.: Eur. Phys. J. D **54**, 451–461 (2009)
117. Di Vita, A.: Eur. Phys. J. D **72**, 7 (2018)
118. Leigh, D.C.: Phys. Fluids **5**, 501 (1962)
119. Lakshmana Rao, S.K.: Phys. Fluids **6**, 598 (1963)
120. Astarita, G.: J. Non-Newtonian Fluid Mech. **2**, 343–351 (1977)
121. Sher Akbar, N., Nadeem, S., Mekheimer, K.S.: J. Egypt. Math. Soc. **24**(1), 138–142 (2016)
122. Sun, T., Meakin, P., Jossang, T.: PRE **49**(6), 4865–4872 (1994)
123. Lorenz, R.D.: Earth system dynamics discussions, 1–13 (2019). https://doi.org/10.5194/esd-2019-73
124. Liu, H.H.: Fluid Flow in the Subsurface. Springer International Publishing (2017). https://doi.org/10.1007/978-3-319-43449-0
125. Waldmann, L.: Zeitschrift fuer Naturforschung A **22**(8), 1269–1280 (1967)
126. Santillan, M., Ares de Parga, G., Angulo-Brown, F.: Eur. J. Phys. **19**, 361–369 (1998)
127. Wurfel, P., Ruppel, W.: J. Phys. C: Solid State Phys. **18**, 2987–3000 (1985)
128. Yoshida, Z., Mahajan, S.M.: Phys. Plasmas **15**, 032307 (2008)
129. Biwa, T., Ueda, Y., Yazaki, T., Mizutani, U.: EPL **60**, 363 (2002)
130. Meija, D., Selle, L., Bazile, R., Poinsot, T.: Proc. Comb. Inst. **35**, 3201–3208 (2014)
131. Hong, S., Shanbhogue, S.J., Ghoniem, A.F.: Impact of the flameholder heat conductivity on combustion instability characteristics. GT2012-70057 Proceedings of ASME Turbo Expo 2012, 11–15 June 2012, Copenhagen, Denmark (2012)
132. Pascale, S., Gregory, J.M., Ambaum, M.H.P., Tailleux, R., Lucarini, V.: Earth Syst. Dynam. **3**, 19–32 (2012)
133. Yang, R., Chong, K.L., Wang, Q., Verzicco, R., Shishkina, O., Lohse, D.: PRL **125**, 154502 (2020)
134. Urban, P., Hanzelka, P., Králik, T., Musilová, V., Skrbek, L.: PRL **128**, 134502 (2022)
135. Zupanovic, P., Juretic, D., Botric, S.: PRE **70**, 056108 (2004)
136. Holyst, R., Maciołek, A., Zhang, Y., Litniewski, M., Knychała, P., Kasprzak, M., Banaszak, M.: PRE **99**, 042118 (2019)
137. Elsgolts, I.V.: Differential Equations and Variational Calculus. Mir, Moscow, URSS (1981)
138. Rebhan, E.: Phys. Rev. A **42**, 781 (1990)
139. Rubin, D.M., Ikeda, H.: Sedimentology **37**, 673–684 (1990)
140. Jinghai, L., Wei, G., Zhongdong, Z., Jie, Y.: Chin. Sci. Bull. **44**(4), 323–327 (1999)
141. Lucia, U., Grisolia, G., Ponzetto, A., Silvagno, F.: J. Theor. Biol. **429**, 181–189 (2017). https://doi.org/10.1016/j.jtbi.2017.06.029
142. Lucia, U., Grisolia, G., Ponzetto, A., Bergandi, L., Silvagno, F., Soc, R.: Open Sci. **7**, 200299 (2020). https://doi.org/10.1098/rsos.200299
143. Malkus, W.V.R., Veronis, G.: J. Fluid Mech. **4**, 225 (1958)
144. Chandrasekhar, S.: Hydrodynamic and Hydromagnetic Stability. Oxford University Press, New York (1961)
145. Welander, P.: J. Fluid Mech. **29**(1), 17–30 (1967)
146. Arango-Restrepo, A., Rubi, J.M., Barragán, D.: J. Phys. Chem. B **123**(27), 5902–5908 (2019)

The Garden of Forking Paths

7

Abstract

We review our results. Linear non-equilibrium thermodynamics is well developed but with a narrow domain of validity. No general-purpose variational principle exists for non-equilibrium thermodynamics. Outside the domain of linear non-equilibrium thermodynamics, all the necessary criteria for stability of far-from-equilibrium relaxed states (both steady and oscillating ones) which are discussed in the book follow from Le Châtelier's principle, provided that the system always satisfies the assumption of local thermodynamic equilibrium.

The fascination and the usefulness of thermodynamics lie in its unique capability in providing predictions concerning the behaviour of macroscopic systems while requiring no explicit solution of the equations of motion of the many microscopic particles these systems are made of. A tenet of thermodynamics is that entropy achieves a maximum at thermodynamic equilibrium. As for steady states far from thermodynamic equilibrium, attempts to derive extremum principles like the Second Principle are so far unconvincing since they often require the introduction of additional hypotheses, which by themselves are less evident than the proved statements. When taking a bird's eye view of the available descriptions of stable steady states far from thermodynamic equilibrium, we see that Gage et al.'s nonexistence theorem [1] stands as a Stone Guest who casts doom over all attempts at writing down a universal, general-purpose criterion for stability in the form of a variational principle (Sect. 5.1). In particular, neither the minimum entropy production principle [2,3] (Sects. 4.1.4 and 4.3.10) nor the least dissipation principle [4] (Sect. 4.1.5) of linear non-equilibrium thermodynamics (LNET, Sect. 4.1.1) are universal, as both rely on Onsager's symmetry [5] (Sect. 4.1.2) which in turn holds for a very limited class of phenomena.[1] This means that even if it exists, any stability criterion which

---

[1] Which includes both Seebeck effect (Sect. 4.1.8), Peltier effect (Sect. 4.1.9), Thomson effect (Sect. 4.1.10), the thermocouple (Sect. 4.1.11) and fluid mixtures of reacting chemical species

© The Author(s), under exclusive license to Springer Nature Switzerland AG 2022   201
A. Di Vita, *Non-equilibrium Thermodynamics*, Lecture Notes in Physics 1007,
https://doi.org/10.1007/978-3-031-12221-7_7

takes the form of an extremum condition can only have limited validity and hold for a well-defined class of problems. In the language of variational calculus, a meaningful variational principle must be subject to constraints that are not universal but depend on the problem. No matter what the variational principle is like, an obvious choice is that these constraints are provided by the conservation equations (of mass, momentum, electric charge...) ruling the particular system of interest in a steady state. Thus, a steady state is a solution of the conservation equation; if, in addition, this solution is also stable, then it satisfies also the variational principle.

This is a paradigm shift with respect to the familiar equilibrium thermodynamics. The latter starts from the Second Principle, which is an extremum condition (maximization of entropy) of stability of thermodynamic equilibrium and draws conclusions of universal validity (e.g. the positiveness of the specific heat at constant volume) concerning thermodynamic equilibrium, regardless of the detailed dynamics of the system. We expect successful non-equilibrium thermodynamics to provide us with stability conditions for the steady-state solutions of the conservation equations in systems far from thermodynamic equilibrium, and we expect it to do the job in a way that is as independent as possible from the detailed dynamics.

To this purpose, it is useful to focus our attention on systems at local thermodynamic equilibrium (LTE, Sect. 3.2), i.e. a configuration where the body is in thermodynamic equilibrium with itself but not with the environment.[2] Through one of its corollaries, Le Châtelier's principle (Sect. 3.1),[3] the Second Principle of thermodynamics makes a body at LTE to counteract the impact of external perturbations in such a way that the absolute value of the modification of a thermodynamic force due to such perturbations is minimum [5].[4] This result is usually restated [6] by assessing that a system at LTE always tries to counteract the impact of a perturbation of a thermodynamic force due to the external world.[5]

Usually, the consequences of Le Châtelier's principle are invoked at full thermodynamic equilibrium only (Sect. 3.3). Remarkably, however, if LTE holds at all times in an arbitrary small mass element of a physical system, then it provides us with information [7] on the evolution of the system (Sect. 3.6), hence on its possible outcome,

---

undergoing diffusion and with chemical affinities $\ll k_B T$ (Sect. 4.3.8), among other things; generally speaking, it does *not* include heat conduction (Sect. 5.2.2) except in particular cases (Sects. 4.1.7 and 4.3.6).

[2] To fix the ideas, let the entropy of a physical system depend on two macroscopic quantities. Correspondingly, we define two thermodynamic forces, say X and Y. Both of them vanish at full thermodynamic equilibrium. At LTE, our freedom when it comes to choose the thermodynamic forces of interest (Sect. 2.5) allows us to identify e.g. $Y = 0$ with the condition of thermodynamic equilibrium among the different parts of the system, namely LTE; as for X, it may remain different from zero.

[3] Which states that the absolute value of a perturbation of X due e.g. to the external world is a minimum whenever Y is kept constant.

[4] Because the value of Y is fixed—it is equal to zero—at LTE by definition, and Le Châtelier's principle ensures that the absolute value of a perturbation of X is minimum in this case.

[5] And the freedom of choice of the thermodynamic forces (Sect. 2.5) allows us to replace the words 'thermodynamic force' with the words 'generic quantity'.

a relaxed state [8], and so does Le Châtelier's principle. We try to derive precisely the consequences of Le Châtelier's principle at LTE concerning relaxed states in many physical problems and to show their agreement with available (mostly empirical) results. To this purpose, we make use of the familiar Gibbs-Boltzmann entropy [5].[6] We try to show how stimulating and far-reaching the perspective of non-equilibrium thermodynamics offered by LTE is.[7]

A droplet of water in a river is an example of a body at LTE: even if it is definitely not at equilibrium with the surrounding environment (the river, the embankment...), we may locally define and measure its temperature, its pressure, its mass density, its enthalpy, etc. just like in thermodynamic equilibrium at all times because the innumerable water molecules the droplet is made of are—locally and instantaneously at least—at thermodynamic equilibrium with each other [15]. The river as a whole is made of many small masses of water, each of them at LTE at all times; we may say that LTE holds everywhere at all times throughout the river.

Since all water droplets in the river are at LTE at all times, Le Châtelier's principle applies to all of them. If we dip a stick into the steady flow of a river, we feel a drag on the stick and we can measure it, e.g. with the help of a dynamometer. Of course, here we may speak of a 'steady state' because we have time-averaged all quantities of interest on time-scales much longer than short time-scales of fast phenomena like turbulent fluctuations and the like. We take for granted neither the existence of a relaxed state of the river[8] nor the resilience of such a relaxed state against perturbations. Rather, we postulate both of them and derive necessary conditions for the stability of this state.[9]

If we hold the stick at a fixed position in spite of this drag, soon the water adjusts itself to a new steady flow, slightly different from the initial one, which remains unchanged until we remove the stick. Correspondingly, the river's answer to the external perturbation (the stick) is to modify the drag, raising it from the initial zero value to some final value, which is the outcome of the contributions of all droplets. Accordingly, a natural choice for the thermodynamic force in Le Châtelier's principle

---

[6] We postulate neither a modification nor a reinterpretation in the definition of entropy like in the approaches of 'extended irreversible thermodynamics' [9] (Sect. 6.1.6), 'steepest ascent' [10] (Sec. 6.1.7), 'information thermodynamics' [11,12] (Sect. 6.1.9) and 'second entropy' [13] (Sect. 6.1.8). We do not even postulate the occurrence of any 'rugged invariant' in the relaxation process, in contrast, e.g. with the approaches reviewed in Ref. [14] (Sects. 6.1.4 and 6.1.5).

[7] Our discussion starts from some consequences of Le Châtelier's principle at LTE described in Sects. 6.2.1, 6.2.11 and 6.2.15.

[8] Where such fluctuations are so small and fast that time-averaging still makes sense.

[9] This is contrast with the 'excess entropy production' approach [16], which leads to a sufficient condition for stability of non-equilibrium steady state [17] (Sect. 6.1.3) and is, therefore, unable to check the stability of an alleged relaxed state.

is the drag itself.[10] Then, Le Châtelier's principle dictates that the final absolute value of the drag is minimum.[11]

The obvious question arises: minimum with respect to what? We are going to answer below; here we stress the point that Le Châtelier's principle tells us nothing about the fluid dynamics of a river: it provides us rather with information about the final configuration the system adapts itself to—if any such relaxation actually occurs. Le Châtelier's principle is, therefore, likely to lead us to write down necessary conditions of stability for relaxed states. Among all possible solutions of the equations of motion which correspond to a final, steady state—if any such solution exists—these necessary conditions provide us with selection rules which point at the particular state which is resilient against perturbations—and is, therefore, more likely to be actually observed. We stress the point that the search of the minimum spans the set of possible solutions corresponding to a steady state and is, therefore, constrained by the boundary conditions relevant to the problem. The necessary conditions we are speaking about are, therefore, always subject to some constraint.

Being all the small droplets at LTE at all times during their evolution, indeed, it is only natural that Le Châtelier's principle provides information concerning the relaxed state of the river,[12] which is the outcome of the evolution of all the droplets. And conversely, if such a relaxed state exists, then its stability relies on the stability of each droplet (including, e.g. the fact that its specific heat at constant volume is positive-definite at all times), which in turn copes ultimately with the Second Principle of thermodynamics and satisfies, therefore, Le Châtelier's principle. The requirement of LTE everywhere at all times replaces the much more restrictive requirement of full thermodynamic equilibrium, the price to be paid being the loss of universality of the results we obtain as both the actual quantities of interest and the explicit form of the related constraints (which include, e.g. the conservation equations in steady state) depend on the system.[13]

In our system made of a river and a stick, the constrained minimization of drag has far-reaching consequences. In the final configuration, indeed, Newton's Third Law of dynamics implies that the stick exerts a force on the water which is equal and opposite to the drag exerted by the water on the stick and opposes the water flow. Through this force, we, the external world, provide the river with an amount of mechanical work per unit time which is basically proportional to the product of the

---

[10] Being the drag a mechanical force which opposes the motion of the fluid, it acts like a thermo-dynamic force which lowers the kinetic energy of the river by raising the production of entropy in the water, i.e. the dissipation.

[11] The fact that we are speaking of 'absolute value' allows the discussion to be independent of the direction of the river's flow.

[12] As for formal consequence of the assumption of LTE for the generic mass element of the system at all times, the so-called 'general evolution criterion', we refer to Ref. [7]. See also Ref. [8] for some examples of applications.

[13] Sometimes, admittedly, self-gravitating systems with time-dependent gravitational field violate the requirement of LTE everywhere at all times [18] (Sect. 3.4.1). Here, we consider static gravitational fields only; however, we recall that even gravitational waves may affect the entropy balance [19] (Sect. 6.2.6).

drag force and of the velocity of the water relative to the stick far away from it.[14] For any reasonable size of the stick,[15] we are allowed to assume that this velocity— which we may interpret as being proportional to the spatially averaged velocity of the river, which in turn is proportional to the water flow of the river—is constant.[16] In particular, we are allowed to assume that its value is the same both before and after dipping the stick—i.e. we may focus our attention on perturbations leaving the water flow unaffected.[17] But then, Le Châtelier's minimization of the drag reduces to constrained minimization of the mechanical work exerted by the external world per unit time on the river with the help of the stick, the constraint being given by the water flow of the river.

Of course, there is nothing special in our stick, which plays the role of a dummy. In particular, the external world does work on the river through many different physical processes, including, e.g. Earth's gravity which drives the river flow towards the sea; but the impact of all of them on the river remains unchanged as we dip our stick and is, therefore, not considered in the discussion above. If we replace, e.g. the dipping of the stick either with a change of shape of the embankment due to the building-up of sediments or with a perturbation of gravity due to tides, we repeat our argument step-by-step and obtain the same result, provided that the river actually relaxes to a final stable state of unchoked flow. Then, we generalize our constrained minimization above from the case of the stick to the total mechanical work done per unit time by the external world on the river.

Being the river in a steady state, if we assume that the temperature is uniform and constant[18] so that the enthalpy of water is also uniform and constant, then conservation of energy requires that the total mechanical work done by the external world per unit time on the river is equal to the total power dissipated inside the fluid because of friction. Then, our constrained minimization is equivalent to constrained minimization of the power dissipated by friction inside the river,[19] at given water flow. Moreover, the actual value of the water flow is uniquely determined by the boundary conditions through mass balance and Navier-Stokes' equation. Accordingly, we may replace the constraint of given water flow with the constraint of fixed boundary conditions.[20]

---

[14] We may convince ourselves if we recall that Galileian invariance ensures that the mechanical power is exactly the same if the water stands still and we move the stick.

[15] I.e. as far as the stick is much smaller than a dam, which stops the river flow altogether.

[16] Here is one of the constraints hinted at above.

[17] This rules out the dam.

[18] Here is another constraint, which we are going to drop below.

[19] To the author's knowledge, Helmholtz is to be credited for being the first to put in evidence the link between the work done by external forces and the energy dissipated in a fluid [20] (Sect. 5.3.10).

[20] Should we be able to rephrase these findings in a more general way, we would say that the relaxed state of a system is stable when the heat produced by dissipation in the bulk of the system is minimum, i.e. when the work done by the environment on the system is minimum, or, in other words, when the environment offers minimum resistance. In particular, whenever a river with a given flow is in a steady, stable state, the external world offers a minimum resistance to it. This is similar to the conclusions a believer in Bejan's 'constructal law' [21] would like to draw (Sect. 6.2.4).

As hinted above, we can repeat the same argument with any other conceivable disturbance. The thermodynamic nature of Le Châtelier's principle ensures that—as far as the constraint of fixed velocity on the boundaries of the system is satisfied—the steady state which the river is supposed to relax onto corresponds to a minimum value of the total power dissipated by friction regardless of the detailed friction mechanism (a stick, the embankment and the bulk viscosity of water). Usually, constant and uniform temperature corresponds to constant and uniform viscosity coefficients.[21] Apart from that, Le Châtelier's principle relies on no assumption on the constitutive relationships concerning viscous dissipation. For example, we do not require that the fluid is Newtonian.[22]

Just like the stick, our river has nothing special. Rather than speaking of a river flowing against the friction due to the embankment and the river bed, we could e.g. speak of either water percolating vertically across a porous layer of rock,[23] or a flow of electrons in a metal or a plasma.[24] The viscous dissipation, the viscosity (tensor) and the constraint of constant flow are replaced by Joule dissipation, resistivity (tensor) and the constraint of constant total electric current, respectively.[25]

Let us focus our attention on the environment (say, the embankment) surrounding our river. Once applied to the boundary between the river and the external world,

---

[21] For a Newtonian fluid with constant viscosity coefficients and subject to fixed boundary conditions, minimization of the total power dissipated by viscosity is just Korteweg-Helmholtz' principle (Sect. 6.2.9) in the limit of low Reynolds number Re where the viscous heating is weak and the assumption of constant and uniform temperature is self-consistent [22]. Even at larger Re, however, such minimization holds both in problems of hydrology [23] and geology [24] (Sect. 5.3.8).

[22] Indeed, generalization of Kortweg-Helmholtz' principle to non-Newtonian fluids with constant coefficients of viscosity is available, provided that the bulk viscosity of an incompressible, Newtonian fluid is replaced by the components of a tensor [25, 26] (Sect. 6.2.9); it has been observed e.g. in human blood [27, 28] (Sect. 5.3.8).

[23] And in this case, we would retrieve the observed minimization of energy expenditure rate per unit volume [29] (Sect. 6.2.10).

[24] Against this analogy, admittedly, one could argue that Korteweg-Helmholtz' principle takes into account the friction between adjacent masses of water, while the electron-electron collisions are usually neglected in a metal. But then thermodynamics turns out to be useful, as it ensures that minimization of dissipated power holds regardless of the detailed friction mechanism, thus making the above distinction irrelevant. Kohler's analysis of Boltzmann kinetic equation [30] provides us with a microscopic justification for the common treatment of viscous and Joule dissipation (Sect. 6.2.5).

[25] So that Kirchhoff's principle of minimum power dissipated by Joule heating [31] replaces Korteweg-Helmholtz' principle of minimum power dissipated by viscosity. Remarkably, Kirchhoff's principle works only if the temperature is uniform, so that resistivity is uniform [32]. In turn, many particular principles stem from Kirchhoff's principle, e.g. Steenbeck's principle for radiative arcs [33]. Together, Korteweg-Helmholtz' and Kirchhoff's principles put constraints on the macroscopic velocity and the magnetic field in viscoresistive, magnetized plasmas with flat temperature gradient and lead [34] to Taylor's principle [35] and Turner's principle [36] in MHD and Hall MHD, respectively (Sect. 6.2.8). Finally, the term which takes into account Joule heating in the entropy balance includes also the effect of emission and absorption of photons between two given energetic levels of an atom; it is, therefore, not surprising that a Kirchhoff-like principle holds for such photons as well [37] (Sect. 6.2.6).

Newton's third law of dynamics ensures that the total amount of mechanical work done by the environment on the river per unit time is equal and opposite to the total amount of work done by the river on the environment. Minimization of the former is, therefore, equivalent to maximization of the latter. A given amount of water flow still provides the constraint, which is equivalent to the constraint of a given amount of power dissipated by friction. Now, a river flowing in a steady state exerts friction on the river bed and the embankment. Eventually, therefore, the corresponding total amount of work done by the river on the environment per unit time is transformed into heat. Such heating raises the entropy of the environment by an amount equal to the ratio of the total amount of heat and the temperature of the environment. Being the environment much larger than the river, the heat capacity of the environment is much larger than the heat capacity of the river, and the heating leaves the temperature of both river banks unaffected. It follows that the rate of entropy growth of the environment is maximized, the constraint being a given amount of water flow.[26] Remarkably, this result does not depend on what occurs inside the bulk of the river; it depends just on the details of friction at the river banks and on the temperature of the banks, i.e. on what occurs both at the boundary of the river and inside the environment regardless of what occurs inside the river bulk. As such, it holds even if the temperature inside the river bulk is not uniform.

Again, our river is just an example. Let us replace it with a fluid flowing between two parallel plates at constant temperatures. The work done by the fluid on the environment is obviously done at the expense of the energy of the fluid, which has to be transported somehow from the bulk towards the plates. Maximization of the amount of energy transported this way per unit time across the fluid is, therefore, required.[27] This result holds regardless of the physical origin of the fluid's energy.[28] In particular, if the temperatures of the two plates are different, then maximization of the heat flow across the fluid from the hotter towards the colder plate follows, with the above constraint of a given amount of dissipated power[29]; being the entropy content of the fluid constant in steady state, this implies maximization of the net

---

[26] Thus, we have retrieved the results of Ref. [38] (Sect. 6.2.18), with no need of the 'quasi-thermodynamic approach' [39] they were originally [40] related to (Sect. 6.1.12): whenever the entropy generation due to a river starts decreasing, the river flow looses stability and a flood follows.

[27] If turbulence rules energy transport across the fluid, this amount increases with increasing amount of kinetic energy stored in the turbulent motions. Maximization of the latter has been postulated in [41], the constraint of given flow being replaced by the equivalent constraint of given spatially averaged velocity across the plates (Sect. 5.3.12).

[28] For example, should a perfect gas be heated by the external world—e.g. with the help of radiofrequency—just at the mid-plane between the two plates, then the relaxed state of the system corresponds to a minimum of the increase of the gas internal energy with the constraint of given heating power [42], in agreement with our maximization of energy transport (Sect. 6.2.16).

[29] Which is equivalent to minimization of temperature gradient in case of natural convection [43,44], with the same constraint (Sect. 6.2.2).

amount of entropy flowing per unit time across the fluid boundaries[30]; as usual by now, the detailed expression of this amount and of the constraints depends on the particular problem of interest.

Being rooted in Le Châtelier's principle, maximization of the net amount of entropy flowing per unit time across the boundaries of the system does not even require that the system is a fluid.[31] The detailed mechanism ruling the entropy outflow is not relevant.[32] Such maximization is sometimes referred to in the literature as 'proof' of a general-purpose maximum entropy production principle [60,61] (Sects. 6.1.9 and 6.1.10).[33] It is often assumed [65] that such principle involves the contribution to entropy production of all irreversible processes, both in the bulk and on the boundaries of the system. However, past attempts to derive the principle under discussion are unconvincing [66] since they require the introduction of additional hypotheses, which by themselves are less evident than the proved statements.[34]

In some cases, this maximization is just a consequence of a particular choice of boundary conditions in systems that could otherwise be described through some minimization procedure [69,70], e.g. for the sake of the reciprocity principle for isoperimetric problems [71].[35] Roughly speaking, the relationship between maximization and minimization procedures resembles the relationship between the minimum entropy production principle and the least dissipation principle in LNET: it is basically a matter of boundary conditions.[36] Generally speaking, indeed, different quantities enjoy different extremum properties in the same relaxed state under the same constraints. No contradiction arises, as these properties refer to different quantities. The choice of the stability criterion to be adopted in a given problem is ultimately a matter of convenience: all criteria hinted at here stem from LTE through Le Châtelier's principle.

---

[30] In problems with no heating occurs inside the system, this conclusion agrees with the results of Refs. [45–49] concerning convection at large Rayleigh number Ra, the plasma edge in a tokamak, shock waves at large Mach number and an analysis of Earth's atmosphere and oceans as a system maximizing the heat flow from the Equator to the Poles, respectively (Sects. 6.2.13, 6.2.17 and 6.2.19). The latter analysis includes no friction and provides no satisfactory description once dissipation in the bulk of the system is taken into account [50–52] (Sect. 5.6.13).

[31] A list of problems with no heating power in the bulk of the system includes dunes [53], metallurgy [54], processes of crystallization [55] and solidification [56]—see Ref. [57] for a review (Sects. 6.2.17 and 6.2.20).

[32] We may, e.g. replace the water flow of our river with the flux of photons emitted by a body at finite temperature, this flux being proportional to the total power lost by the body through radiation. Let the system be at LTE, i.e. let us deal with a black body. Then, of all bodies which radiate the same quantity of power, a black body is that which produces the largest amount of entropy per unit time [58] (Sect. 6.2.12). A case with non-vanishing heating power is the maximization of entropy production at given heating power in a network of resistors [59] (Sect. 6.2.14).

[33] Which is often invoked by researchers in ecology, in loose connection [62] with the so-called 'Gaia hypothesis' [63]. However, the latter fails to meet general consensus [64].

[34] The arguments of [65] in favour of this principle have been criticized in Refs. [67,68] (Sect. 6.1.9).

[35] For example, this principle allows reformulation of the problem of Ohmic conduction in terms of a maximization principle [72] (Sect. 6.2.8).

[36] This is in agreement with the discussions in Refs. [66,69,73].

We recall that our 'steady' state is just a time-average. If the relevant quantities in the relaxed state of the system keep on oscillating at a given frequency, the time-average is to be taken on a time-scale much longer than the period of oscillation. Should the relaxed state be an oscillatory one, we would perform the time-average on a time-scale much longer than the period of oscillation. Should the system of interest undergo a spontaneous transition (a bifurcation [74]) from a steady state to an oscillatory state, or from an oscillatory state to another oscillatory state with different periods, the necessary criterion of stability relevant to the problem would provide us with a selection rule: the system selects the configuration which corresponds to a minimum (a maximum) of the quantity minimized (maximized) in the criterion. This selection rule is the thermodynamic counterpart of a bifurcation between different limit cycles, in full agreement with the original suggestion of Ref. [7], and confirms our extension of the concept of 'relaxed configuration' from steady to periodically oscillating states (Sect. 5.8.1).[37] In particular, maximization is basically a maximization of some time-average and leads, therefore, to a prescription of the phases of the quantities involved in the oscillation.[38]

Admittedly, and with the exception of both the results based on Onsager's symmetry (Sects. 4.1.4, 4.3.10 and 4.1.5) and Glansdorff and Prigogine's excess entropy production (which is a just a sufficient condition, Sect. 6.1.3), the only stability criteria rigorously proven to date starting from well-known first principles of dynamics are due to Kirchhoff (Joule heating, Hermann [31]), Korteweg and Helmholtz (viscous heating [22]), Kohler (the microscopic foundation of Joule and viscous heating, [30]), Chandrasekhar [43] and Busse [44] (convection just above threshold), Schechter [25] and Astarita [26] (non-Newtonian fluids), Kroell [37] (absorption and emission of photons), Santillan et al. [58] (black body) and Rebhan [48] (shock waves at large Mach number). Apart from that, our list of necessary criteria of stability of relaxed

---

[37] The experiments described in Refs. [75] (Sect. 5.8.2), [76] (Sect. 5.8.3) and [77] (Sect. 5.8.4) provide us with excellent examples.

[38] For example, if a spontaneous oscillation in the subsonic motion of a viscous fluid moved by natural convection in a closed, loop-like pipe occurs, then the perturbation of the amount of heat supplied by the external world to a unit volume of the fluid oscillates in phase with the perturbation of fluid temperature due to the exchange of heat between the fluid and the walls of the pipe [78] (Sect. 5.8.11). In the outer layer of a Cepheid pulsating star, maximization of the total flux of entropy—produced by the nuclear fusion reactions in the star core—implies maximization of the correlation between oscillations of temperature and of heat flux [79] (Sect. 5.8.12). If a fluid does work on the environment through acoustic waves, then maximization of the total amount of mechanical work done by the system on the external world per unit time implies that the development of vibrations in an oscillating system evolves towards those relationships involving amplitudes and phases which maximize the amount of acoustic energy irradiated from the region where combustion occurs [80] (Sect. 5.8.7). Finally, if the system is an inviscid fluid with uniform entropy density and is endowed with an internal source of heat, then it can be shown that this result is equivalent to the maximum correlation between oscillations of pressure and oscillations of the amount of heat released per unit time in a unit volume of fluid, in agreement with Rayleigh's criterion of thermoacoustics [81] (Sect. 5.8.8) as well as with Rijke's (Sect. 5.8.9) and Sondhauss' (Sect. 5.8.10) experiments.

states far from thermodynamic equilibrium includes either more or less reasonable postulates or empirical observations and is definitely far from complete; here, we have just picked up some examples which could be explained with our example of a river and a stick.

However, our discussion shows how versatile Le Châtelier's principle at LTE can be as a tool for searching necessary conditions of stability. Moreover, all the results listed above have been either proven or postulated by different authors independently; Le Châtelier's principle at LTE provides us with a unifying perspective. The Holy Grail of a general-purpose variational principle which generalizes the Second Principle to all stable, steady states far from thermodynamic equilibrium may be fading, but we are left with an unexpected web of intertwined, surprising results, a never-ending garden of forking paths with the exciting perspective of finding useful results and with LTE as the guiding compass.

Let us conclude with Galileo's poetic words [82]: *Ora, perché é tempo di por fine ai nostri discorsi, mi resta a pregarvi, se nel riandar piú posatamente le cose da me arrecate incontraste delle difficoltá o dubbi non ben resoluti, scusiate il mio difetto, sí per la novitá del pensiero, sí per la debolezza del mio ingegno, sí per la grandezza del suggetto, e sí finalmente perché io non pretendo né ho preteso da altri quell'assenso ch'io medesimo non presto a questa fantasia.*[39]

References

1. Gage, D.H., Schiffer, M., Kline, S.J., Reynolds, W.C.: The non-existence of a general thermokinetic variational principle. In: Donnelly, R.J. (ed.) Non-equilibrium Thermodynamics: Variational Techniques and Stability. University of Chicago Press, Chicago, IL (1966)
2. Gyarmati, I.: Non-equilibrium Thermodynamics. Springer, Berlin (1970)
3. Glansdorff, P., Prigogine, I.: Physica **20**, 773–780 (1954) (in French)
4. Ichiyanagi, M.: Phys. Rep. **243**, 125–182 (1994)
5. Landau, L.D., Lifshitz, E.: Statistical Physics. Pergamon, Oxford, UK (1960)
6. Fermi, E.: Thermodynamics. Dover Publications (1956)
7. Glansdorff, P., Prigogine, I.: Physica **30**, 351 (1964)
8. Di Vita, A.: Phys. Rev. E **81**, 041137 (2010)
9. Jou, D., Casas-Vázquez, J., Lebon, G.: Rep. Prog. Phys. **51**, 1105 (1988)
10. Beretta, G.P.: Steepest entropy ascent model for far-non-equilibrium thermodynamics. unified implementation of the maximum entropy production principle. arXiv:1409.6672v1 [cond-mat.stat-mech]. Accessed 23 Sept 2014
11. Jaynes, E.T.: Phys. Rev. **106**, 620–630 (1957)
12. Jaynes, E.T.: Phys. Rev. **108**, 171–190 (1957)
13. Attard, P.: J. Chem. Phys. **122**, 154101 (2005)
14. Brown, M.R.: J. Plasma Phys. **57**(1), 203–229 (1997)

---

[39] Now, since it is time to end our arguments, the last thing I have to do is to beg you, if, in reconsidering with more attention what I have presented, you will find difficulties, or doubts not well resolved, to pardon my deficiency, due to the novelty of the ideas, the weakness of my understanding, the difficulty of the topic and the fact that I do not pretend nor I ever pretended from others the certitude that I myself do not lend to this imagination.

15. DeGroot, S.R., Mazur, P.: Non-equilibrium Thermodynamics. North Holland, Amsterdam (1962)
16. Glansdorff, P., Prigogine, I.: Physica **46**, 344–366 (1970)
17. Keizer, J., Fox, R.: Proc. Nat. Acad. Sci. USA **71**(1), 192–196 (1974)
18. Velazquez, L.: J. Stat. Mech.: Theory Exp. **2016**(3), 033105 (2016)
19. DeWitt, C.M. (ed.): Conference on the Role of Gravitation in Physics at the University of North Carolina, Chapel Hill, March 1957; WADC Technical Report 57-216 (Wright Air Development Center, Air Research and Development Command, United States Air Force, Wright Patterson Air Force Base, Ohio). https://edition-open-sources.org/sources/5/34/index.html
20. von Helmholtz, H.: Wissenschaftliche Abhandlungen, Bd. 1, (1882) (in German). http://echo.mpiwg-berlin.mpg.de/ECHOdocuViewfull?url=/mpiwg/online/permanent/einstein_exhibition/sources/QWH2FNX8/index.meta&start=231&viewMode=images&pn=237&mode=texttool
21. Bejan, A., Lorente, S.: J. Appl. Phys. **100**, 041301 (2006)
22. Lamb, H.: Hydrodynamics. Cambridge University Press, Cambridge, UK (1906)
23. Sun, T., Meakin, P., Jossang, T.: PRE **49**(6), 4865–4872 (1994)
24. Lorenz, R.D.: Earth system dynamics discussions, pp. 1–13 (2019). https://doi.org/10.5194/esd-2019-73
25. Schechter, R.S.: Chem. Eng. Sci. **17**, 803–806 (1962)
26. Astarita, G.: J. Non-Newtonian Fluid Mech. **2**, 343–351 (1977)
27. Thompson, D.W.: On Growth and Form. Cambridge, New York (1945)
28. Murray, C.D.: Proc. Natl. Acad. Sci. (Physiol.) **12**, 207–214 (1926)
29. Liu, H.H.: Fluid Flow in the Subsurface. Springer International Publishing (2017). https://doi.org/10.1007/978-3-319-43449-0
30. Kohler, M.: Z. Phys. **124**, 772–789 (1948) (In German)
31. Hermann, F.: Eur. J. Phys. **7**, 130 (1986)
32. Jaynes, E.T.: Ann. Rev. Phys. Chem. **31**, 579 (1980)
33. Steenbeck, M.: Wissenschaftlichen Veroeffentlichungen aus den Siemens Werke **1**, 59 (1940). (in German)
34. Di Vita, A.: Eur. Phys. J. D **54**, 451–461 (2009)
35. Taylor, J.B.: PRL **33**, 1139–1141 (1974)
36. Turner, L.: IEEE Trans. Plasma Sci. **PS-14**(6), 849–857 (1986)
37. Kroell, W.: J. Quant. Spectrosc. Radiat. Transf. **7**, 715–723 (1967)
38. Lucia, U., Buzzi, P., Grazzini, G.: Irreversibility in river flow. Intl. J. Heat Technol. **34**(Special Issue 1), S95–S100 (2016)
39. Lavenda, B.H.: Phys. Rev. A **9**(2), 929–942 (1974)
40. Lucia, U.: Atti della Accademia Peloritana dei Pericolanti **94**(1), A4 (2016)
41. Malkus, W.V.R.: J. Fluid Mech. **1**, 521 (1956)
42. Holyst, R., Maciołek, A., Zhang, Y., Litniewski, M., Knychała, P., Kasprzak, M., Banaszak, M.: PRE **99**, 042118 (2019)
43. Chandrasekhar, S.: Hydrodynamic and Hydromagnetic Stability. Oxford University Press, New York (1961)
44. Busse, F.H.: J. Fluid Mech. **30**(4), 625–649 (1967)
45. Ozawa, H., Shimokawa, S., Sakuma, H.: Phys. Rev. E **64**, 026303 (2001)
46. Malkus, W.V.R., Veronis, G.: J. Fluid Mech. **4**, 225 (1958)
47. Yoshida, Z., Mahajan, S.M.: Phys. Plasmas **15**, 032307 (2008)
48. Rebhan, E.: Phys. Rev. A **42**, 781 (1990)
49. Paltridge, G.W.: Nature **279**, 630 (1979)
50. Nicolis, C.: Q.J.R. Meteorol. Soc. **125**, 1859–1878 (1999)
51. Marston, J.B.: Ann. Rev. Cond. Matter Phys. **3**, 285–310 (2012)
52. Bartlett, S., Virgo, N.: Entropy **18**, 431 (2016)
53. Rubin, D.M., Ikeda, H.: Sedimentology **37**, 673–684 (1990)
54. Kirkaldy, J.S.: Metall. Trans. A **16A**(10), 1781 (1985)
55. Hill, A.: Nature **348**, 426 (1990)

56. Sekhar, J.A.: J. Mater. Sci. **46**, 6172 (2011)
57. Martyushev, L.M., Seleznev, V.D., Kuznetsova, I.E., Eksp, Z.: Theor. Fiz. **118**, 149 (2000)
58. Santillan, M., Ares de Parga, G., Angulo-Brown, F.: Eur. J. Phys. **19**, 361–369 (1998)
59. Zupanovic, P., Juretic, D., Botric, S.: PRE **70**, 056108 (2004)
60. Martyushev, L.M.: Phil. Trans. R. Soc. B **365**, 1333–1334 (2010)
61. Ziegler, H.: ZAMP **34**, 832–844 (1983)
62. Kleidon, A.: Phil. Trans. R. Soc. A **370**, 1012–1040 (2012)
63. Lovelock, J.E., Margulis, L.: Tellus **XXVI**, 1–2 (1974)
64. Tyrrell, T.: New Scientist, pp. 30–31 (2013)
65. Dewar, R.D.: J. Phys. A **38**, L371 (2005)
66. Martyushev, L.M., Seleznev, V.D.: Phys. Rep. **426**, 1–45 (2006)
67. Grinstein, G., Linsker, R.J.: Phys. A: Math. Theor. **40**, 9717 (2007)
68. Polettini, M.: Eur. Phys. Lett. **97**, 30003 (2012)
69. Reis, A.H.: Ann. Phys. **346**, 22?27 (2014)
70. Sree Harsha, N.R.: Eur. J. Phys. **40**, 033001 (2019)
71. Elsgolts, I.V.: Differential Equations and Variational Calculus. Mir, Moscow, URSS (1981)
72. Christen, T.: J. Phys. D: Appl. Phys. **39**, 4497–4503 (2006)
73. Kleidon, A., Malhi, Y., Cox, P.M.: Phil. Trans. R. Soc. B **365**, 1297–1302 (2010)
74. Kuznetsov, Y.A.: Elements of Applied Bifurcation Theory. Springer, New York (2004)
75. Biwa, T., Ueda, Y., Yazaki, T., Mizutani, U.: EPL **60**, 363 (2002)
76. Meija, D., Selle, L., Bazile, R., Poinsot, T.: Proc. Comb. Inst. **35**, 3201–3208 (2014)
77. Hong, S., Shanbhogue, S.J., Ghoniem, A.F.: Impact of the Flameholder heat conductivity on combustion instability characteristics. In: Proceedings of ASME Turbo Expo 2012, 11–15 June 2012, Copenhagen, Denmark (2012). https://doi.org/10.1115/GT2012-70057
78. Welander, P.: J. Fluid Mech. **29**(1), 17–30 (1967)
79. Eddington, A.S.: The pulsation theory of cepheid variables. Observatory **40**, 290–293 (1917)
80. Rauschenbach, B.V.: Vibrational Combustion State Editions of Physico-Mathematical Literature. URSS, Moscow (1961) (in Russian)
81. Rayleigh, J.W.S.: Nature **18**, 319–321 (1878)
82. Galilei, G.: Dialogo dei massimi sistemi, Firenze (1632)

Appendix

A.1 Proof of the General Evolution Criterion

The relationships listed in Sect. 3.6 lead to the following chain of relationships, which eventually reduce to GEC:

$$
\Psi = -\frac{d}{dt}\left(\frac{1}{T}\right)\frac{dp}{dt} + \frac{d}{dt}\left(\frac{1}{T}\right)\frac{d(\rho h)}{dt} - \rho\frac{d}{dt}\left(\frac{\mu_k^0}{T}\right)\frac{dc_k}{dt} + \frac{\rho}{T}\frac{dp}{dt}\frac{d}{dt}\left(\frac{1}{\rho}\right) - \frac{d}{dt}\left(\frac{1}{T}\right)h\frac{d\rho}{dt} =
$$

$$
= -\frac{d}{dt}\left(\frac{1}{T}\right)\frac{dp}{dt} + \frac{d}{dt}\left(\frac{1}{T}\right)\frac{d(\rho h)}{dt} - \frac{\rho}{T}\left[\left(\frac{\partial\mu_k^0}{\partial p}\right)_{T,c_1\ldots}\right]\frac{dc_k}{dt}\frac{dp}{dt} - \rho\left[\frac{\partial}{\partial T}\left(\frac{\mu_k^0}{T}\right)\right]_{p,c_1\ldots}\frac{dc_k}{dt}\frac{dT}{dt} +
$$

$$
-\frac{\rho}{T}\left(\frac{\partial\mu_k^0}{\partial c_j}\right)_{T,p,\ldots}\frac{dc_j}{dt}\frac{dc_k}{dt} + \frac{\rho}{T}\frac{dp}{dt}\frac{d}{dt}\left(\frac{1}{\rho}\right) - \frac{d}{dt}\left(\frac{1}{T}\right)h\frac{d\rho}{dt} =
$$

$$
= -\frac{d}{dt}\left(\frac{1}{T}\right)\frac{dp}{dt} + \frac{d}{dt}\left(\frac{1}{T}\right)\frac{d(\rho h)}{dt} - \frac{\rho v_k}{T}\frac{dc_k}{dt}\frac{dp}{dt} +
$$

$$
-\rho\left[\frac{\partial}{\partial T}\left(\frac{\mu_k^0}{T}\right)\right]_{p,c_1\ldots}\frac{dc_k}{dt}\frac{dT}{dt} - \frac{\rho}{T}\left(\frac{\partial\mu_k^0}{\partial c_j}\right)_{T,p,\ldots}\frac{dc_j}{dt}\frac{dc_k}{dt} + \frac{\rho}{T}\frac{dp}{dt}\frac{dv}{dt} - \frac{d}{dt}\left(\frac{1}{T}\right)h\frac{d\rho}{dt} \leq
$$

$$
\leq -\frac{d}{dt}\left(\frac{1}{T}\right)\frac{dp}{dt} + \frac{d}{dt}\left(\frac{1}{T}\right)\frac{d(\rho h)}{dt} - \frac{\rho v_k}{T}\frac{dc_k}{dt}\frac{dp}{dt} +
$$

$$
-\rho\left[\frac{\partial}{\partial T}\left(\frac{\mu_k^0}{T}\right)\right]_{p,c_1\ldots}\frac{dc_k}{dt}\frac{dT}{dt} + \frac{\rho}{T}\frac{dp}{dt}\frac{dv}{dt} - \frac{d}{dt}\left(\frac{1}{T}\right)h\frac{d\rho}{dt} =
$$

$$
= -\frac{d}{dt}\left(\frac{1}{T}\right)\frac{dp}{dt} + \rho\frac{d}{dt}\left(\frac{1}{T}\right)\frac{dh}{dt} - \frac{\rho v_k}{T}\frac{dc_k}{dt}\frac{dp}{dt} - \rho\left[\frac{\partial}{\partial T}\left(\frac{\mu_k^0}{T}\right)\right]_{p,c_1\ldots}\frac{dc_k}{dt}\frac{dT}{dt} + \frac{\rho}{T}\frac{dp}{dt}\frac{dv}{dt} =
$$

$$
= -\frac{d}{dt}\left(\frac{1}{T}\right)\frac{dp}{dt} + \rho\frac{d}{dt}\left(\frac{1}{T}\right)\frac{dh}{dt} - \frac{\rho v_k}{T}\frac{dc_k}{dt}\frac{dp}{dt} - \frac{\rho}{T}\left(\frac{\partial\mu_k^0}{\partial T}\right)_{p,c_1\ldots}\frac{dc_k}{dt}\frac{dT}{dt} + \frac{\rho\mu_k^0}{T^2}\frac{dc_k}{dt}\frac{dT}{dt} + \frac{\rho}{T}\frac{dp}{dt}\frac{dv}{dt} =
$$

$$
= -\frac{d}{dt}\left(\frac{1}{T}\right)\frac{dp}{dt} + \rho\frac{d}{dt}\left(\frac{1}{T}\right)\frac{dh}{dt} - \frac{\rho v_k}{T}\frac{dc_k}{dt}\frac{dp}{dt} + \frac{\rho s_k}{T}\frac{dc_k}{dt}\frac{dT}{dt} + \frac{\rho\mu_k^0}{T^2}\frac{dc_k}{dt}\frac{dT}{dt} + \frac{\rho}{T}\frac{dp}{dt}\frac{dv}{dt} =
$$

$$
= -\frac{d}{dt}\left(\frac{1}{T}\right)\frac{dp}{dt} + \rho\frac{d}{dt}\left(\frac{1}{T}\right)\frac{dh}{dt} - \frac{\rho v_k}{T}\frac{dc_k}{dt}\frac{dp}{dt} + \frac{\rho h_k}{T^2}\frac{dc_k}{dt}\frac{dT}{dt} + \frac{\rho}{T}\frac{dp}{dt}\frac{dv}{dt} =
$$

A. Di Vita, *Non-equilibrium Thermodynamics*, Lecture Notes in Physics 1007, https://doi.org/10.1007/978-3-031-12221-7

$$= -\frac{d}{dt}\left(\frac{1}{T}\right)\frac{dp}{dt} - \frac{\rho}{T^2}\frac{dT}{dt}\frac{dh}{dt} + \frac{\rho h_k}{T^2}\frac{dc_k}{dt}\frac{dT}{dt} + \frac{\rho}{T}\left(\frac{\partial v}{\partial p}\right)_{T,c_1,\ldots}\left(\frac{dp}{dt}\right)^2 + \frac{\rho}{T}\left(\frac{\partial v}{\partial T}\right)_{p,c_1,\ldots}\frac{dp}{dt}\frac{dT}{dt} =$$

$$= -\frac{d}{dt}\left(\frac{1}{T}\right)\frac{dp}{dt} - \frac{\rho}{T^2}\left(\frac{\partial h}{\partial p}\right)_{T,c_1,\ldots}\frac{dp}{dt}\frac{dT}{dt} - \frac{\rho}{T^2}\left(\frac{\partial h}{\partial T}\right)_{p,c_1,\ldots}\left(\frac{dT}{dt}\right)^2 +$$

$$+ \frac{\rho}{T}\left(\frac{\partial v}{\partial p}\right)_{T,c_1,\ldots}\left(\frac{dp}{dt}\right)^2 + \frac{\rho}{T}\left(\frac{\partial v}{\partial T}\right)_{p,c_1,\ldots}\frac{dp}{dt}\frac{dT}{dt} =$$

$$= \frac{1}{T^2}\frac{dT}{dt}\frac{dp}{dt} - \frac{\rho}{T^2}\left[v - T\left(\frac{\partial v}{\partial T}\right)_{p,c_1,\ldots}\right]\frac{dp}{dt}\frac{dT}{dt} - \frac{\rho}{T^2}\left(\frac{\partial h}{\partial T}\right)_{p,c_1,\ldots}\left(\frac{dT}{dt}\right)^2 +$$

$$+ \frac{\rho}{T}\left(\frac{\partial v}{\partial p}\right)_{T,c_1,\ldots}\left(\frac{dp}{dt}\right)^2 + \frac{\rho}{T}\left(\frac{\partial v}{\partial T}\right)_{p,c_1,\ldots}\frac{dp}{dt}\frac{dT}{dt} =$$

$$= -\frac{\rho}{T^2}\left(\frac{\partial h}{\partial T}\right)_{p,c_1,\ldots}\left(\frac{dT}{dt}\right)^2 + \frac{\rho}{T}\left(\frac{\partial v}{\partial p}\right)_{T,c_1,\ldots}\left(\frac{dp}{dt}\right)^2 + 2\frac{\rho}{T}\left(\frac{\partial v}{\partial T}\right)_{p,c_1,\ldots}\frac{dp}{dt}\frac{dT}{dt} =$$

$$= -\frac{\rho}{T^2}\left\{\left(\frac{\partial u}{\partial T}\right)_{v,c_1,\ldots} - T\left[\left(\frac{\partial v}{\partial T}\right)_{p,c_1,\ldots}\right]^2\left[\left(\frac{\partial v}{\partial p}\right)_{T,c_1,\ldots}\right]^{-1}\right\}\left(\frac{dT}{dt}\right)^2 +$$

$$+ \frac{\rho}{T}\left(\frac{\partial v}{\partial p}\right)_{T,c_1,\ldots}\left(\frac{dp}{dt}\right)^2 + 2\frac{\rho}{T}\left(\frac{\partial v}{\partial T}\right)_{p,c_1,\ldots}\frac{dp}{dt}\frac{dT}{dt} \le$$

$$\le \frac{\rho}{T}\left[\left(\frac{\partial v}{\partial T}\right)_{p,c_1,\ldots}\right]^2\left[\left(\frac{\partial v}{\partial p}\right)_{T,c_1,\ldots}\right]^{-1}\left(\frac{dT}{dt}\right)^2 + \frac{\rho}{T}\left(\frac{\partial v}{\partial p}\right)_{T,c_1,\ldots}\left(\frac{dp}{dt}\right)^2 + 2\frac{\rho}{T}\left(\frac{\partial v}{\partial T}\right)_{p,c_1,\ldots}\frac{dp}{dt}\frac{dT}{dt} =$$

$$= \frac{\rho}{T}\left[\left(\frac{\partial v}{\partial p}\right)_{T,c_1,\ldots}\right]^{-1}\left[\left(\frac{\partial v}{\partial T}\right)_{p,c_1,\ldots}\frac{dT}{dt} + \left(\frac{\partial v}{\partial p}\right)_{T,c_1,\ldots}\frac{dp}{dt}\right]^2 \le 0$$

A.2 Euler-Lagrange Equations

We focus our attention on the following variational principle [1]:

$$\int_{t_1}^{t_1} L\,dt = \text{extremum} \; ; \quad q\,(t_1) = c_1 \; ; \quad q\,(t_2) = c_2$$

where $L = L\left(q, \frac{dq}{dt}, t\right)$ and $q = q\,(t)$ are dubbed 'Lagrangian' and 'Lagrangian coordinate' respectively and c_1, c_2 are constant quantities. The extremum property implies that if we perturb the Lagrangian coordinate, i.e. replace $q\,(t)$ with $q\,(t) + \delta q\,(t)$ with $\delta q\,(t_1) = \delta q\,(t_2) = 0$ (so that $q\,(t_1)$ and $q\,(t_2)$ remain unaffected), then $\frac{dq(t)}{dt}$ is replaced by $\frac{dq(t)}{dt} + \frac{d\delta q(t)}{dt}$ but $\delta\int_{t_1}^{t_1} L\,dt = 0$. This means that:

$$0 = \delta\int_{t_1}^{t_2} L\,dt = \int_{t_1}^{t_2} L\left[q + \delta q, \frac{d\,(q + \delta q)}{dt}, t\right]dt - \int_{t_1}^{t_1} L\left(q, \frac{dq}{dt}, t\right)dt =$$

$$= \int_{t_1}^{t_2}\delta q\frac{\partial L}{\partial q}dt + \int_{t_1}^{t_2}\delta\frac{dq}{dt}\frac{\partial L}{\partial\left(\frac{\partial q}{\partial t}\right)}dt = \int_{t_1}^{t_2}\delta q\frac{\partial L}{\partial q}dt + \int_{t_1}^{t_2}\frac{d\delta q}{dt}\frac{\partial L}{\partial\left(\frac{\partial q}{\partial t}\right)}dt =$$

$$= \int_{t_1}^{t_2} \delta q \frac{\partial L}{\partial q} dt + \int_{t_1}^{t_2} \frac{d}{dt} \left[\delta q \frac{\partial L}{\left(\frac{\partial q}{\partial t} \right)} \right] dt - \int_{t_1}^{t_2} \delta q \frac{d}{dt} \frac{\partial L}{\left(\frac{\partial q}{\partial t} \right)} dt =$$

$$= \int_{t_1}^{t_1} \delta q \left[\frac{\partial L}{\partial q} - \frac{d}{dt} \frac{\partial L}{\left(\frac{\partial q}{\partial t} \right)} \right] dt + \left[\frac{\partial L}{\left(\frac{\partial q}{\partial t} \right)} \delta q \right]_{t_1}^{t_2} = \int_{t_1}^{t_1} \delta q \left[\frac{\partial L}{\partial q} - \frac{d}{dt} \frac{\partial L}{\left(\frac{\partial q}{\partial t} \right)} \right] dt$$

for arbitrary δq, hence the solution (which is dubbed 'extremal') $q(t)$ of the variational principle satisfies the ('Euler-Lagrange') equation:

$$\frac{\partial L}{\partial q} - \frac{d}{dt} \frac{\partial L}{\left(\frac{\partial q}{\partial t} \right)} = 0$$

with the boundary conditions $q(t_1) = c_1$ and $q(t_2) = c_2$. Remarkably, understanding if the extremal is actually a maximum or a minimum is usually no trivial task—see Ref. [2].

Generalization to both the case of a Lagrangian depending on higher-order derivatives of the Lagrangian coordinate, to the case of a Lagrangian coordinate depending on either both time and space or on spatial coordinates only and to the case of many Lagrangian coordinates (which is the case the text refers to) is cumbersome but straightforward, as it follows exactly the same line of reasoning outlined above—see, e.g. [3,4].[1]

A.3   Lagrange Multipliers

Let $f(t)$ and $g(t)$ be functions of t. We want to identify the extrema of $f(t)$ which satisfy the constraint $g(t) = 0$. To this purpose, we introduce the quantity λ ('Lagrange multiplier') and solve the following system of two equations in two unknown quantities:

$$\frac{d}{dt} [f(t) + \lambda g(t)] = 0 \quad ; \quad \frac{d}{d\lambda} [f(t) + \lambda g(t)] = 0$$

The second equation is equivalent to the original constraint $g(t) = 0$.

Generalization is straightforward. Let, e.g. both L and g be functions of $q(t)$, $\frac{dq(t)}{dt}$ and t. We want to find the extremal of the variational principle $\int_{t_1}^{t_1} L dt = $ extremum which satisfies the constraint $\int_{t_1}^{t_1} g dt = 0$ (with similar boundary conditions $g(t_1) = c_1, g(t_2) = c_2$). To this purpose, we introduce the Lagrange multiplier $\lambda(t)$ and solve

---

[1] If L depends on both time and space it is sometimes referred to as 'Lagrangian density'.

the following systems of two Euler-Lagrange equations in two unknown functions of t, namely $q(t)$ and $\lambda(t)$:

$$\frac{\partial (L + \lambda g)}{\partial q} - \frac{d}{dt} \frac{\partial (L + \lambda g)}{\left(\frac{\partial q}{\partial t}\right)} = 0$$

$$\frac{\partial (L + \lambda g)}{\partial \lambda} - \frac{d}{dt} \frac{\partial (L + \lambda g)}{\left(\frac{\partial q}{\partial t}\right)} = 0$$

Generalization to more Lagrangian coordinates, higher-order derivatives etc. ... is trivial. If $q = q(x_j)$ then $\lambda = \lambda(x_j)$ too. Further constraints require further Lagrange multipliers, and—correspondingly—further Euler-Lagrange equations.

A. 4   Proof of Gage et al.'s Theorem

Following Ref. [5], let us investigate the possible existence of a universal variational principle in non-equilibrium thermodynamics, whose Euler-Lagrange equations (Sect. A. 2) are equivalent to the conservation equations in steady state.[2] In particular, let $\Theta \equiv \int_a^b \Phi\left(\Gamma_i, L_{jk}, X_i, y\right) dy$, $\Gamma_i = \Gamma_i(y)$ Lagrangian coordinates, $X_i = \frac{\partial \Gamma_i}{\partial y}$, $L_{jk} = L_{jk}(\Gamma_i, y)$ and $J_i = L_{ij} X_j$, with $a \leq y \leq b$ and $i, j, k, \ldots = 1, \ldots N$. We perturb $\Gamma_i \to \Gamma_i + \delta \Gamma_i$ with $\delta \Gamma_i (y = a) = \delta \Gamma_i (y = b) = 0$ and require that $\Theta = $ extremum. No quantities other than the Γ_i's, the L_{jk}'s and the X_i's—and in particular no Lagrange multiplier—are involved.

The Euler-Lagrange equations are:

$$\frac{\partial \Phi}{\partial \Gamma_i} + \frac{\partial \Phi}{\partial L_{jk}} \frac{\partial L_{jk}}{\partial \Gamma_i} - \frac{d}{dy}\left(\frac{\partial \Phi}{\partial X_i}\right) = 0$$

They are equivalent to:

$$\frac{\partial \Phi}{\partial \Gamma_i} + \frac{\partial \Phi}{\partial L_{jk}} \frac{\partial L_{jk}}{\partial \Gamma_i} = \frac{d}{dy}\left(\frac{\partial \Phi}{\partial X_i}\right) = \frac{\partial}{\partial X_i}\left(\frac{d\Phi}{dy}\right) =$$

$$= \frac{\partial}{\partial X_i}\left(\frac{\partial \Phi}{\partial \Gamma_j} \frac{\partial \Gamma_j}{\partial y} + \frac{\partial \Phi}{\partial X_j} \frac{\partial X_j}{\partial y} + \frac{\partial \Phi}{\partial L_{jk}} \frac{\partial L_{jk}}{\partial y} + \frac{\partial \Phi}{\partial y}\right) =$$

$$= \frac{\partial}{\partial X_i}\left(\frac{\partial \Phi}{\partial \Gamma_j} X_j + \frac{\partial \Phi}{\partial X_j} \frac{\partial X_j}{\partial y} + \frac{\partial \Phi}{\partial L_{jk}} \frac{\partial L_{jk}}{\partial y} + \frac{\partial \Phi}{\partial y}\right) =$$

---

[2] Our discussion is concerned with one-dimensional systems only. Generalization is straightforward.

$$= \left(\frac{\partial^2 \Phi}{\partial X_i \partial \Gamma_j}\right) X_j + \left(\frac{\partial^2 \Phi}{\partial X_i \partial X_j}\right) \frac{\partial X_j}{\partial y} + \left(\frac{\partial^2 \Phi}{\partial X_i \partial L_{jk}}\right) \frac{\partial L_{jk}}{\partial y} + \frac{\partial^2 \Phi}{\partial X_i \partial y}$$

where we have taken into account that all Lagrangian coordinates are perturbed independently from both each other and their derivatives.[3] We rewrite the expression above—which is perfectly equivalent to the original Euler-Lagrange equations—as follows:

$$\left(\frac{\partial^2 \Phi}{\partial X_i \partial X_j}\right) \frac{\partial X_j}{\partial y} = -\left(\frac{\partial^2 \Phi}{\partial X_i \partial L_{jk}}\right) \frac{\partial L_{jk}}{\partial y} + \frac{\partial \Phi}{\partial L_{jk}} \frac{\partial L_{jk}}{\partial \Gamma_i} + \left[\frac{\partial \Phi}{\partial \Gamma_i} - \left(\frac{\partial^2 \Phi}{\partial X_i \partial \Gamma_j}\right) X_j - \frac{\partial^2 \Phi}{\partial X_i \partial y}\right]$$

We want this relationship—hence, the Euler-Lagrange equations—to be equivalent to the steady-state conservation equations $\frac{\partial J_i}{\partial y} = 0$, i.e. $0 = \frac{\partial}{\partial y}\left(L_{ij} X_j\right)$, or, equivalently:

$$L_{ij} \frac{\partial X_j}{\partial y} = -X_j \frac{\partial L_{ij}}{\partial y}$$

Formally, both the relationship which the Euler-Lagrange equations are equivalent to and the relationship which the conservation equations are equivalent to are linear systems of N equations in the N quantities $\frac{\partial X_j}{\partial y}$. Since the two relationships above are postulated to be equivalent, we must have an invertible matrix a_{ij} such that $\left(\frac{\partial^2 \Phi}{\partial X_i \partial X_j}\right) \frac{\partial X_j}{\partial y} = a_{ik} L_{kj} \frac{\partial X_j}{\partial y}$. Generally speaking, $a_{ij} = a_{ij} \left(\Gamma_k, X_k, L_{kl}, y\right)$. Since $L_{ij} \frac{\partial X_j}{\partial y} = -X_j \frac{\partial L_{ij}}{\partial y}$, we obtain $\left(\frac{\partial^2 \Phi}{\partial X_i \partial X_j}\right) \frac{\partial X_j}{\partial y} = -a_{ik} X_j \frac{\partial L_{kj}}{\partial y}$, so that the Euler-Lagrange equations reduce to[4]:

$$-\frac{\partial \Phi}{\partial L_{kj}} \frac{\partial L_{kj}}{\partial \Gamma_i} + \left(\frac{\partial^2 \Phi}{\partial X_i \partial L_{kj}} - a_{ik} X_j\right) \frac{\partial L_{kj}}{\partial y} - \left[\frac{\partial \Phi}{\partial \Gamma_i} - \left(\frac{\partial^2 \Phi}{\partial X_i \partial \Gamma_j}\right) X_j - \frac{\partial^2 \Phi}{\partial X_i \partial y}\right] = 0$$

Since the L_{kj}'s are arbitrary, $\frac{\partial L_{jk}}{\partial \Gamma_i}$ and $\frac{\partial L_{kj}}{\partial y}$ are independent, and their coefficients must vanish identically everywhere:

$$\frac{\partial \Phi}{\partial L_{kj}} \equiv 0 \quad ; \quad \text{and} \quad \frac{\partial^2 \Phi}{\partial X_i \partial L_{kj}} \equiv a_{ik} X_j \quad ;$$

---

[3] Admittedly, we have ignored the fact that $\frac{\partial X_i}{\partial X_j} = 1$ for $i = j$. However, nothing changes in the following if we introduce the corresponding correction to our discussion.

[4] Remarkably, no Onsager symmetry is invoked (all the same, dummy indices may be swapped at will).

Since $\frac{\partial^2 \Phi}{\partial X_i \partial L_{kj}} = \frac{\partial}{\partial X_i}\left(\frac{\partial \Phi}{\partial L_{kj}}\right)$, the relationship on the left implies $\frac{\partial^2 \Phi}{\partial X_i \partial L_{kj}} \equiv 0$, and the relationship on the right implies therefore $a_{ik} X_j \equiv 0$. Being a_{ik} invertible, this implies

$$X_j \equiv 0$$

and as a consequence

$$J_i = L_{ij} X_j \equiv 0$$

In the familiar thermodynamic interpretation, the X_j's and the $J_i's$ are thermodynamic forces and fluxes; both disappear simultaneously and identically at full thermodynamic equilibrium only. In non-equilibrium thermodynamics, therefore, no variational principle $\Theta =$ extremum ever exists, at least as far as no other quantity different from thermodynamic forces and fluxes, Onsager's coefficients and spatial coordinates is involved in the computation—e.g. as Lagrange multipliers (Sect. A. 3). This fact suggests that extremum conditions in non-equilibrium thermodynamics may only hold if suitably constrained.

References

1. Landau, L.D., Lifshitz, E.: Fluid Mechanics Pergamon, Oxford (1960)
2. Elsgolts, I.V.: Differential Equations and Variational Calculus. Mir, Moscow, URSS (1981)
3. Gyarmati, I.: Non-equilibrium Thermodynamics. Springer, Berlin (1970)
4. Landau, L.D., Lifshitz, E.: The Classical Theory of Fields. Pergamon, Oxford (1971)
5. Gage, D.H., Schiffer, M., Kline, S.J., Reynolds, W.C.: The non-existence of a general thermokinetic variational principle. In: Donnelly, R.J. (ed.) Non-Equilibrium Thermodynamics: Variational Techniques and Stability. University of Chicago Press, Chicago (1966)

Index

© The Editor(s) (if applicable) and The Author(s), under exclusive license to
Springer Nature Switzerland AG 2022
A. Di Vita, *Non-equilibrium Thermodynamics*, Lecture Notes in Physics 1007,
https://doi.org/10.1007/978-3-031-12221-7

nted in the United States
aker & Taylor Publisher Services